GLOBAL DEFORESTATION

Forests cover 41 billion hectares or 31% of the global land surface, yet 8,000 years ago they covered nearly 50% of the global land surface. While the rate of deforestation has decreased since the 1990s, the current rate is still 13 million hectares per year, and this has widespread environmental impacts. For example, forest cover loss affects the hydrological functioning of watersheds, biogeochemical cycling and the availability of nutrients, carbon dioxide sequestration, and climate change. The adverse impacts of deforestation are recognized as a serious issue by governments globally.

Global Deforestation provides a concise but comprehensive examination of the variety of ways in which deforestation modifies environmental processes, as well as the societal implications of these changes. The book stresses how forest ecosystems may be prone to nearly irreversible degradation. To prevent the loss of important biophysical and socio-economic functions, forests need to be adequately managed and protected against the increasing demand for agricultural land and forest resources. The book:

- describes the spatial extent of forests, including methods used to detect forest cover and its current and historical changes, leading to an understanding of the past and present drivers of deforestation;
- presents a theoretical background to understand the impacts of deforestation on biodiversity, hydrological functioning, biogeochemical cycling, and climate;
- bridges the physical and biological sciences with the social sciences by examining economic impacts and socioeconomic drivers of deforestation;
- reviews the rich body of literature on deforestation and synthesizes information across disciplines, allowing readers to learn about deforestation on an interdisciplinary level without having to consult multiple texts and journal articles.

This book will appeal to anyone in search of a comprehensive yet concise reference on deforestation, including graduate and undergraduate students, researchers and policymakers in environmental science, ecology, forestry, hydrology, geography, biogeochemistry, plant science, ecohydrology, and environmental economics.

Christiane Runyan is a lecturer in Hydrology and Water Resources in the Advanced Academic Program of the Zanvyl Krieger School of Arts and Sciences at Johns Hopkins University. Her research examines how deforestation affects the dynamics of hydrological and biogeochemical processes, and includes modeling the control that vegetation has on the soil's physical and biogeochemical conditions.

Paolo D'Odorico is Ernest H. Ern Professor of Environmental Sciences at the University of Virginia. His research focuses on the role of hydrological processes in the functioning of terrestrial ecosystems and societies. He was awarded the Sustainability Science Award from the Ecological Society of America in 2009, and he was made a Fellow of the John Simon Guggenheim Memorial Foundation in 2011. He edited *Dryland Ecohydrology* (2006, Springer), co-authored *Noise-Induced Phenomena in the Environmental Sciences* (2011, Cambridge University Press), and *Elements of Physical Hydrology, Second Edition* (2014, Johns Hopkins University Press).

GLOBAL DEFORESTATION

CHRISTIANE RUNYAN
Johns Hopkins University

PAOLO D'ODORICO
University of Virginia

CAMBRIDGE
UNIVERSITY PRESS

32 Avenue of the Americas, New York NY 10013

Cambridge University Press is part of the University of Cambridge.

It furthers the University's mission by disseminating knowledge in the pursuit of education, learning, and research at the highest international levels of excellence.

www.cambridge.org
Information on this title: www.cambridge.org/9781107135260

© Christiane Runyan and Paolo D'Odorico 2016

This publication is in copyright. Subject to statutory exception and to the provisions of relevant collective licensing agreements, no reproduction of any part may take place without the written permission of Cambridge University Press.

First published 2016

Printed in the United States of America

A catalog record for this publication is available from the British Library.

Library of Congress Cataloging in Publication Data
Names: Runyan, Christiane, 1982–, author. | D'Odorico, Paolo, 1969– author.
Title: Global deforestation / Christiane Runyan, Johns Hopkins University, Paolo D'Odorico, University of Virginia.
Description: New York, NY: Cambridge University Press, 2016. |
Includes bibliographical references and index.
Identifiers: LCCN 2015042438 | ISBN 9781107135260 (hardback)
Subjects: LCSH: Deforestation.
Classification: LCC SD418.R86 2016 | DDC 634.9–dc23
LC record available at http://lccn.loc.gov/2015042438

ISBN 978-1-107-13526-0 Hardback

Cambridge University Press has no responsibility for the persistence or accuracy of URLs for external or third-party Internet Web sites referred to in this publication and does not guarantee that any content on such Web sites is, or will remain, accurate or appropriate.

Contents

Preface		*page* ix
1.	Introduction: *Patterns and Drivers*	1
	1.1 Definitions and Classifications of Forest Ecosystems	1
	1.1.1 Biogeography of Forest Ecosystems	3
	1.2 Spatial and Temporal Trends in Forest Cover Change	10
	1.2.1 Current (1990–Present) Spatial and Temporal Trends in Deforestation	10
	1.2.2 Historical Spatial and Temporal Trends in Deforestation	13
	1.2.3 Historical Patterns of Global Deforestation	14
	1.2.4 Reforestation/Afforestation	21
	1.3 Historical, Current, and Future Drivers of Deforestation	23
	1.3.1 Agricultural Production	23
	1.3.2 Logging	24
	1.3.3 Shifting Cultivation	25
	1.3.4 Biofuels	26
	1.4 Projected Geographic Changes in Deforestation	26
	1.5 Remote Sensing Methods Used to Quantify and Map Deforestation	33
	1.5.1 Optical Remote Sensing	34
	1.5.2 Radar	35
	1.5.3 Laser (i.e., LiDAR)	37
	1.6 Concluding Comments	38
2.	Hydrological and Climatic Impacts	39
	2.1 Introduction	39
	2.2 Precipitation and Forest Canopies	42
	2.2.1 Canopy and Litter Interception	42
	2.2.2 Canopy Condensation and Occult Precipitation	43
	2.3 Infiltration and Runoff Generation	45
	2.4 Effects of Deforestation on the Hydrologic Response	47
	2.4.1 Effects on Flood Dynamics at the Event Timescale	47
	2.4.2 Effects on Water Yields	47

		2.4.3 Effect on Snowmelt	50
		2.4.4 Effect of Fire	50
	2.5	Forest Effects on Groundwater	52
	2.6	Effect of Deforestation on Wetlands	52
	2.7	Evaporation and Transpiration	54
		2.7.1 The Effect of Climate Change on Evapotranspiration from Landmasses	58
	2.8	Effect of Forest Vegetation on Precipitation	59
		2.8.1 Effect of Deforestation on Precipitation Recycling	60
		2.8.2 Impact on Surface Energy Balance and Boundary Layer Dynamics	61
		2.8.3 Effect of Forest Vegetation on Cloud Microphysics	62
		2.8.4 The Effect of Mesoscale Circulations Induced by "Small-Scale" Canopy Gaps	63
	2.9	Effect of Forest Vegetation on Microclimate	63
	2.10	Effects of Deforestation on Large-Scale Climate	68
	2.11	Summary	69
3.	Biogeochemical Impacts		71
	3.1	Carbon Cycle	71
		3.1.1 Carbon Cycle in Undisturbed Forests	71
		3.1.2 Global Estimates of C Pools, Emissions, and Uptake in Terrestrial Ecosystems	76
		3.1.3 Changes in the Carbon Cycle as a Result of Deforestation	78
	3.2	Nitrogen Cycle	85
		3.2.1 Nitrogen Cycle in Undisturbed Forests	85
		3.2.2 Location of N-Limited Forests	88
		3.2.3 Change in the N Cycle Following Deforestation	90
	3.3	Phosphorus Cycle	93
		3.3.1 Phosphorus Cycling in Undisturbed Forests	93
		3.3.2 Location of P-Limited Forests	96
		3.3.3 P Losses after Deforestation	99
	3.4	Conclusion	101
4.	Irreversibility and Ecosystem Impacts		103
	4.1	Background on Irreversibility and Bistability in Deforested Ecosystems	103
	4.2	Feedbacks That Modify Resource Availability	106
		4.2.1 Precipitation-Vegetation	106
		4.2.2 Canopy Deposition	113
		4.2.3 Soil Moisture	117
		4.2.4 Water Table	118
		4.2.5 Permafrost	121
		4.2.6 Nutrient Cycling	125
	4.3	Feedbacks That Modify the Disturbance Regime	129
		4.3.1 Landslides	129
		4.3.2 Fire	132

	4.3.3 Exposure to Freezing Events (Climate-Air-Temperature)	135
	4.3.4 Salinity	138
4.4	Noise Induced Transitions	142
4.5	Leading Indicators of State Shifts	143
4.6	Concluding Comments	144

5. Economic Impacts and Drivers of Deforestation — 145

5.1	Background	145
5.2	Economic Uses of Forested Land	148
5.3	Factors Driving Deforestation	149
	5.3.1 Proximate Causes	149
	5.3.2 Underlying Causes	152
5.4	Modeling Frameworks to Examine Deforestation	163
	5.4.1 Microeconomic models	164
	5.4.2 Regional models	164
	5.4.3 Macroeconomic	164
5.5	Economic Effects of Deforestation	165
	5.5.1 Benefits	165
	5.5.2 Costs	166
5.6	Alternative Policies Aimed at Deriving Value from Forested Lands (i.e., REDD)	170
5.7	Conclusion	171

6. Synthesis and Future Impacts of Deforestation — 173

6.1	Benefits of Preserving Forests	173
6.2	Ecohydrological and Climate Impacts of Deforestation	174
6.3	Effect of Forest Loss on Biogeochemical Processes	174
6.4	Economic Impacts of Deforestation	176
6.5	Irreversible Changes Induced by Deforestation	177
6.6	Biodiversity Loss	178
	6.6.1 Role of Biodiversity in Ecosystem Processes	183
	6.6.2 Societal Impacts of Biodiversity Loss	184
	6.6.3 Strategies for Protecting Biodiversity	185
6.7	Impact of Deforestation on Human Health	186
6.8	Food Security as a Major Future Driver of Deforestation	187
	6.8.1 Reduce Food Losses	188
	6.8.2 Intensification versus Extensification	188
	6.8.3 Mitigating the Environmental Impacts of Intensification	189
	6.8.4 Crop Selection and Sequencing	191
	6.8.5 Location of Agricultural Expansion	192
6.9	Concluding Comments	193

References — 195
Index — 249
Color Figures — 255

Preface

Deforestation disrupts hydrological processes, climate, biogeochemical cycling, and socioenvironmental dynamics. It can lead to irreversible losses of biodiversity, natural capital, and rural livelihoods, while favoring an unsustainable use of natural resources and enhancing unbalanced relationships between private benefits and public losses associated with land clearance. Deforestation is a *disturbance* because it leads to biomass losses over timescales much shorter than those needed for forest regeneration. In some cases recovery is not possible because the disturbance induces a shift in forest ecosystems to a permanently deforested state by impacting the availability of resources and environmental conditions that are necessary for forest regeneration.

According to the 2010 Food and Agriculture Organization (FAO) Forest Resource Assessment, forests cover 41 billion hectares, or 31% of the global land surface, yet used to cover nearly 50% of the global land surface 8,000 years ago. While the current rate of deforestation has decreased since the 1990s from 16 million ha yr^{-1} to 13 million ha yr^{-1}, it remains relatively high. Deforestation alters the coupled natural and human systems with important impacts on the potential for forests to regenerate. Understanding these impacts is also important in light of international programs that seek to provide financial incentives for reduced deforestation and have an estimated market potential of U.S. $10 billion.

This book is motivated by the need for a comprehensive cross-disciplinary analysis of the existing literature on global deforestation. We review the geography of deforestation, analyze the major drivers and effects of forest loss, and examine theories as well as empirical evidence on how forests affect their natural environment. We stress how forest removal may cause the loss of important ecosystem functions, leading to a permanent and nearly irreversible shift to a treeless state. We investigate the biotic-abiotic feedbacks that determine the stability and resilience of forest ecosystems and analyze the socioeconomic processes underlying current patterns of deforestation. While doing so, we review a large number of recent studies on this body of literature and synthesize information across disciplines, thereby bridging the physical and biological sciences with the social sciences.

This analysis addresses a broad readership of ecologists, hydrologists, economists, biogeochemists, geographers, resource analysts, and policy makers whose work is related to deforestation. As such, it was written with the goals of readability and accessibility by both social and natural scientists. While providing a relatively thorough synthesis of research that is currently spread across a diverse and broad body of literature, this book is not intended to be a comprehensive treatise on deforestation; this is a fast-moving research field that produces new important contributions every day. It would not be possible to contain in this volume a complete analysis of this growing body of literature.

This book would have not been possible without the help, motivation, and support of our colleagues, families, and institutions. We are grateful to Deborah Lawrence (University of Virginia) for her unfailing support through years of continued collaboration and companionship. We are truly indebted to her for drawing us into this research field and inspiring this work. Christiane Runyan thanks her husband, Joshua, daughter, Georgiana and son, Waylon, for the support they have provided during the time it has taken to write this book. We are grateful to the University of Virginia, Department of Environmental Sciences, for providing the academic environment that stimulated our work. We also thank Michelle Faggert and Kailiang Yu of the University of Virginia for thier assistance with formatting and artwork. We acknowledge the support of the Vice President for Research Office at the University of Virginia and the National Social-Environmental Synthesis Center (SESYNC) of the University of Maryland.

1

Introduction: *Patterns and Drivers*

1.1 Definitions and Classifications of Forest Ecosystems

Forests, which can be defined as woody plant communities in areas that are large enough to modify the local environment and microclimate (Chang, 2002), currently cover 3.8 billion hectares, roughly 30% of the Earth's land surface (FAO, 2010). Approximately 25% of the world's forested area is located in Europe, followed by South America (21%), North and Central America (17%), Africa (17%), Asia (15%), and Oceania (5%) (Table 1.1). The most extensive forest biomes are tropical, boreal, and temperate (Figure 1.1). Tropical forests cover approximately 1.76 billion hectares (or 42% of the world's forested area), followed by boreal forests with their roughly 1.37 billion hectares (or ~33% of the world's forests), and temperate forests (1.04 billion hectares, or 25% of the world's forested areas) (IPCC, 2000).

Forest ecosystems play a fundamental role in the dynamics of the Earth system and provide services of great environmental, societal, and economic value. They are a major determinant of the regional and global climate (Chapter 2), modulate water and nutrient cycling (Chapter 3), and provide invaluable resources and services (Chapter 5) that have played a crucial role for the social, economic, and cultural development of several civilizations. Depending on how resources derived from forests are used, they can be either renewable (i.e., not depleted) or nonrenewable (Chang, 2002). Moreover, the rise and fall of several civilizations in human history have been determined by their use and overuse of forests (Box 1.1).

This book is concerned with the ongoing phenomenon of global deforestation (Box 1.2). This Introduction and the following chapters will discuss the major drivers along with the environmental and societal implications of deforestation. It will also analyze social-environmental processes that affect the stability and resilience of forest ecosystems and their ability to recover after deforestation.

Table 1.1. *Forest area by region in 2010 from the Forest Resources Assessment (FAO, 2010): Global tables in million hectares (Mha) while the forest carbon stock represents the carbon contained in living forest biomass (in million tonnes)*

Region	Land area (Mha)	Primary forest (Mha)	Regenerated forest (Mha)	Permanent forest estate (Mha)	Total forest area (Mha)	Forest carbon stock
Africa	2978	48	437	241	674	55,736
Asia	4362	110	359	417	593	34,891
Europe	983	262	669	301	1,005	45,009
North and Central America	2137	280	385	422	705	37,457
Oceania	849	35	151	37	191	3,902
South America	1755	624	180	350	864	93,270
World	13064	1,359	2,182	1,767	4,033	270,265

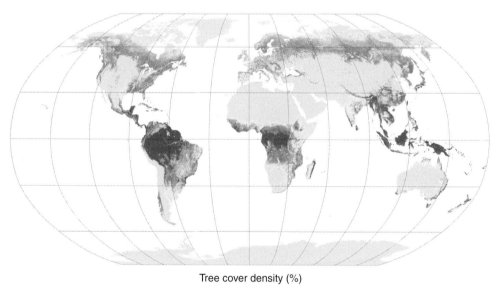

Figure 1.1. Distribution of the World's Forest by Climatic Domain (FAO, 2010). See also color figures at the end of the book.
Source: Available at: http://www.fao.org/forestry/fra/80298/en/.

Box 1.1 Deforestation and the Collapse of Past Civilizations

Geographers and anthropologists have often related the decline of past civilizations to environmental degradation resulting from deforestation. Deforestation might have triggered the collapse of the Viking, Maya, Anasazi, and Rapa Nui civilizations (Diamond, 2005; Turner and Sabloff, 2012). In most of these cases, deforestation enhanced soil erosion, thereby leading to the permanent loss of soil resources. In Chapter 4, we will discuss some positive feedbacks that could prevent forest regeneration. One of such processes is associated with soil erosion due to the loss of the sheltering effect of vegetation (e.g., Runyan et al., 2012a). Because tree establishment and growth require adequate soil resources, nutrient depletion after prolonged deforestation and erosion may impede forest regrowth. Soils are the ultimate foundation of a society's livelihood, and soil loss has undermined several civilizations' ability to thrive (Montgomery, 2008).

The historical case of Easter Island is particularly instructive because of its remoteness (more than 2,000 km away from other inhabited islands) and the high level of cultural and technological development of the Rapa Nui civilization. Rolett and Diamond (2004) showed that Easter Island was particularly prone to permanent deforestation when compared to 69 other Pacific islands between Hawaii in the north and New Zealand in the south. Their analysis showed that islands more likely to remain permanently deforested were those with slower plant growth rates (i.e., at high latitudes and with relatively low annual precipitation), lower rates of nutrient replenishment by atmospheric deposition, smaller area (hence smaller diversity and resilience (MacArthur and Wilson, 1967; Walker and Salt, 2006)), and higher isolation (i.e., fewer emigration opportunities). Easter Island was the third island in the South Pacific most likely to be permanently deforested after Necker and Nihoa, which were also completely deforested. The theory of an island-wide environmental collapse has been challenged by recent archaeological evidence, suggesting that the decline in land use prior to the European arrival might have been due to environmental limitations rather than degradation (Stevenson et al., 2015).

The history of this island is often considered as a good metaphor of the fate of planet Earth and the human race, should the world's natural resources be used unsustainably (Diamond, 2005). As with the Rapa Nui people, we will have no other place to go once we have exhausted the Earth's resources (Pointing, 1991).

1.1.1 Biogeography of Forest Ecosystems

The global distribution of forests is strongly controlled by climate, with energy and water playing key roles for woody plant establishment, survival, and growth. Regions that either are too arid or have too-short frost-free seasons are generally not suitable for woody plants. Water limitations constrain tree growth because they induce stomatal closure in order to reduce water losses and prevent hydraulic failure (i.e., when

> **Box 1.2 Definitions**
>
> The following definitions are from the FAO (2001), except where otherwise indicated.
>
> - **Deforestation**: the conversion of forest to another land use or the long-term reduction of the tree canopy cover below a 10% threshold. Deforestation implies the long-term or permanent loss of forest cover and its transformation into another land use.
> - **Reforestation**: the establishment of forest plantations on temporarily unstocked lands that are considered forest.
> - **Afforestation**: the establishment of forest plantations on land that previously was not classified as forest. It implies a transformation from nonforest to forest.
> - **Primary forest**: forest that has never been logged and has developed following natural disturbances and under natural processes, regardless of its age (Convention on Biological Diversity, 2001).
> - **Secondary forest**: forests regenerated largely through natural processes after significant human or natural disturbance of the original forest vegetation (Chokkalingam, 2001).
> - **Modified natural forest**: all secondary forests and forests significantly disturbed by human activity (Putz and Redford, 2010; Wright, 2010).
> - **Planted forests:** forests predominantly composed of trees established through planting and/or deliberate seeding of native or introduced species (Carle and Holmgren, 2008).
> - **Degraded forest**: a secondary forest that has lost, through human activities, the structure, function, species composition, or productivity normally associated with a natural forest type expected on that site (Convention on Biological Diversity, 2001).
> - **Frontier forests**: large, ecologically intact, and relatively undisturbed forests that support the natural range of species and forest functions (WRI, 1997).
> - **Woodland**: the type of land cover characterized by trees and shrubs: "other wooded land." Other wooded land is defined by the most recent global forest resource assessment (i.e., FRA-2010) by the FAO as land not classified as "forest," spanning more than 0.5 hectare; with trees higher than 5 meters and a canopy cover of 5%–10%, or trees able to reach these thresholds in situ; or with a combined cover of shrubs, bushes, and trees above 10%.

water transport is disrupted in a large number of embolized vessels [cavitation], resulting in the desiccation of plant tissues; Urli et al., 2013, see Chapter 2, Box 2.2). Thus, other plant types are better suited for growth in hyperarid climates with scarce, unreliable, and strongly variable precipitation patterns (Noy-Meir, 1973). Likewise, climate zones with low temperatures and short growing seasons are not favorable for woody plants (Box 2.3) because carbon assimilation rates may be too low to maintain a positive carbon budget and sustain growth (e.g., Tranquillini, 1979). Cold stress may also inhibit new cell production and tissue differentiation (i.e., the development or maturity of tissue cells to have a more distinct form and function) (Körner, 1998). Moreover low temperatures reduce the regeneration capacity of woody plants (i.e., seed production and germination) and induce frost damage and winter frost desiccation (Tranquillini, 1979; Körner, 1998; Box 2.3). Thus, the biogeography of forest ecosystems is strongly controlled by precipitation and temperature.

Forests are typically found in regions with annual precipitation greater than 400–500 mm and frost-free periods longer than 14–16 weeks (Chang, 2002). Their occurrence is mainly limited by temperature regime at high latitudes and altitudes, and precipitation in tropical, subtropical, and midlatitude lowland regions. Other factors such as aspect (i.e., position relative to the Sun), topography, exposure to wind, water table depth, and soil thickness also play a role in determining the presence or absence of trees (e.g., Sveinbjörnsson, 2000). The northern latitudinal limit of forests (or "arctic tree line") roughly coincides with the 10–12°C maximum temperature isotherms for the warmest month of the year (i.e., July) (Figure 4.19). This tree "line" is a latitudinal band stretching between the forest limit to the south and the cold tolerance limit for tree survival to the north (Epstein et al., 2004).

Temperature regimes also strongly control tree distribution along altitudinal gradients. As a general rule, the change in temperature along a 1,000 m elevation gradient is similar to that observed over 5° of latitude (*Hopkins's biogeoclimatic law*) (Perry et al., 2008; see also Hopkins, 1938). Thus, some of the vegetation changes found along mountain slopes (e.g., the transition from broadleaf to coniferous species) may resemble those observed along latitudinal gradients. As mean annual temperatures decrease with elevation, conditions become increasingly harsh for the survival and establishment of trees. Thus, trees are not found above a certain elevation. In the transitional zone between alpine forests and treeless, high-elevation meadows (known as the *alpine tree line* or *kampfzone*), trees are typically stunted. Similarly to its arctic counterpart, the alpine tree line occurs around the 10°C summer isotherm (Daubenmire, 1954) – though other authors refer to a mean growing season temperature of 6–7°C (Körner & Paulsen, 2004). The elevation of this tree line depends on aspect, soil type, and latitude. It is found at about 680 m in northern Sweden (68°N), 950 m in southern Alaska (63°N), 1,300 m in the Craigieburn Mountains of New Zealand (43°S), 1,800–2,100 m in the Alps (47°N), 2,359 m in the Hida Mountains of Japan (36°N), 3,000–3,500 m in the equatorial Andes (8°N–3°S; Bader et al., 2007), and 3,950 m on Mount Kilimanjaro in Tanzania (3°S) (Körner, 1998). Despite the important role played by aspect, topography, soil thickness, exposure to winds, and cloudiness, the major factor determining the location of alpine tree lines is temperature (Smith et al., 2003). Both the arctic and alpine tree lines have fluctuated in the past 10,000 years in response to climate variability (Kullman, 1988; Grace, 1989).

Tree lines associated with the cold intolerance of woody plants can also be found away from arctic and alpine regions (D'Odorico et al., 2013). For instance, low-energy, intertidal environments in tropical regions are dominated by trees known

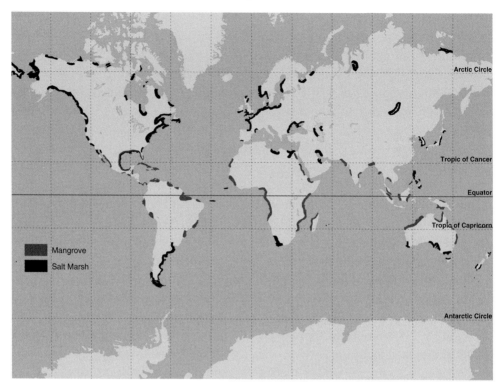

Figure 1.2. Global distribution of mangrove forests and salt marshes. See also color figures at the end of the book.
Source: D'Odorico et al., 2013. Redrawn after Chapman, 1975.

as *mangroves*. These salt and flood tolerant trees are cold sensitive and cannot survive in midlatitude intertidal landscapes (Figure 1.2). Thus, low-energy extratropical coastal ecosystems are dominated by salt marshes (i.e., with herbaceous life-forms) because there are no woody species that are both cold and salt tolerant enough to fill that niche. Similarly, temperature regimes can also play a crucial role in grassland to shrubland transitions in deserts (D'Odorico et al., 2010a).

Despite these exceptions, the geographic distribution of midlatitude and tropical forests is strongly controlled by precipitation (Figure 1.3). In the tropics, mean annual precipitation is at a maximum at the equator ("wet tropics") and decreases while moving toward the subtropics ("dry tropics"), consistent with the Hadley circulation pattern (i.e., an atmospheric circulation pattern characterized by warm air that rises near the equator; cools as it travels toward the subtropics, where it sinks; and warms as it travels back toward the equator). Moreover rainfall becomes more seasonal, with a well-defined alternation between dry and rainy seasons determined by the seasonal migration of the intertropical convergence zone (i.e., the area near the equator where the trade winds converge, causing air masses to rise) between the Northern (e.g., May–August) and

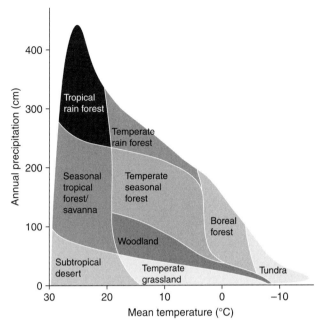

Figure 1.3. Distribution of plant communities as a function of temperature and precipitation.
Source: Redrawn from Ricklefs, 2008.

Southern (e.g., November–February) Hemispheres. The rainfall patterns within the intertropical region partly explain the vegetation gradients from tropical rain forests closer to the equator – where the climate is hot and wet throughout the year – to seasonal dry forests, scrubs, and savannas in the dry subtropics.

Tropical rain forests occupy a broad band across the Earth's warm, moist equatorial regions, including Central Africa (the Congo Basin), the Amazon, and Southeast Asia (Lieth and Werger, 1989). They are dominated by evergreen or semideciduous broadleaf species, and their most common type is dry ground (*terra firme*) lowland rain forests (Malhi et al., 1999). Montane forests, flooded forests, and mangroves, however, can also be found in this region. Rain forests are characterized by tall stature (usually exceeding 30 m), a tightly closed canopy, and very high diversity (200–300 tree species per hectare) (Malhi et al., 1999). The amount of aboveground biomass reported in tropical forests varies substantially across location. Lewis et al. (2013) analyzed data from 260 forest plots across old-growth African forest and found that African rain forests have a mean aboveground biomass of 395.7±14.3 Mg dry biomass ha^{-1}, which increases to 429 Mg dry biomass ha^{-1} in Central Africa. This is much higher than the mean value of 289 Mg ha^{-1} reported for Amazonia, and comparable with the mean value of 445 Mg ha^{-1} reported for Borneo (Malhi et al., 2013; Table 1.2).

Tropical dry forests occur in frost-free areas experiencing a 4– to 7-month period with limited or no rainfall (Janzen, 1988). The mean annual precipitation may

Table 1.2. *Stand characteristics for different forest biomes; values are the mean ± the standard deviation of the observed values for 513 forest sites. The leaf area index (LAI) is the total foliage area in the canopy per unit ground area (see Chapter 3). Evergreen is abbreviated as Ev, deciduous as De. and Temperate as Temp*

	Boreal humid		Boreal semiarid		Temp. humid		Temp. semiarid	Mediterranean warm	Tropical humid
	Ev.	Ev.	De.	Ev.	De.	Ev.	Ev.	Ev.	Ev.
Latitude (°)	58 ± 7	59 ± 5	61 ± 5	44 ± 8	44 ± 9	44 ± 2	40 ± 4	14 ± 8
Max LAI ($m^2\,m^{-2}$)	4.1 ± 3.0	3.4 ± 1.8	3.5 ± 1.5	7 ± 2.9	6.1 ± 3.5	1.8 ± 1.0	3.5 ± 1.2	5.2 ± 1.2
Tree height (m)	14 ± 7	8 ± 2	19 ± 5	20 ± 12	19 ± 7	10 ± 5	12 ± 8	28 ± 9
Basal area ($m^2\,ha^{-1}$)	28 ± 12	26 ± 10	28 ± 4	42 ± 24	31 ± 15	8 ± 2	24 ± 14	23 ± 13
Tree density (number ha^{-1})	3767 ± 5652	4230 ± 3018	1451 ± 720	1399 ± 1985	1723 ± 2439	506 ± 326	2136 ± 2815	385 ± 221
Aboveground biomass (g C m^{-2})	5761 ± 3708	4766 ± 2498	7609 ± 2438	14934 ± 13562	10882 ± 5670	6283 ± 5554	5947 ± 1808	11389 ± 5824
Belowground biomass (g C m^{-2})	1388 ± 836	1604 ± 925	1352 ± 645	4626 ± 4673	2565 ± 2609	2238 ± 1728	3247 ± 2212	2925 ± 2284

Source: Luyssaert et al., 2007.

broadly range between 250 and 2,000 mm with annual potential evapotranspiration typically exceeding annual precipitation (Murphy and Lugo, 1986). Tropical dry forests are located in Central and South America, northern Australia, Southeast Asia, Africa, and India (Janzen, 1988) and contribute about 42% of the tropical forest cover (Murphy and Lugo, 1986). They are known with different names, depending on the region, including "monsoon forests" in Asia, "caatinga" in Brazil, and "bosque tropical caducifolio" in Mexico (Pennington et al., 2009). Studies on the structure and functional characteristics of tropical dry forests suggest that trees are shorter and have a smaller basal area than those in tropical rain forests (Murphy and Lugo, 1986). Aboveground phytomass (i.e., aboveground plant biomass) varies from 28 to 269 Mg ha^{-1} with 9%–50% of the total phytomass allocated to roots (Martinez-Yrizar et al., 1995). In seasonal dry forests, most species are deciduous, with the litterfall occurring mainly during the dry season. The degree of deciduousness tends to increase in areas with lower mean annual rainfall (Pennington et al., 2009), though not all dry forests are predominantly deciduous. Annual litter production varies from 2 to 13 Mg ha^{-1} yr^{-1} with leaves constituting on average 70% of total litterfall (Martinez-Yrizar et al., 1995).

Temperate forests occupy many of the zones adjacent to tropical forests at latitudes between 25° and 50° in both hemispheres. They are found in midlatitude regions with a well-defined winter season, at least 4–6 frost-free months, relatively long growing seasons, and mean annual precipitation that exceeds average annual evapotranspiration. Principal genera include pines (*Pinus*), oaks (*Quercus*), beeches (*Fagus*), maples (*Acer*), and eucalypts (*Eucalyptus*) (Malhi et al., 1999).

In these regions the winter months are generally not too cold for broadleaved angiosperms (Perry et al., 2008). Temperate forests are typically dominated by deciduous broadleaf trees (e.g., eastern North America, western Europe, and northeast Asia), though they more broadly range from deciduous forests in areas with moist, warm summers and more frost-free days in winter (Röhrig and Ulrich, 1991); to broadleaf evergreen forests in moist regions with mild, nearly frost-free winters (Ovington, 1983); coniferous forests in mountain regions with warm summers and cold winters; mixed coniferous evergreen and broadleaf forests in "Mediterranean" regions (e.g., the Mediterranean Basin and the Pacific coast of North America) with mild, wet winters and dry summers; and *sclerophyllous forests* (i.e., scrub and forest vegetation with small, thick, and leathery or waxy leaves that minimize water losses) in drier regions with Mediterranean climate. Areas with a Mediterranean climate are found across western North America, the Mediterranean Basin, temperate Asia, and Australia. Wetter regions of Pacific North America with winter precipitation are dominated by giant, long-lived coniferous species such as redwoods, Douglas fir, hemlock, and Sitka spruce (Perry et. al., 2008). Wet, frost-free temperate regions with precipitation evenly distributed throughout the year typically exhibit *temperate (broadleaved) rain forests*. These forests can be found in small patches in the southeastern United

States, China, and Japan and are more abundant in the Southern Hemisphere (Chile, New Zealand, and Tasmania).

The *boreal forest* occupies a circumpolar belt in high northern latitudes, between circumpolar tundra and temperate forests and grasslands (Larsen, 1980). The transition between temperate and boreal forests is where cold-tolerant conifers replace hardwoods as the dominant forest species (Perry et al., 2008). Thus, this transition is determined by climate conditions unsuitable (i.e., too cold) for angiosperms (i.e., hardwoods) with minimum winter temperatures dropping below −40 °C, which is the temperature at which xylem water in hardwoods freezes (see Box 2.3). These conditions occur at roughly 50°–60° latitude. The northernmost limit of the boreal forest is found at about 70 °N, which is where the growing season becomes too short for conifer growth. Boreal forests are found in two bands: one between Newfoundland and Alaska and the other between Scandinavia and the Pacific coast of Siberia (e.g., Chang, 2002). They are characterized by a limited number of conifer genera, particularly spruce (*Picea*), pine (*Pinus*), larch (*Larix*), and fir (*Abies*), and few deciduous genera such as birch (*Betula*) and poplar (*Populus*) (Malhi et al., 1999).

1.2 Spatial and Temporal Trends in Forest Cover Change

In this section, we examine historical, current, and projected temporal and spatial patterns of deforestation. In addition, we assess the areas where afforestation and reforestation are occurring globally.

1.2.1 Current (1990–Present) Spatial and Temporal Trends in Deforestation

Between 1990 and 2005, there was a net reduction in global forest area of 66.4 million ha (Mha), or 1.7% (FAO, 2010). This reduction in forest area occurred although the global rate of deforestation had decreased from 16 Mha yr^{-1} to 13 Mha yr^{-1} since the 1990s (FAO, 2010). Much of the net forest loss (i.e., gross forest loss minus gross forest gain) between 1990 and 2005 was concentrated in South America, followed by Africa (Figure 1.4). Over the period 2000–2012, Hansen et al. (2013) found that tropical areas experienced the largest rates of forest loss of four climate domains (i.e., tropical, subtropical, temperate, and boreal; Figure 1.5). Tropical forests accounted for 58% (86 Mha) of net global forest loss, followed by boreal (27%; 40 Mha), temperate (8%; 12 Mha), and subtropical (8%; 11 Mha) biomes. Tropical dry forests of South America had the highest rate of tropical forest loss. Gross forest cover loss was high in the boreal zone largely as a result of natural processes such as fire, which accounted for nearly 60% of forest loss (Potapov et al., 2008). The remaining 40% of boreal forest loss was attributable to logging and natural processes such as insect- and disease-related forest mortality (Royama, 1984; Royama et al., 2005). For example,

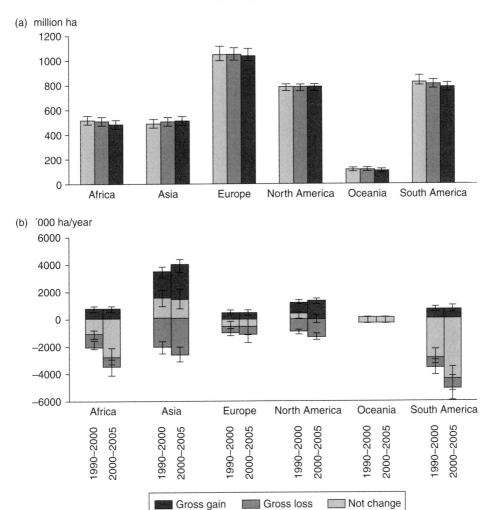

Figure 1.4. (a) Forest area by region, 1990, 2000, and 2005, and (b) gross gains and losses and net changes in forest area by region, 1990–2000 and 2000–2005.
Source: Food and Agriculture Organization of the United Nations (FAO) and JRC, 2012, and available at: http://www.fao.org/docrep/017/i3110e/i3110e00.htm.

by 2006, the mountain pine beetle infestation severely impacted western Canadian forests, infesting a cumulative area greater than 10 Mha (Westfall, 2004).

Notably, there was a substantial reduction in forest loss rates observed in Brazil (0.13 Mha yr^{-1}) with forest loss being less than 2 Mha yr^{-1} in 2010–2011, which is substantially less than the nearly 3 Mha yr^{-1} observed in the Brazilian Amazon during the mid-1990s (INPE, 2014). In contrast, Indonesia had the largest increase in forest loss rates (0.10 Mha yr^{-1}), with rates reaching 2 Mha yr^{-1} in 2011–2012 (double the forest loss observed during the early 2000s). The top five countries with the highest amounts of primary forest loss – Brazil, Papua New Guinea, Gabon, Indonesia, and

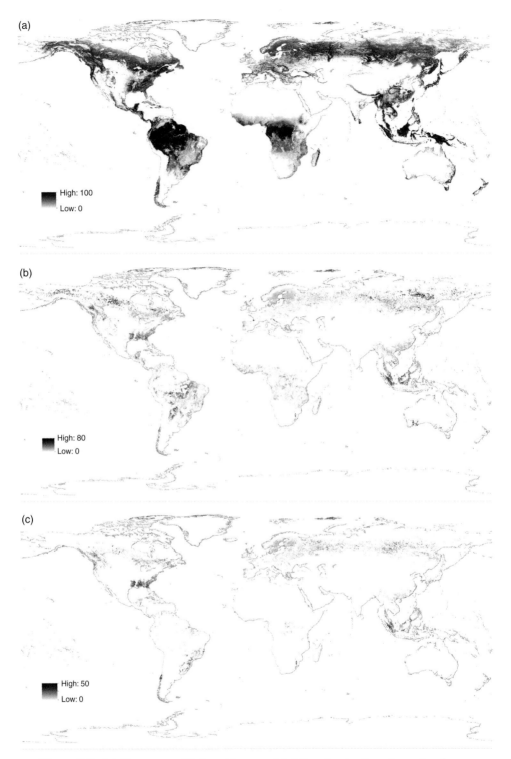

Figure 1.5. (a) Tree cover, (b) forest loss, and (c) forest gain. See also color figures at the end of the book.
Source: Redrawn using data sets from Hansen et al., 2013, for the period 2000–2012.

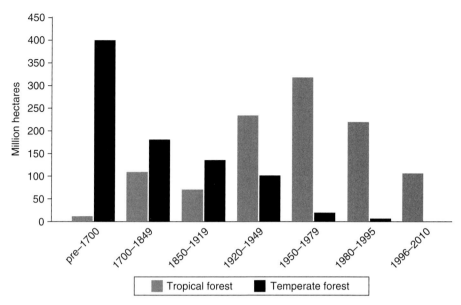

Figure 1.6. Estimated deforestation by forest type and period.
Source: FAO, 2012, and available at: http://www.fao.org/docrep/016/i3010e/i3010e00.htm.

Peru – lost a total of 3.6 Mha of primary forests annually over the past decade (Rudel et al., 2009b; Gibbs et al., 2010).

1.2.2 Historical Spatial and Temporal Trends in Deforestation

Over the last 5,000 years, there has been a cumulative loss of approximately 1.8 billion hectares of forested land (Williams, 2002). It is estimated that 8,000 years ago, forests covered 6.2 billion hectares of the planet – about 47% of Earth's land surface (Billington et al., 1996). Forest area decreased by roughly 15% from 5,000–6,200 Mha in 1700 to 4,300–5,300 Mha in 1990 (Lambin et al., 2003). Williams (1990) estimated that globally between 12% and 28% of the cleared land in 1978 had been cleared before 1750. Between 1850 and 1980, the area of forests and woodlands decreased by 11% (603 Mha) worldwide (Houghton, 1996). Approximately 84% (508 Mha) of this loss was in the tropics, 15% (91 Mha) in the temperate zone, and <1% (4 Mha) in the boreal zone (Houghton, 1996).

Until the early twentieth century, the highest rates of deforestation occurred in temperate forests in Asia, Europe, and North America, resulting primarily from the expansion of agricultural production (FAO, 2012; Figure 1.6). This pattern changed during the twentieth century with deforestation essentially coming to a halt in the world's temperate forests (Figure 1.6). Between 1850 and 1950, rates of net temperate forest clearance were fairly constant at about 1 Mha yr^{-1}, but they have subsequently declined to about zero (Houghton, 1996). In contrast, deforestation increased rapidly in the world's tropical forests during this same period. In the tropics, clearance rates ranged between 1 and 3 Mha yr^{-1} between 1850 and 1930, but have since accelerated rapidly, reaching a peak in the early to mid-1990s possibly as high as 12.9 Mha yr^{-1} (Malhi and Grace, 2000).

1.2.3 Historical Patterns of Global Deforestation

We examine and review historical factors driving widespread deforestation across different continents to shed light on similar factors that have driven and may continue to drive deforestation. In reviewing these histories of deforestation, we also observe that prior drivers of deforestation do not necessarily continue to drive deforestation today. However, these lessons can be extrapolated to provide a better understanding of future drivers of deforestation because they show that while deforestation can be controlled by similar historic factors, there are often shifting drivers controlling forest cover and deforestation rates.

1.2.3.1 Latin America

European settlers began to harvest exotic timber in South America selectively around 1600, and cattle ranching began after Spanish and Portuguese colonization, yet significant changes in forested area did not occur until the mid- to late twentieth century (Lambin and Geist, 2003). Of the forests standing in 1850, about one-third had been cleared by the mid-1980s (Houghton et al., 1991). Large-scale frontier colonization, occurring in the arc of deforestation on the southern fringe of the Amazon Basin, started in the 1960s (Lambin and Geist, 2003). Deforestation in the Brazilian Amazon accelerated after 1970, and Skole et al. (1994) suggested that 90% of the deforested area in 1988 was created after 1970. Since the late 1970s, state subsidies and massive infrastructure development have led to large-scale forest conversion and colonization for cattle ranching in the Brazilian Amazon (Lambin et al., 2003; Houghton et al., 1991; Fearnside, 2007, Rudel, 2005). In the 1990s, expanding world markets, improved access to local credit, and government incentives such as tax exemptions, funding of agricultural research, and improved marketing channels and infrastructure rapidly encouraged the expansion of export crops, thereby driving higher rates of deforestation (Valdes, 2006; Brown et al., 2004; Barbier, 2004; Madi, 2004). Deforestation of the Brazilian moist Atlantic forest started much earlier than in the Amazon. This forest includes coastal evergreen and interior semideciduous tree communities (Oliveira-Filho and Fontes, 2000; Brannstrom, 2002) stretching from the coast to about 3,000 km inland. Its deforestation started early in the history of Portuguese colonization (16th century) and continued until the mid 20th century with the greatest losses of forest cover occurring in the state of Sao Paulo during the early 1900s (e.g., Viana and Tabanez, 1996).

Prior to European arrival in Central America, human populations were concentrated in the Mayan lowlands of the Yucatan Peninsula, Guatemala, and Honduras. Forest cover in the Mayan lowlands was substantially transformed by the fifteenth century from agricultural cultivation that slowly removed the original forest (Myers and Tucker, 1987). The arrival of Spanish conquistadores in Central America in the 1500s influenced Central America's forest cover because it carried epidemic diseases and led to the early enslavement of native Indian populations (Myers and Tucker, 1987). These factors rapidly reduced native populations by as much as 90% in many parts of the region (Myers and Tucker, 1987). As a result, large areas of previously cultivated land were abandoned and reverted to forest. This trend lasted for about 300 years until populations began to recover (Myers and Tucker, 1987). The independence of Central American countries by the early 1820s eventually led to large-scale plantation cropping, which displaced large areas of forest (Myers and Tucker, 1987). By the late nineteenth century, a series of revised land laws converted communal land to private ownership (Skidmore and Smith, 1984). This led to the expansion of plantations for cultivation of coffee and later bananas as the region's primary export crops, driving the conversion of forests in tropical lowlands. Since 1960, however, the most substantial source of forest clearing resulted from the expansion of cattle ranching with Central America experiencing its highest rates of deforestation between the 1960s and 1980s (Redo et al., 2012).

1.2.3.2 South Asia

In South Asia, the percentage of land occupied by forest was relatively low in 1880 because it was already actively cropped at this time, yet decreased 38% over the next hundred years from 110 Mha to 68 Mha (Figure 1.7; Flint, 1994). Similarly, it is estimated that 45% of China was initially forested (Houghton, 2002); yet by 1800, primary forests were largely nonexistent in China (excluding the Northeast and Southwest, i.e., lower-elevation areas in Tibet) as a result of timber and fuel harvesting (Vermeer, 1998). In mainland Southeast Asian countries (i.e., Burma, Thailand, Cambodia, Laos, and Vietnam), a much lower percentage of the land was cropped in 1880, and thus the potential for agricultural expansion was much greater (Flint, 1994). Tropical Asia is estimated to have lost 26% of its initial forest cover before 1850 and another 33% after 1850 (Houghton, 2002). Between 1880 and 1980, a major driver for deforestation was the conversion of land to agricultural uses. Across South and Southeast Asia, the total cultivated area increased during this period by 86%, leading to a decline in forest area of 137 Mha (Flint, 1994). Rubber was an important permanent crop, but food grains such as rice dominated in mainland Southeast Asia (Flint, 1994). For example, the area under rubber cultivation in Indonesia, Malaysia, and Thailand expanded from 0.3 Mha in 1910 to almost 7 Mha in 1990, mostly at the expense of forest cover (Byerlee and Rueda, 2015). The highest rates of conversion to agriculture occurred in insular Southeast Asia, where the net cultivated area more

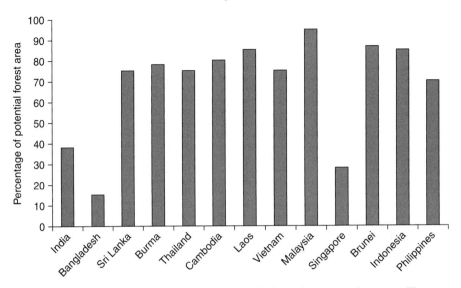

Figure 1.7. Percentage of potential forest area within each country that was still occupied in 1880 by forest. Between 1880 and 1980, total forest cover dropped in South Asia (i.e., India, Bangladesh, and Sri Lanka) by 38% (42 Mha), in mainland Southeast Asia (i.e., Burma, Thailand, Cambodia, Laos, and Vietnam) by 28% (43 Mha), and in insular Southeast Asia (i.e., Malaysia, Singapore, Brunei, Indonesia, and the Philippines) by 25% (50 Mha).
Source: Redrawn from Flint, 1994.

than quadrupled over a century (Flint, 1994). The development of insular Southeast Asia as a supplier of export crops to the world economy was a critical factor accelerating the permanent transformation of forests to agricultural land uses (Flint, 1994). More efficient and powerful logging techniques and increased world demand for tropical Asian timber also pressured forest resources. Particularly after 1950, commercial timbering claimed millions of hectares of forest (Poffenberger, 1989). In addition, the population requiring forest biomass for subsistence uses (i.e., firewood, charcoal, and small timber) continued to increase. Combined, these factors led to tropical Asia's losing half of its forested area by 1980 (Houghton, 2002).

1.2.3.3 United States of America

At the time of European settlement, forest covered approximately half of the present contiguous 48 states (Meyer, 1995). Although forest cover had been altered by Native American land use practices, a continuous decline in forest area began at the onset of European settlement and lasted until the early twentieth century (Meyer, 1995). Clearance resulted from the conversion of forest to farmland and from timber harvesting for fuel, timber, and other wood products. From an estimated 365 Mha in 1850, the forested area of the entire United States reached a low point of 243 Mha around 1920 (Greeley, 1925; Figure 1.8). Farm abandonment in much of the eastern United States

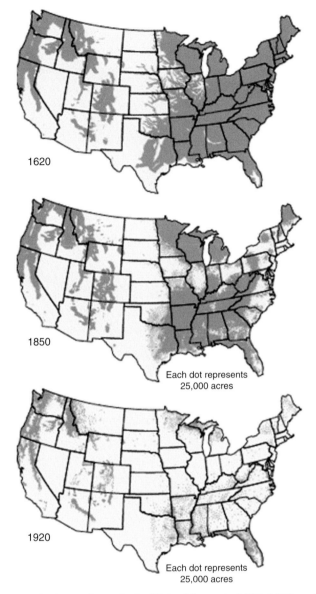

Figure 1.8. Area of primary forest in the United States in 1620, 1850, and 1920. Note that the shaded regions represent primary forests and not total tree cover resulting from regrowth of deforested areas.
Source: Meyer, 1995. Redrawn from Greeley, 1925.

led to forest recovery beginning in the mid- to late nineteenth century (Meyer, 1995). Historically, timber harvesting followed a regular pattern of harvesting one region's timber and then moving on to the next; the once-extensive old-growth forest of the Great Lakes, the South, and the Pacific Northwest represented successive and overlapping harvest frontiers (Meyer, 1995). After about 1930, frontier-type exploitation gave way to the permanent management of stands by timber companies. Forested area increased slowly through the postwar decades, largely through abandonment of cropland and regrowth on cleared areas, but around 1960, another modest decline occurred as a result of settlement expansion and higher rates of mechanized timber extraction (Meyer, 1995). More recently the trend in forest area has been upward; between 2000 and 2010, there was an increase in forested area of 383,000 ha yr^{-1} (FAO, 2010).

1.2.3.4 Europe

Since the establishment of the first agricultural societies in Europe in the mid-Holocene (Price, 2000), humans have substantially altered the European landscape. In general, western and central Europe were minimally deforested at 1000 BC (Berglund, 2006); however, by 300 BC well-established agriculture and the rise of classical civilizations led to population increases and significant deforestation throughout central and western Europe (Kaplan et al., 2009). High forest clearance rates began in Greece (roughly 1000 BC to 300 BC) and several centuries later in Italy (roughly 300 BC to 200 AD) (Kaplan et al., 2009). However, the fall of both the Greek and Roman Empires were marked by periods of afforestation, as populations declined and land was abandoned (Kaplan et al., 2009). During the period from 400 AD to 750 AD, war, plagues, invasions from the East, and climate deterioration resulted in stagnating or declining populations across much of Europe (Wanner et al., 2008). In most regions of Europe, this period was marked by stable forest cover or afforestation (Kaplan et al., 2009). The next gradual decline in forest cover occurred during the 11th–13th centuries and coincided with the development of feudal societies; the emergence of western European nation-states in Spain, France, and England; the establishment of cities across Europe; and climatic amelioration (Wanner et al., 2008; Figure 1.9). Eastern Europe experienced lower levels of deforestation because it was subject to political and cultural unrest during the medieval period (Kaplan et al., 2009). Deforestation ended across Europe when the Black Death devastated populations around 1350 AD (Darby, 1956; Williams, 2000). As populations began to recover 100 years later, the Renaissance and later the ages of discovery and Enlightenment brought about an advancement of ideas, science, trade, and religion throughout all of Europe, which led to higher forest loss rates (Kaplan et al., 2009). Most deforestation in Europe occurred before the Industrial Revolution (Kaplan et al., 2009), but the amount of forest is now increasing since the land is no longer being used for agricultural purposes, which is a trend that began in the mid-twentieth century (FAO, 2011b) and is expected to continue over the next few decades (van Vuuren et al., 2006).

1.2.3.5 Tropical Africa

Approximately 3,500 years ago, human populations settled forested areas in Central Africa and began using the first forms of slash and burn agriculture (Oslisly et al., 2013). Population densities began to increase substantially, with a negative impact

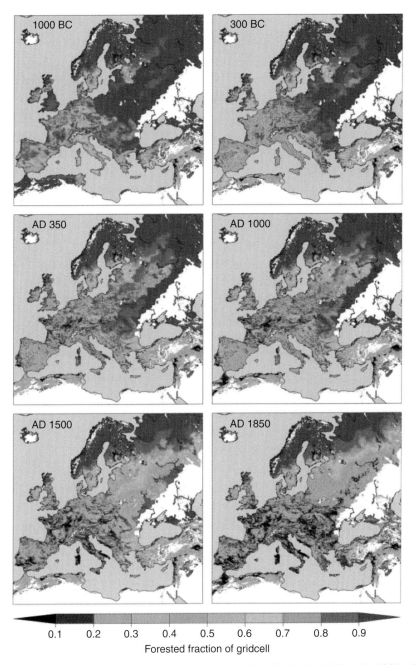

Figure 1.9. Historical forest clearance maps for 1000 BC, 300 BC, 350 AD, 1000 AD, 1500 AD, and 1850 AD. See also color figures at the end of the book.
Source: Kaplan et al., 2009.

on forest cover after the spread of iron tools with the Bantu people beginning around 400 BC (Oslisly et al., 2013). Iron Age settlement peaked around 100 AD and the substantial increase of iron tools was likely to have profoundly affected forest cover as a result of widespread use of slash and burn agriculture (Oslisly et al., 2013). However, around 400–1000 AD there was an extensive population crash, possibly as the result of widespread disease, leaving the Atlantic coast of Central Africa almost devoid of people (Oslisly et al., 2013). As the population crashed, the forest likely expanded and recovered. Around 1400 AD, the region suffered another population crash possibly resulting from the Atlantic slave trade and the direct removal of people, which subsequently led to a period of forest regrowth (Malhi et al., 2013). From the late sixteenth century onward, European settlers began to establish colonial plantations in coastal West Africa (Lambin and Geist, 2003). Merchants transported timber and other products on rivers to the coast then shipped these goods to Europe, reaching the Congo Basin as late as the 1890s (Lambin and Geist, 2003). Beginning in the late 1930s, colonial road construction began to provide access to remote forests. Infrastructure development together with the post–World War II economic boom triggered rapid agricultural expansion, leading to the conversion of old-growth forests to cocoa, coffee, oil-palm, rubber, and banana plantations (Lambin and Geist, 2003).

After political independence in the mid-twentieth century, two divergent pathways of deforestation resulted in this area. For countries in the Congo Basin (i.e., Cameroon, Central African Republic, Democratic Republic of the Congo, Gabon), state attention shifted away from inland agriculture after the discovery of oil as well as poor maintenance of transportation infrastructure (Lambin and Geist, 2003). Congo Basin countries, which account for almost 90% of Africa's rain forest, have large extractive oil and mineral industries (Oslisly et al., 2013). The extraction of oil and minerals triggers economic booms that can induce economic side effects such as high labor costs; less competitive agricultural exports; more food imports, thereby slowing agricultural expansion, accelerating urbanization, and leading to reduced rates of deforestation in remote, rural regions (Mayaux et al., 2013; Rudel, 2013). In contrast, industrial-scale agriculture grew in the Guinean zone of West Africa (i.e., Gambia, Ghana, Guinea, Liberia, Nigeria, and Sierra Leone), triggering the world's highest rate of contemporary tropical deforestation, affecting mainly the coastal forests (Lambin and Geist, 2003). Timber extraction became an important component of national economies in this portion of the region and by the end of the twentieth century the coastal rain forests of West Africa had largely been deforested (Lambin and Geist, 2003).

1.2.3.6 Australia

Australia has undergone substantial land use changes since human settlement, including transformation of the land cover by Aborigines going back as much as

75,000 years ago (Flannery, 2002; Rasmussen et al., 2011). It is estimated that ~30% of Australia's land area was covered by forest at the time of first European colonization in the late 18th century (Barson et al., 2000). After the first permanent European settlement was established in Sydney Cove in 1788, forest clearing for agriculture followed almost immediately (Bradshaw, 2012). The highest clearance rates occurred in areas with soils best suited for agriculture (Braithwaite, 1996) and generally in coastal areas. In 1861, the newly formed government of Australia passed the Crown Lands Alienation Act, which was designed to open up the colony to settlement (Bradshaw, 2012). Over the following century, that act effectively guaranteed unrestricted settlement and, in turn, the rapid clearing of vegetation because it penalized entitled landholders for failing to develop their lands (Braithwaite, 1996). Thus, most land clearing occurred in southeastern Australia from the turn of the 19th century to the mid-20th century (Bradshaw, 2012). In New South Wales most deforestation occurred between 1892 and 1921, resulting from the rapid proliferation of wheat and sheep industries (Norton, 1996). Afterward, emphasis shifted to southwestern Western Australia, which experienced its most rapid deforestation as a result of the expansion of wheat production in 1920–1980 (Deo, 2011). In more recent decades, deforestation has been concentrated in Queensland and northern New South Wales (Bradshaw, 2012). For instance, most (>80%) of the 1.2 Mha cleared in Australia between 1991 and 1995 was in Queensland (Barson et al., 2000; Wilson et al., 2002). Nonetheless, since the mid-1940s, forest clearing has been less extensive than during the 19th century (Braithwaite, 1996). Australia's native forests now cover 147.4 Mha, or 19% of the total land area, which represents a total loss of ~38% since European settlement (Australian Bureau of Rural Sciences, 2010).

1.2.4 Reforestation/Afforestation

Although global cropland area has increased globally by 12% since 1961, certain regions have witnessed the opposite trend. For instance, over the last 50 years, cropland areas in North America, Europe, and China have stabilized and even decreased (Ramankutty and Foley, 1999). This decrease in cropland has led to increased rates of afforestation in certain areas. For instance, around the mid-twentieth century, China made the planting of trees a national objective with a goal of restoring 30% of the country to forest (Houghton, 2002). Subsequently, over the period 2000–2005, China experienced an increase in forest cover (FAO, 2009). Forest cover in North America and Europe has stabilized and cropland abandonment in the eastern United States and Europe has allowed for the regrowth of forests (Figure 1.8). Between 1990 and 2005, there was an increase of 12 Mha of forested area in Europe (FAO, 2009). A very low net loss of ~2 Mha of forest is projected in the United States between 1997 and 2020, resulting from the conversion of forest land to other uses such as urban and suburban development, as well as afforestation and natural reversion of abandoned crop and

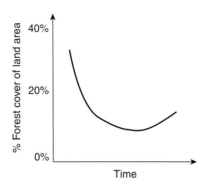

Figure 1.10. The Forest transition curve.
Source: Redrawn from Rudel et al., 2005.

pasture to forest (U.S. Forest Service, 2008). Similarly, deforestation rates are so low in Canada that it is estimated it would take 40 years for Canada to lose 1% of its forested area (Canadian Council of Forest Ministers, 2006). Globally, gains in forest area through afforestation and natural forest expansion were 6.8 Mha yr^{-1} between 1990 and 2000 and 7.3 Mha yr^{-1} between 2000 and 2005 (FAO, 2010). In fact, a shift from a net decrease in forest cover to an increase in forest cover has been documented in many countries globally (Rudel et al., 2005; see also Chapter 5).

The idea that forested area changes in relatively predictable ways as societies undergo economic development, industrialization, and urbanization is termed a forest transition (Mather, 1992; Walker, 1993; Figure 1.10). Thus, a forest transition occurs when a large decline in forest cover is followed by a national-scale long-term shift to an expansion in forest cover (Rudel, 1998; Figure 1.10). There are two primary theories to explain when forest cover begins to expand: 1) *"The economic development path"* when farm workers leave the land for better paying nonfarm jobs that make agricultural enterprises unprofitable by raising wages of the remaining workers, thereby leading to the abandonment of agricultural land; or 2) *"the forest scarcity path"'* that occurs in places with stable or growing populations and little ability to import forest products. In these areas, continued declines in forest cover spur increases in the prices of forest products, and these price increases induce landowners to invest in trees instead of crops or pasture grasses. Rudel et al. (2005) used data from a recent FAO survey of forest cover (i.e., the FRA-2000) to examine when forest transitions occur and global drivers of these transitions. Nations that gained forest cover during the 1990s had an average GNP per capita of $8,453 in 1990 compared with $1,614 among nations that lost forest cover (Rudel et al., 2005). They found that labor scarcities induced an increase in forest cover in European countries (i.e., Greece, Ireland, and Portugal) that recently shifted to a net gaining state while the scarcity of forest products explained the shift in forest cover in South and East Asian countries (i.e., Bangladesh, India, and China) during the period 1990–2000. Notably, earlier forest transitions that triggered forest recoveries do

Introduction

not always lead to sustained increases in forest cover when poverty, war, and expanding markets drive the decline in forest cover (Rudel et al., 2005).

1.3 Historical, Current, and Future Drivers of Deforestation

In this section, we broadly review large-scale historical and current drivers of deforestation as well as discussing potential future drivers of deforestation.

1.3.1 Agricultural Production

Clearing of forests and woodlands has historically been driven by the demand for crops and agricultural production. Ramankutty and Foley (1999) estimated a net loss of 1,140 Mha of forests/woodlands (with approximately half of this clearing occurring since 1850) while Matthews (1983) and Williams (1990) estimated a net loss of 910 Mha and 740–800 Mha, respectively, since preagricultural times. In the period 1850–1990, approximately 27% (603 Mha) of the increase in the area of land under cultivation globally was attributable to forest clearance (Houghton, 1996). These losses have resulted from the substantial conversion to pasture and cropland from grassland and forest. Pongratz et al. (2008) reported that 4,070 Mha of natural vegetation was put under agricultural use between 1700 and 2000 compared to 490 Mha between 800 and 1700 AD. The area of cropland has increased globally from an estimated 300–400 Mha in 1700 to 1,500–1,800 Mha in 1990, a 4.5- to 5-fold increase in three centuries and a 50% net increase in the twentieth century alone (Ramankutty and Foley, 1999; Goldewijk, 2001). Yet much of this expansion occurred during the beginning of the mid-twentieth century as the area of global cropland only grew by 12% between 1961 and 2009 although agricultural production expanded by 150% (FAO, 2009). Goldewijk et al. (2011) estimated that the total global area of cropland doubled almost every century after the sixteenth from 300 Mha in 1700 to 420 Mha in 1800, 850 Mha in 1900 and 1,530 Mha in 2000. The area under pasture increased from 320 Mha in 1700 to 510 Mha in 1800, then accelerated to 1,290 Mha in 1900, finally reaching 3,410 Mha in 2000 (Goldewijk et al., 2011).

Prior to 1700, Europe along with South and Southeast Asia already had extensive crop cover (Ramankutty and Foley, 1999). Disturbed forests were most frequently converted in forest-poor areas such as South Asia, where most of the intact forests had been cleared prior to 1700 (Ramankutty and Foley, 1999). After this time, cropland continued to expand rapidly in Europe, followed by North America and the former Soviet Union (Ramankutty and Foley, 1999). This expansion occurred at the expense of forests/woodlands, particularly in the eastern United States and Europe, where forest cover declined by 45% and 24%, respectively, between 1700 and 1900 (Figure 1.11). Rapid intensification of cropland in Southeast Asia between 1930 and 1992 resulted primarily from forest/woodland clearing, leading to an estimated 21%

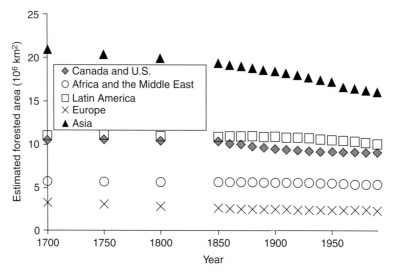

Figure 1.11. Estimated forested area, 1700–1990, for a given geographic region using data from Ramankutty and Foley (1999). Asia includes the former Soviet Union, and Latin America includes Mexico as well as Central and South America. Ramankutty and Foley (1999) estimated historical changes in cropland area using cropland data from 1992 along with a land use change model calibrated against national and subnational data (i.e., state, province) to reconstruct global 5 minute resolution data of croplands from 1700 to 1990. Historical cropland data were overlaid onto a potential vegetation data set to examine the extent to which vegetation types such as forest have been altered since 1700. These spatial and temporal trends in land use change were consistent with the history of human settlement and patterns of economic development.
Source: Data from Ramankutty and Foley, 1999.

loss of forested area (Ramankutty and Foley, 1999; Figure 1.11). Agricultural expansion in forest-rich regions of Latin America, Central Africa, and Southeast Asia relied predominantly on clearing intact forests for new agricultural land (Gibbs et al., 2010). Goldewijk et al. (2011) estimated that over the period 1700–1990 there was a 10-fold expansion in the area of pasture, with substantial conversion to pasture taking place in East Asia (442 Mha) followed by Oceania (401 Mha) and South America (398 Mha). While pasture expansion often results from conversion of grasslands, savannas, and steppes (e.g., Pongratz et al., 2008), it can also result from loss of forested area. For instance, cattle production is estimated to have caused nearly 80% of Amazonian deforestation (Fearnside, 2005; Nepstad et al., 2009).

1.3.2 Logging

Historically, deforestation has partially been driven by logging and the demand for wood and paper products. Most logging over the past century has been in the temperate and

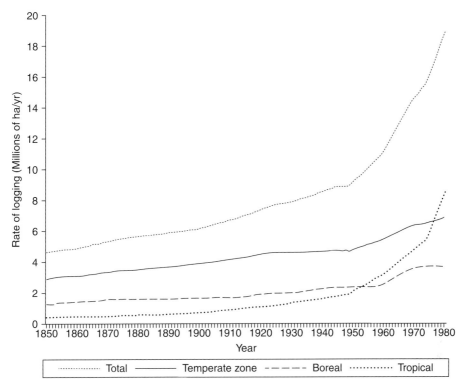

Figure 1.12. Annual rates of logging in 1850–1980 in tropical, temperate, and boreal zone forests.
Source: Houghton, 1996.

boreal zones (Figure 1.12). Rates of logging of boreal forests have steadily increased from about 1 Mha year^{-1} in 1850 to 3.5 Mha year^{-1} in 1980, and in temperate forests from 3 Mha year^{-1} in 1850 to 6 Mha year^{-1} in 1980 (Houghton 1996; Figure 1.12). In contrast, logging of tropical forests was below 0.5 Mha year^{-1} in 1850, less than 2 Mha year^{-1} in 1950, but accelerated to 8 Mha year^{-1} in 1980, overtaking temperate forest logging in the mid-1970s (Houghton, 1996). Overall, 1,069 Mha of forests were logged between 1850 and 1990, an area that was 77% larger than the area of forest converted to agriculture (Houghton, 1996). Once logged, however, forests are often left to recover either spontaneously (secondary forest) or by planting trees (Malhi et al., 1999). For instance, forest plantations are approximately 5% of the total global forest area, and the spread of planted forests has been accelerating, from 3.7 Mha annually in the 1990s to 4.9 Mha annually the following decade (FAO, 2001).

1.3.3 Shifting Cultivation

Shifting cultivation is one type of deforestation that consists of clearing forests by slash and burn, using the land for cropping, and leaving the land fallow so the forest regenerates. This sequence of land use change is then repeated over multiple cycles. Historically, shifting cultivation cleared (including reclearing) more than ten times the amount of land that wood harvesting cleared during the 1700s and 1800s, with this amount declining to about twice as much land as wood harvesting by 2000 (Hurtt et al., 2006). Hurtt et al. (2006) used existing historical reconstructions of land use to estimate that 600–1,000 Mha of tropical land (forest and nonforest) were currently in shifting cultivation. Global forest resource assessments (FRA-2000) by the FAO (FAO, 1996; 2001) reported about 200 Mha in short (i.e., a mosaic of young secondary forest, various stages of natural regrowth, and cultivated areas with cultivated areas covering between 30% and 50% of the total area) and long (i.e., a mosaic of mature forest, secondary forest, various stages of natural regrowth, and cultivated areas with cultivated areas covering between 5% and 30% of the total area) fallow forests and another 200 Mha in fragmented forest (i.e., patches of forest separated by nonforested lands). It is estimated that approximately 0.45 billion people are currently engaged in shifting agriculture at a mean per capita clearing rate of 0.17 ha yr^{-1} (Rojstaczer et al., 2001). Slightly more than half of all shifting cultivation occurs in tropical forest (largely tropical secondary forest) while slightly less than half of all shifting cultivation occurs in grassland (Rojstaczer et al., 2001).

1.3.4 Biofuels

In recent years, both the United States and the European Union have adopted bioenergy policies that mandate a certain degree of reliance on biofuels (Hermele, 2014). As a result, the human pressure on agricultural land has increased, sometimes leading to the emergence of competition for natural resources between food production and the bioenergy sector (e.g., Hermele, 2014). As discussed in Chapter 5, the production of biofuel crops has been associated with negative impacts on the environment, including land use change and deforestation (Fargione et al., 2008; Fitzherbert et al., 2008; Lima et al., 2011).

1.4 Projected Geographic Changes in Deforestation

Although it is difficult to project where deforestation will increase because of the many temporally variable environmental and socioeconomic factors affecting deforestation rates (see Chapter 5), in this subsection we focus on potential future drivers of deforestation. Specifically, we consider how population growth, urbanization, and the demand for bioenergy are expected to result in an increase in agricultural production, which could influence future patterns of deforestation.

Population. Food production will need to expand as a result of increasing population (Figure 1.13) and increasing per capita demand for agricultural products. World

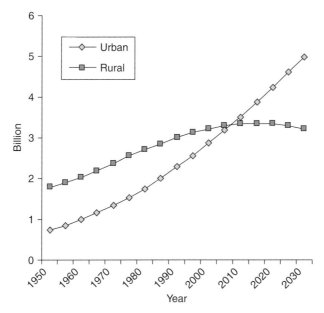

Figure 1.13. Rural and urban world population (1950–2030).
Source: Pretty, 2008.

population is likely to continue increasing until approximately 2040–2050 and then stabilize or fall as a result of changes in fertility patterns (Pretty, 2008; Lee, 2011). Per capita agricultural production has outpaced population growth (Hazell and Wood, 2008): For each person today, there is an additional 25% more food than in 1960, which has notably saved millions from being malnourished (Pretty, 2008) but also reflects an increasing overconsumption or reliance on resource intensive diets in more affluent societies, as discussed in the following sections. Future projections suggest that by 2050, the world's average daily calorie demand could rise to 3,070 kcal per person, an 11% increase over its level in 2005/2007 (i.e., 2,770 kcal per person) (Alexandratos and Bruinsma, 2012; Davis et al., 2014). Combined, the projected 50% increase in global population and 11% increase in food consumption per person are projected to increase global food demand 60% by 2050. In turn, this is projected to result in the conversion of roughly 70 Mha of intact and disturbed forested land by 2050 (Alexandratos and Bruinsma, 2012).

Agricultural production. Worldwide demand for agricultural products is projected to increase 60% from 2005/2007 to 2050 (Alexandratos and Bruinsma, 2012), and evidence suggests that similarly to what occurred in 1980–2000, land in tropical countries will be needed to meet much of this increased demand (Gibbs et al., 2010; Figure 1.14). Although overall demand for agricultural products is expected to grow at only 1.1% per year from 2005/2007 to 2050, down from 2.2% per year in the past four decades (Alexandratos and Bruinsma, 2012), if agricultural expansion were to

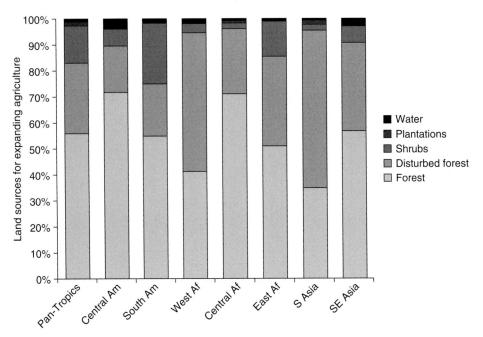

Figure 1.14. The origin of new agricultural land in the tropics between 1980 and 2000. *Source:* Gibbs et al., 2010. Copyright (2008) National Academy of Sciences, U.S.A.

double again as it did during the period 1961–1996, there would be an 18% increase in cropland (Tilman, 1999). This 18% increase would require the loss of 268 Mha of nonagricultural ecosystems worldwide comparable in size to cultivating all of the currently forested land of the United States (Tilman, 1999; Figure 1.14).

Projections suggest that 80% of future production increases need to result from intensification (i.e., increased agricultural production on the same area of land; FAO, 2002). Agricultural intensification is crucial to meeting the increasing human demand for agricultural products while protecting forests and other ecosystems (Foley et al., 2011; Godfray et al., 2010). Yet, it does not appear that the same efficiency gains in agriculture (e.g., via fertilizer and irrigation) that occurred during the Green Revolution can be achieved (Ray et al., 2013). Annual growth in crop yields is expected to be half the rate that occurred during the Green Revolution: 0.8% per annum from 2005/2007 to 2050, versus 1.7% per annum from 1961 to 2007 (Alexandratos and Bruinsma, 2012). For instance, Ray et al. (2013) assessed global average rates of yield increase for the top four crops (which account for nearly two-thirds of global agricultural calories) using 2.5 million agricultural statistics across ~13,500 political units. They found global average rates of yield increase to be 1.6%, 1.0%, 0.9%, and 1.3% per year for maize, rice, wheat, and soybean, respectively. In contrast, a ~2.4% annual rate of yield gain is needed to double crop production by 2050. Other studies have looked at the global distribution of yield gaps (i.e., the difference between potential

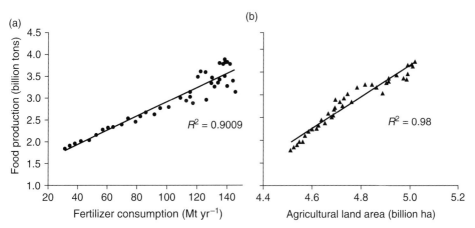

Figure 1.15. Relationship between world plant food production (including cereals, coarse grains, pulses, roots and tubers, and oil crops) and a) all fertilizers applied and b) agricultural land over the period 1961–2002 using FAO (2005).
Source: Data from Pretty, 2008.

and actual crop yields (Mueller et al., 2012)) and shown that the closing of such gaps would dramatically increase food availability around the world but would not be sufficient to meet the increasing human demand (e.g., projections for 2050) unless substantial changes in consumption habits were to occur (Davis et al., 2014).

Over the last half-century, the doubling of world food production resulted from i) the development of higher-yielding strains of crops, ii) increased use of herbicides for weed control and insecticides and fungicides for pest control, iii) marked increases in the amounts of nitrogen and phosphorus fertilizers applied worldwide annually, iv) an increase in the proportion of arable land that was irrigated, and v) an increase in the total amount of land worldwide cultivated annually (Tilman, 1999). In fact, 70% of the increase in crop production over the last four decades has occurred from intensification through high-yield seed varieties, synthetic fertilizer, other chemical inputs, irrigation, mechanization, multiple cropping, and shorter fallow periods (FAO, 2002; Figure 1.15). Since the early 1960s, the area under irrigation and number of agricultural machines have grown approximately twofold and the consumption of all fertilizers fourfold (nitrogen fertilizers sevenfold; Pretty, 2008). The use of pesticides in agriculture has also increased dramatically and now amounts to ~2.56 billion kg yr^{-1} (Pretty, 2008). Notably, these efficiency gains have been at the expense of many environmental problems such as increased nutrient loading to waterways and ecosystems and drawdown of groundwater and surface water resources (e.g., Tilman, 1999).

Continued crop productivity gains from these means may not be possible. For instance, the easiest and greatest gains from crop breeding programs may have already occurred because once most of the original genetic variance preserved within crop landraces[1] and remaining wild relatives has been exploited, future breeding-based yield gains are likely to be small or difficult to obtain (Tilman, 1999). Moreover, reductions in the projected yield of a crop are expected because the yield is a saturating function of the rate of supply of its limiting resource (e.g., water, nitrogen, phosphorus) (Ray et al., 2012). Thus, adding fertilizer to already-well-fertilized areas, such as productive croplands in developed nations that produce the majority of the world's cereals, will have little impact on yield. However, significant regional gains could be achieved in many developing countries (where greater yield gaps exist) by fertilizing croplands not currently receiving fertilizer. Introducing modern technology (e.g., irrigation, fertilizers, mechanization) to countries where small-scale farming has not been reached by the innovation of the Green Revolution will require investments in agriculture that, to date, smallholders in developing countries have been unable to afford (D'Odorico and Rulli, 2013). Thus, beside physical constraints on the increase in food production achievable at yield gap closure, there are also major economic and institutional limitations.

It is projected that agricultural land will need to expand by 70 Mha, including an expansion of about 120 Mha (12%) in developing countries being offset by a decline of 50 Mha (8%) in developed countries (Alexandratos and Bruinsma, 2012). Future expansion of agricultural land into forested areas may be high because forests and woodlands, largely in the tropics, are the only remaining biomes where large tracts of land are available for additional expansion of agricultural production (DeFries and Rosenzweig, 2010). For instance, it is currently estimated that 80% of new croplands replacing forests are located in the tropics (Gibbs et al., 2010). The lower productivity of land in these areas will require more clearing because a unit of cleared land in the tropics produces less than half the crop yield of temperate regions (West et al., 2010). Interestingly, because of the globalization of food through international trade (D'Odorico et al., 2014), important teleconnections exist between the increasing demand for agricultural products and deforestation. Thus, forest loss in a given country will likely be contributed to by consumer demand in other regions of the world, a phenomenon known as *global displacement of land use* (Meyfroidt et al., 2013).

Augmented demand for certain types of crops will also drive forest loss. For instance, two forms of agriculture drive deforestation in the Brazilian Amazon: cattle ranching and industrial soybean farming. Growth of the number of cattle (e.g., Kaimowitz et al., 2004; Ferraz et al., 2005; Barreto et al., 2006) and increasing extent of soybean farms in the Amazon have both been directly linked to forest loss (Ewers et al., 2008). Soares-Filho et al. (2006) estimate that if expansion of agricultural production in the Amazon continues unabated, its current area of 530 Mha (2003, 85% of the original area) will decrease to 320 Mha (53%) by 2050. In Southeast Asia,

---

[1] A crop landrace is the dynamic population(s) of a cultivated plant that has historical origin, has distinct identity, and lacks formal crop improvement, as well as often being genetically diverse, locally adapted, and associated with traditional farming systems (Camacho Villa et al., 2005).

oil palm agriculture has expanded rapidly over the last several decades (Koh and Wilcove, 2008; Koh et al., 2011). The global extent of oil palm cultivation increased from 3.6 Mha in 1961 to 13.2 Mha in 2006, and the total cultivated area of oil palm accounts for nearly one-tenth of the world's permanent cropland (FAO, 2007; WRI, 2007). Oil palm is set to continue expanding in part because of its low production cost and low price relative to those of major alternative plant and vegetable oils (Carter et al., 2007). Indonesia and Malaysia are the world's top producers of palm oil (≈43 million Mg/yr), accounting for 87% of global production (USDA, 2010). More than half the recent oil palm plantation expansion in Indonesia and Malaysia occurred at the expense of forests (Koh and Wilcove, 2008; Naylor, 2011; Miyake et al., 2012). Future projections of land use change in that region indicate that oil palm plantations are unlikely to expand further into existing cropland because of the decreasing availability of cropland; instead their expansion will result primarily from the conversion of primary forests (Bauen et al., 2010). In contrast to that in Southeast Asia and the neotropics, 60% of deforestation in Africa has been driven by subsistence or smallholder agriculture, and the extraction of primary products such as wood fuel, timber, and charcoal for domestic use (Fisher, 2010). Notably, this trend may change because large-scale agriculture could become a major driver of future deforestation in Africa (FAO, 2009).

Bioenergy. Pressures to expand agriculture will also result from increased production of bioenergy crops (Melillo et al., 2009). Projected world primary energy (i.e., energy that has not been subjected to any conversion or transformation process) demand is expected to be in the range of 600 to 1,000 exajoules/year (EJ/yr) by 2050 compared to about 540 EJ in 2010 (Popp et al., 2014). Of the 540 EJ consumed at the primary energy level in 2010 (IEA, 2012), 80% was provided by fossil fuels, and 10% by bioenergy, mainly from wood combustion (accounting for roughly 80% of the energy derived from renewable sources) (IEA, 2009). Traditional biomass is a major source of energy in developing countries and is used primarily for heating and cooking. Heating accounted for the majority of biomass use, contributing an estimated 6%–7% of total global primary energy demand (Popp et al., 2014). Modern forms of bioenergy being used in 2011 amounted to 23.6 EJ as heat, biofuel, and electricity, while 31.4 EJ of traditional biomass was used for cooking/heating in poor rural areas, mainly in Africa (REN21, 2013). Recently published estimates of the global primary bioenergy potential in 2050 range from 30 to 1,300 EJ/year (Haberl et al., 2010; 2013). However, when accounting for factors such as water limitation, biodiversity protection, and food demand, a more plausible range is approximately 200–500 EJ/year by the year 2050, which is nearly four- to ninefold higher than 2010 levels (Dornburg et al., 2010).

Transport biofuels are currently the fastest growing bioenergy sectors, representing 3%–4% of total road transport fuel and only 5% of total current bioenergy consumption (Balat and Balat, 2009). The term "biofuels" is generally used for liquid fuels in the transportation sector and can be broadly classified as biodiesel and bioethanol (Balat and Balat, 2009). Almost all commercially available biofuels are produced from either starch- or sugar-rich crops (for bioethanol), or oilseeds (for biodiesel). Currently, bioethanol accounts for around 80% of the global production of liquid biofuels (Balat and Balat, 2009). In 2012, global fuel ethanol production reached 86 billion liters, while global biodiesel production amounted to 20 billion liters (Popp et al., 2014). Global production of bioethanol increased from 17.25 billion liters in 2000 (Balat, 2007) and could exceed 125 billion liters by 2020 (REN21, 2008). The United States is the world's largest producer of bioethanol fuel, at 60% (i.e., 51 billion liters) of total global supply, while Brazil is the second largest producer, at 25% (i.e., 21 billion liters) (Lane, 2013). In the United States, the production and use of biofuels, particularly bioethanol made from corn, started in the early 1980s, largely to revitalize the farming sector during a time of oversupply of agricultural products (Jull, 2007). Since this time, the United States has remained the largest bioethanol producer in the world (RFA, 2010). In contrast to the United States, which uses starch from corn in bioethanol production, Brazil utilizes sugarcane (Linde et al., 2008).

The proportion of global cropland used for biofuels is currently some 2.5% (40 Mha); it is projected to rise to 65 Mha by 2030, and 105 Mha by 2050 (IEA, 2011). Two types of land use change can result from bioenergy crop production: direct land use change, which is the change in land use on a site used for bioenergy crop production, and indirect land use change, which comprises the unintended effects that result from displacing existing crops, often in countries outside the original country (Berndes et al., 2010; IEA, 2009). For instance, Lapola et al. (2010) projected that sugarcane ethanol and soybean biodiesel production in Brazil would push the rangeland frontier into the Amazon forest, accounting for 40% and 60%, respectively, of indirect deforestation by 2020. This is the case because most of Brazil's soybean plantations (i.e., 90%) and sugarcane expansion since 2010 occurred on land previously used as rangeland (Nassar et al., 2008; Camargo, 2008; Lapola et al., 2010). In turn, this causes deforestation at the forest frontier because it pushes cattle ranches and others (i.e., the area previously occupied as rangeland) into forested area (Hermele, 2014). As explained in Chapter 5, this effect of displacement of land use associated with environmental policies – such as those mandating an increasing reliance on renewable energy – is known as *leakage* (Meyfroidt et al., 2011).

Urbanizantion. Tropical deforestation has been positively related to urban population growth (DeFries et al., 2010). Urbanization raises consumption levels and increases demand for agricultural products. Urban consumers generally eat more processed foods and animal products than rural consumers, thereby inducing increased commercial production of crops and livestock (DeFries et al., 2010). For instance, 62% of global crop production (on a mass basis) is allocated to human food while 35% is allocated to animal feed (Foley et al., 2011). Seto et al. (2012) assessed how urban land cover will change through 2030 using global land cover data, projections of

Introduction 33

urban population, and gross domestic product (GDP) growth in a probabilistic model of urban land change. In total, they project 120 Mha of high probability (>75%) new urban land globally by 2030. Moreover, they estimate the immediate aboveground biomass carbon losses associated with land clearing from these new urban areas in the pantropics (i.e., tropical regions across all continents) to be 1.38 Pg (i.e., petagram) C between 2000 and 2030. Nearly half of the global increase in high-probability urban expansion is forecasted to occur in Asia, with China and India absorbing 55% of the regional total. The rate of increase in urban land cover is predicted to be highest in Africa, at 590% over 2000 levels.

1.5 Remote Sensing Methods Used to Quantify and Map Deforestation

Being able to quantify the amount of biomass present as well as to monitor deforestation, reforestation, and forest regrowth on a global basis is important not only in terms of improving our understanding of where deforestation is occurring and how forests respond after disturbance, but also for understanding the effectiveness of policies and guiding policy makers' decisions. For instance, REDD (i.e., Reducing Emissions from Deforestation and Forest Degradation) is one program under the United Nations Framework Convention on Climate Change that requires documenting the extent of changes in forest area and the carbon stock associated with those changes resulting from the implementation of REDD activities. REDD is a global scale program that seeks to provide financial incentives to developing countries for reducing deforestation by creating a value for the amount of carbon stored in forests. Thus, being able to document the amount of deforestation that has occurred requires global scale estimates of forest carbon stocks and relatively frequent measurements of these stocks. In this section, we review several of the more common types of methods used to estimate forest loss at the global scale.

There are two broad types of sensors used to map forest biomass and quantify changes in the areal extent of forests: passive and active sensors. Passive sensors depend on an external energy source such as the Sun, while active sensors send out a known quantity of energy and measure the amount of that energy reflected back to the sensor (Figure 1.16). The amount of energy received or reflected back to the sensor can then be related to characteristics of the land surface and forest canopy. In turn, aboveground biomass (AGB) can be mapped from satellite observations by calibrating satellite measurements to in situ (i.e., field based) estimates of AGB. Methods based on radar or laser are examples of active remote sensing, while methods relying on remote sensing in the reflected solar portion of the spectrum (often known as "optical remote sensing") are passive.

Alternatively, remote sensing methods are classified depending on the wavelength at which they operate, particularly on whether they use the solar portion of the spectrum (400–2500 nm) (sometimes termed the "optical window") or the thermal

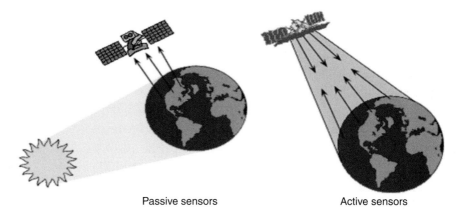

Figure 1.16. Difference between passive and active sensors: Passive sensors depend on an external source of energy (typically the Sun) whereas active sensors have their own source of energy, send out a signal, and measure the amount of energy reflected back to the sensor.
Source: Ashraf et al., 2011.

infrared (IR) (~4–6 μm) (sometimes called the mid-IR) and microwave, ~ 10 μm, which are the wavelengths of the Earth's long wave radiation. Some of the remote sensing products used in land cover change research rely on both reflected solar and thermal infrared and microwave spectral bands.

1.5.1 Optical Remote Sensing

Optical remote sensing uses passive sensing in the reflected solar portion of the spectrum (400–2500 nm), including the visible, near-infrared (NIR, 700–1,400 nm), shortwave infrared I (SWIR I, 1,400–1,800), and shortwave infrared II (SWIR II, 1,800–2,500). Measurements of (passive) reflectance from the Earth are used to map features of the global land surface. Optical measurements have been widely used to estimate and map deforestation and AGB from satellite observations. Estimates of AGB are based on the sensitivity of the optical reflectance to variations in canopy structure. Optical satellite data classify forest change as occurring when a pixel or polygon changes from one land cover class to another. Optical methods work well when large areas of forest are cleared, but can perform poorly when small-scale deforestation or forest degradation is present because of the binary nature of detection (Mitchard et al., 2013). Using optical data to detect degradation is difficult because the loss of canopy cover caused by degradation tends to be small and short-lived (Lu et al., 2005). Cloud cover can also be problematic when working with optical remotely sensed data sets in humid tropical forest environments (Hoekman, 1997; Asner, 2001; Hansen et al., 2008, 2009). For instance, although there is regular acquisition of annual cloud-free imagery over the "arc of deforestation" in Brazil (Fuller, 2006), Indonesia does not have a seasonally cloud-free window, thus requiring more data intensive methods to overcome persistent cloud cover (Broich et al., 2011).

These systems are operational at the global scale, and some satellite systems (e.g., MODIS, Landsat, and AVHRR) have provided globally consistent records for the last three decades (Gibbs et al., 2007). These data are also routinely collected (i.e., MODIS, Moderate Resolution Imaging Spectroradiometer, takes daily global measurements) and often freely available at the global scale (Gibbs et al., 2007). The MODIS product has a spatial resolution varying from 250 to 1000 m and uses seven bands in the reflected solar spectrum (visible, NIR, and the two SWIRs) as well as several other bands in the thermal IR and some microwave. Likewise, the Landsat products have both thermal IR bands and bands in the reflected solar spectrum. Landsat data are used to detect forest cover and its spatiotemporal changes at relatively high spatial resolution (30 m). These data were used to quantify and map forest loss (230 Mha) and gain (80 Mha) in the years 2000–2012 (Hansen et al., 2013). Miettinen et al. (2011) used MODIS data to analyze deforestation rates in insular Southeast Asia between 2000 and 2010. Results showed an overall 1.0% yearly decline in forest cover in insular Southeast Asia primarily occurring in plantations and secondary vegetation. Regardless of the coarse resolution of optical sensors such as MODIS, they do generate consistent and frequent measurements over large areas, making them ideal for identifying locations of rapid change deforestation (GOFC-GOLD, 2011). For instance, the Brazilian Institute for Space Research (INPE) has an operational near-real-time warning system (DETER) to map large deforestation events (>25 ha) in the Amazon using MODIS data (Shimabukuro et al., 2007).

1.5.2 Radar

Since the 1960s, synthetic aperture radar (SAR) has been used to produce images of the Earth's surface based on the principles of radio detection and ranging (RADAR, often used as a synonym for SAR) and has been widely used to map AGB (Kasischke et al., 1997; Tatem et al., 2008). SAR systems are active, meaning they transmit microwave energy and measure the amount of that energy reflected back to the sensor (Goetz et al., 2009). SAR sensors can operate day or night in addition to penetrating through haze, smoke, and clouds. The transmitted microwave energy penetrates into forest canopies, with the amount of backscattered energy largely dependent on the size and orientation of canopy structural elements, such as leaves, branches, and stems (Goetz et al., 2009). The radar signal returned from the ground and tops of trees is then used to estimate tree height, which can be converted to forest carbon stock estimates using allometric relationships between simple plot-level measurements of canopy height and/or depth and AGB, where the latter is determined from trees that have been dissected, oven-dried, and weighed.

The sensitivity of SAR sensors to different AGB components is a function of the wavelength of the sensor, with shorter wavelengths (X and C band) being sensitive to smaller canopy elements (e.g., leaves and small branches) and longer wavelengths (L and P band) being sensitive to large branches and stems (Goetz et al., 2009). A decrease in the sensitivity of backscattered energy to biomass is referred to as the backscatter saturation effect (Minh et al., 2014). Extensive analyses with existing SAR sensors, mostly L-band, suggest the sensitivity of radar backscatter "saturates" around 100–150 tons ha^{-1} (Kasischke et al., 1997). At the P band, the decrease of sensitivity can occur at biomass values higher than 300 tons ha^{-1} (e.g., those of many dense tropical forests) (Minh et al., 2014). Thus, longer wavelength data (i.e., L band) have the highest accuracies, with greater ability to detect deforestation and reasonably assess the quantity of carbon lost in lower biomass areas (Mitchard et al., 2013). Shorter wavelength data (i.e., C band) can also reasonably detect the presence of forest loss, especially in areas with a strong dry season, but they cannot be relied on to detect the quantity of carbon lost (Mitchard et al., 2013). Another limiting factor is associated with ground effects that lead to mountainous or hilly conditions' increasing errors (Minh et al., 2014). As a consequence, terrain topography or ground moisture status can lead to variations in the observed signal that are not due to forest biomass (Minh et al., 2014). More recent satellites such as the ALOS-PALSAR, which was launched in 2006, have increased the potential to use radar to measure biomass because this is the first long-wavelength (L-band) SAR satellite sensor with the capability of collecting cross-polarized (HV, horizontal-send, vertical receive) data in addition to horizontal-send, horizontal-receive (HH) data (Mitchard et al., 2011). This is advantageous for detecting biomass because HV will only detect scattering elements that change the polarization of the incoming electromagnetic radiation, so complex three-dimensional structures such as trees will produce a strong response, but soil moisture, which does not change the polarization of the incoming radiation, will not be detected (Mitchard et al., 2011).

There are a number of radar satellites currently operating at different wavelengths being used to monitor and assess deforestation (e.g., RADARSAT 1/2 (C band), TerraSAR-X (X band), Tandem-X (X band InSAR), MAPSAR (L Band), and Cosmo/SkyMed (X Band)) (Goetz et al., 2009). Data from these satellites have become increasingly available as a result of recent initiatives, such as the JAXA Kyoto and Carbon Initiative, which provide preprocessed radar data free of cost. Many studies have examined the appropriateness of using these data for mapping biomass and deforestation at both global and regional scales. For instance, studies have found that ALOS-PALSAR data have provided improved estimates of carbon stocks across the tropics for degraded or young forests but have been less useful for mature, higher biomass forests (Rosenqvist et al., 2003; Shimada et al., 2005). Shimada et al. (2014) used these data, which had an average of three observations globally (70 km × 70 km for each location) over the 5-year period of observations, 2006–2011, to produce global maps of forest/nonforest cover. Future satellite missions that are scheduled will improve upon errors existing with current radar satellites in operation. For instance, BIOMASS is a mission slated to launch in 2020 that will measure aboveground forest biomass from 70° N to 56° S at a spatial scale of 100–200 m, with error not exceeding ±20% or ±10 t ha^{-1} and forest height with error of ±4 m (Le Toan et al., 2011).

1.5.3 Laser (i.e., LiDAR)

LiDAR is based on the concept of actively sensing vegetation using a pulse of energy, in this case, from a laser operating at optical wavelengths (Dubayah and Drake, 2000; Patenaude et al., 2004). As with all optical systems, LiDAR does not penetrate clouds but has the unique capability of measuring the three-dimensional vertical structure of vegetation in great detail (Dubayah and Drake, 2000). The height and vertical structure of forests are directly estimated by measuring the signal return time and converting these measurements to estimates of forest carbon stocks by applying tree height–carbon relationships (Hese et al., 2005). However, these relationships can be problematic in tropical forests because trees reach their maximum height relatively quickly but continue to accumulate carbon for many decades. Nonetheless, LiDAR has been useful for characterizing fine scale spatial variability in terrestrial carbon stocks across large spatial extents and establishing baseline deforestation data (Goetz and Dubayah, 2011). For instance, Hudak et al. (2012) used data from repeat LiDAR surveys in 2003 and 2009 across ~20,000 ha of an actively managed, mixed conifer forest landscape in northern Idaho to quantify changes in forest cover. Repeat LiDAR surveys accurately quantified high-resolution, spatially explicit biomass and carbon dynamics in these conifer forests. However, this method is limited because aircraft data are not ideal for routinely and repeatedly acquiring systematic observations over very large (i.e., continental to global) spatial extents. Additionally, there are few LiDAR instruments currently operating from satellite platforms, and none currently in operation designed specifically for vegetation characterization. A currently operating satellite LiDAR sensor originally designed for monitoring ice dynamics, the Geoscience Laser Altimetry System (GLAS) onboard ICESAT, has been used for vegetation analysis despite having limited spatial coverage and a relatively large ground footprint (Saatchi et al., 2011). Saatchi et al. (2011) used these data (in combination with other remote sensing databases and 4,079 in situ inventory plots) to map the total carbon stock in live above- and belowground biomass in tropical regions across three continents. They estimated the total biomass carbon stock of forests in the study region to be 247 Gt C (Latin America accounted for 49% of the total, sub-Saharan Africa 25%, and Southeast Asia 26%), with 78% stored aboveground and 22% stored belowground in roots. Future satellite missions using LiDAR could greatly improve our ability to measure global forest carbon stocks by better understanding the spatial arrangement of tree canopies. For instance, NASA's GEDI (scheduled to be completed in 2018) will carry a trio of specialized lasers, creating a total swath width of about four miles that will sample all of the land between latitudes 50° N and 50° S, covering nearly all tropical and temperate forests. These measurements will provide enough detail to measure tree height with an accuracy of ~1 meter.

1.6 Concluding Comments

In this chapter, we have reviewed the global distribution of forests, including environmental factors that control this range. In addition, we have examined both historical and current rates of deforestation as well as large-scale drivers of deforestation and historical factors that have influenced patterns and rates of deforestation. We also discussed potential future drivers of deforestation and areas that may be susceptible to deforestation. Finally, we discussed remote sensing methods to describe the spatial and temporal changes in forest cover. In future chapters, we build on this understanding of forests and deforestation to address the socioeconomic and environmental (i.e., hydroclimatic, biogeochemical, and ecological) impacts of deforestation.

2
Hydrological and Climatic Impacts

2.1 Introduction

Forest ecosystems strongly affect the water cycle through their impact on evapotranspiration, precipitation, infiltration, runoff, and, consequently, soil erosion and streamflow (see following sections). The removal of forest vegetation leads to an increase in water yields (e.g., Bosch and Hewlett, 1982; Section 2.4) and a shift in the predominant mechanism of runoff generation (Dunne and Black, 1970a; Dunne, 1978; Section 2.3). It also enhances snowpack accumulation and shortens the snowmelt season (Section 2.4). Moreover, in deforested watersheds, evapotranspiration is strongly reduced (Section 2.7). The impact on precipitation is more complex: Large-scale (i.e., $>10^5$ km^2) deforestation is expected to reduce regional precipitation (e.g., Bonan, 2008a), though the effect of forest removal also depends on synoptic patterns of atmospheric circulation and geographic setting (e.g., latitude, location with respect to mountain ranges and oceans). The deforestation of small watersheds (<10 km^2) is not expected to have a substantial impact on precipitation, whereas the clearing of intermediate sized areas (15,000–50,000 km^2 [Lawrence and Vandecar, 2015]) could increase local precipitation (see also Chapter 4). Deforestation can also affect the hydrologic conditions and microclimate of nearby (downwind) ecosystems (Ray et al., 2006).

Landmasses receive water as precipitation and lose it either as water vapor fluxes into the atmosphere (evapotranspiration) or as surface and subsurface flows in the liquid phase (runoff). In recent years, these two fluxes have been named *green water* and *blue water* flows, respectively, to stress the fact that evapotranspiration receives a strong contribution from vegetation (transpiration) (Falkenmark and Rockstrom, 2004; Figure 2.1). As explained in the following sections, the overall effect of deforestation on the hydrologic cycle is a decrease in water vapor fluxes and increase in runoff. Thus, green water flows decrease and blue water flows increase. This means that more water is likely to become available for societal withdrawals (but also for environmental uses) in areas located downstream from the cleared watershed. In turn, forest management

40 Global Deforestation

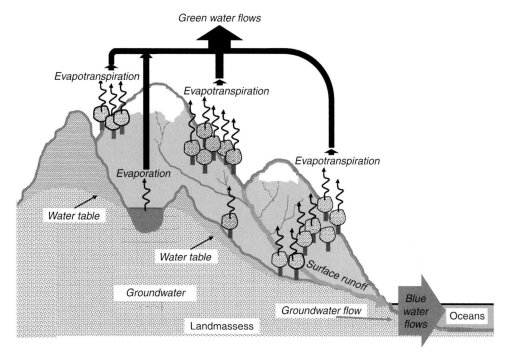

Figure 2.1. Water fluxes out of landmasses. Water vapor fluxes (or *green water fluxes*) are mostly mediated by vegetation (transpiration). Fluxes in the liquid phase (*blue water fluxes*) are due to surface runoff and (to a smaller extent) groundwater flow. See also color figures at the end of the book.

can strongly impact the water resources of a watershed, and sometimes the thinning of woody vegetation has been proposed as an option to increase water availability in semiarid areas (e.g., Ingebo, 1971; Griffin and McCarl, 1989). Such an approach, however, has only limited applicability because forest removal and the consequent increase in overland flow have the effect of increasing sediment yields and soil erosion rates, thereby damaging the landscape, often irreversibly within human timescales.

At regional to global scales, the decrease in green water flows resulting from deforestation are expected to slow the water cycle. In fact, a portion of regional precipitation is contributed by regional evapotranspiration, a process known as *precipitation recycling* that will be discussed in greater detail in Sections 2.8 and 4.2 (e.g., Eltahir and Bras, 1996). This impact of forest cover on the global water cycle was highlighted by Fraedrich et al. (1999), who used a coupled climate-biosphere model to simulate two extreme scenarios of a *green planet* (with all the nonglaciated land covered by trees) and a *desert planet* (with vegetation removed from the whole planet). By comparing these two scenarios, they found that in the presence of forest vegetation, precipitation and evapotranspiration over land increase by about 90% and 250%, respectively, while over the ocean the differences between the two scenarios were almost negligible. Thus, the global biosphere is responsible for sustaining a

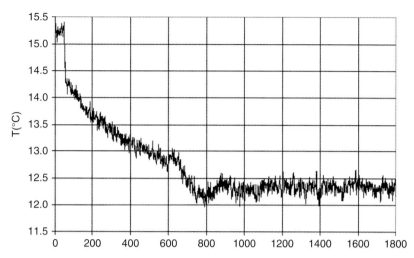

Figure 2.2. Expected changes in temperature induced by global deforestation. Forest vegetation was replaced with grassland 50 years from the beginning of this time series. The long-term effect of global deforestation was a decrease in average global temperature by 3 °C.
Source: Renssen et al., 2003.

much more intense terrestrial water cycle than in a desert planet. In the green planet scenario, runoff is expected to be 24% smaller than in the desert planet. Despite the clear exaggeration inherent to the choice of these scenarios, Fraedrich et al. (1999) highlighted the important role played by the terrestrial biosphere in the global water cycle. Already there is evidence that deforestation has caused a decrease in global water vapor flows from the land to the atmosphere. Gordon et al. (2005) compared water vapor flows from potential vegetation with flows from the current extent of vegetation using a GIS model of the land surface on Earth and found that deforestation has decreased global vapor flows from land by 4% (3,000 km^3/yr). The decreases in evapotranspiration and precipitation associated with current or projected scenarios of forest conversion to cropland or rangeland are expected to have a weaker but still important impact on the water cycle that will be discussed in the following sections.

Likewise, Renssen et al. (2003) used model simulations to investigate the effect of global deforestation. They run a coupled vegetation-climate model with preindustrial radiative forcings (AD 1750) until the model reached equilibrium. All forests on Earth were replaced with grassland and the model was run for another 1,750 years. They found a generalized cooling effect over the whole world (Figure 2.2), particularly in the temperate and boreal regions. Cooling results from the higher albedo of unvegetated land surfaces. Studies on specific regions (e.g., the Amazon), however, suggest that air temperature should increase in response to regional deforestation (e.g., Bonan, 2008a), consistent with the expected decrease in evapotranspiration. In other words, because of the decrease in the latent heat fluxes (and hence,

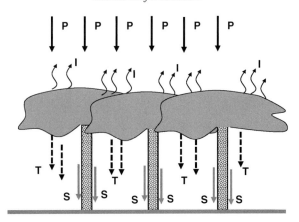

Figure 2.3. Partitioning of precipitation (P) into canopy interception (I), throughfall (T), and stem flow (S).

evapotranspiration), more energy is available for surface heating. Thus, in areas with relatively high rates of evapotranspiration (Section 2.7), the effect of deforestation on temperature is dominated by the impact of reduced evapotranspiration (Section 2.10).

2.2 Precipitation and Forest Canopies

As precipitation reaches the forest canopy (either as rainfall or as snowfall), part of it is retained and temporarily stored by the canopy, while the rest moves through the foliage to the forest floor as throughfall or stem flow (Figure 2.3). Precipitation trapped by the canopy eventually evaporates (or sublimates in the case of snow) or drips down to the ground, contributing to additional throughfall. The fraction of precipitation that evaporates or sublimates from the canopy is known as *canopy interception* (Figure 2.3). The partitioning of precipitation into throughfall and interception depends on canopy structure and rainfall intensity, while stem flow is typically only a small fraction (<5%) of precipitation (e.g., Lee, 1980).

2.2.1 Canopy and Litter Interception

Interception typically ranges between 10% and 40% of precipitation, with the higher fractions corresponding to rainfall events of smaller magnitude and droplet size (Lee, 1980). Vegetation properties determining interception and canopy storage capacity include the amount of foliage (expressed in terms of leaf area index, LAI, i.e., the total foliage area per unit ground area), type of leaves (interception is greater with needle than with broad leaves), and whether the forest is deciduous or evergreen (Table 2.1). Other factors such as leaf angle, roughness, and wettability of the leaf surface also play a role in determining canopy water retention, storage capacity, and interception (Rosado and Holder, 2012).

Table 2.1. *Median values of canopy interception expressed as a percentage of total precipitation at the seasonal or annual timescales*

Forest	Median canopy interception (percentage of total precipitation)
Deciduous forest	13
Coniferous forest	
-Rain and snow interception	28
-Rainfall interception only	22
European forests	35
North American forests	27

Source: Values from Dunne and Leopold, 1978.

Once throughfall and stem flow water reaches the forest floor, part of it evaporates after having been retained by the litter layer, while the rest percolates to the underlying soil and eventually either infiltrates or generates runoff. Known as *litter interception*, the additional evaporative losses from forest litter strongly depend on its thickness, rainstorm frequency, and potential evapotranspiration. Overall, litter interception is believed to be a small fraction (<5%) of annual precipitation (Lee, 1980).

Deforestation has an obvious effect on interception and throughfall: After the removal of the forest canopy, interception losses by residual vegetation or by replacement of the forest by crops and pastures are strongly reduced. Thus, *ceteris paribus*, more water is available for infiltration and runoff in deforested watersheds than in their forested counterparts.

2.2.2 Canopy Condensation and Occult Precipitation

In areas with moist air, low clouds, or fog, forest canopies may favor either the condensation of moisture (dew formation) or the deposition of fog and cloud droplets onto leaf surfaces. Thus, the canopy traps atmospheric moisture by scavenging water out of the air moving through the foliage. This moisture eventually drips down to the forest floor, thereby increasing (unlike interception) the supply of water to forested watersheds. Often known as *horizontal* or *occult precipitation* (e.g., Chang, 2002), this phenomenon may provide an important input of water to montane cloud forests or other ecosystems. Occult precipitation is a major contributor to the water budget of coastal regions, particularly those close to cold ocean surfaces that are typically affected by fog, low clouds, or high relative humidity (Table 2.2). Notable examples include the redwood forests along the coast of California; these ecosystems exhibit a "Mediterranean climate" (i.e., winter rainfall) and may partly rely on fog deposition as a moisture source during the growing season (Dawson, 1998). Occult precipitation and cloud immersion are some of the main defining attributes of tropical montane cloud forests. These ecosystems are

Table 2.2. *Occult precipitation as a function of mean annual precipitation (MAP)*

Location	Elevation (m)	Mean annual rain (mm yr^{-1})	Occult precipitation (mm yr^{-1})	Occult precipitation (percentage)
Colombia/Venezuela	815–3100	450–1125	72–796	3.5–48.3
Costa Rica	–	72–435	0–42	0–37
Costa Rica	1500	3191	886	21.7
Guatemala	2100	2559	23	<1
Guatemala	2550	2559	203	7.4
Hawaii	981–3397	300–2449	134–832	2.6–61.2
Mexico	1330–2425	215–1082	0–339	0–50.7
Panama	500–1270	1495–6763	138–2299	2.3–60.6
Puerto Rico	930–1015	3204–4001	0–435	0–26.2
Venezuela	1750–2150	828–1009	354–592	26–41.7
Australia	1000	1350	727	35
California	47–191	1315	447	25

Source: Based on case studies reported in Dawson, 1998; Bruijnzeel, 2002; and Holder, 2003.

important water sources because they intercept moisture from orographic clouds. The removal of these forests results in an overall decline in water inputs and stream flow, which could limit forest regeneration. This positive deforestation feedback (Wilson and Agnew, 1992) is discussed in more detail in Chapter 4.

It is still unclear how climate warming might affect canopy condensation and fog or cloud deposition (e.g., Foster, 2001). Condensation requires air cooling to the dew point and the presence of surfaces upon which water molecules can condensate, in this case, leaves. Canopy condensation is enhanced if climate change increases the chances for air temperature to drop below dew point. These conditions correspond to an increase in the relative humidity, RH, to values close to or slightly above 100% (i.e., when the dew point temperature is equal to or greater than the air temperature). RH is the ratio between actual and saturation vapor pressure; an increase in temperature increases evapotranspiration and, consequently, the concentration of atmospheric water vapor and actual vapor pressure. At the same time, however, the saturation vapor pressure increases too (Clausius-Clapeyron equation). Therefore, it is unclear how relative humidity is responding to climate warming (e.g., Katul et al., 2012). The situation, however, is more complicated as it also depends on the effect of climate change on cloudiness and cloud base height. Recent studies in the Andes-Amazon transition zone in southwestern Amazonia have shown an overall decrease in cloudiness in lowland areas adjacent to tropical montane cloud forests. This change in cloudiness was likely an effect of increasing sea surface temperatures in the tropical North Atlantic (Halladay et al., 2012).

Cloud forests are particularly vulnerable to changes in microclimate because they are typically adapted to a habitat with moist air, low clouds, and occult precipitation, which depends on a narrow range of climate conditions (Pounds et al., 1999). The ability of these species to adapt to climate change (e.g., through uphill migration) depends on the warming rate as well as on topographic and geographic constraints (Foster, 2001).

Deforestation and land use change may strongly affect the microclimate and modify the narrow range of environmental conditions that are suitable for montane cloud forests. The removal of forest vegetation has the effect of increasing diurnal temperatures and lowering the dew point temperature, thereby decreasing the relative humidity and the chance for the occurrence of condensation. Moreover, deforestation in lowland and premontane areas can increase the cloud base height in downwind highland regions (Lawton et al., 2001; Nair et al., 2003; Fairman et al., 2011). Model simulations in the Monteverde region of Costa Rica have shown how deforestation leads to an increase in cloud base height: Orographic clouds form at higher elevations, and in some scenarios they would not intersect the mountaintop (Ray, et al., 2006). According to these models, the conversion from a pristine situation to the current forest cover would have led to a 5%–13% decrease in the montane forest area covered with fog, while further deforestation in lowland and premontane areas could lead to an additional 15% decrease. In this region, deforestation has the effect of increasing the cloud base by 50–300 m, which would result in the shrinking or disappearance of highland forest areas that depend on cloud immersion, a major defining characteristic of cloud forests (Ray et al., 2006).

2.3 Infiltration and Runoff Generation

Throughfall water reaching the soil surface partly infiltrates and partly contributes to overland flow. Infiltration, the flux of water into the ground, depends both on soil properties and on the rate at which water reaches the soil surface as throughfall or snow melt. The soil infiltration capacity is the maximum infiltration rate allowed for by the soil properties. It depends on soil texture, moisture content, soil profile, and vegetation cover. Vegetation shelters the soil surface, thereby preventing or limiting soil compaction (and erosion) by rain splash, while plant roots provide preferential pathways for water flow into the ground. Macropore formation resulting from the decay of dead roots and bioturbation by other soil organisms further enhances the soil infiltration capacity in forested watersheds.

Runoff generation can be due either to limited soil infiltration capacity, when the rainfall intensity exceeds the soil infiltration capacity (*infiltration-excess runoff* or *Hortonian overland flow*), or to limited soil storage capacity (*saturation-excess runoff*), when the soil reaches saturation and rain falls onto saturated areas (Dunne, 1978; Figure 2.4). In terms of the preceding discussion, forest soils have a relatively

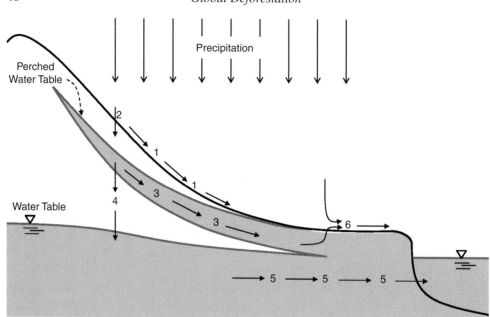

Figure 2.4. Main mechanisms contributing to runoff generation: 1, Hortonian overland flow in areas with low infiltration capacity; 2, infiltration; 3, subsurface storm flow in the presence of higher permeability shallow soil; 4, groundwater recharge; 5, groundwater return flow; 6, saturation-excess runoff in saturated areas.

high soil infiltration capacity and therefore are unlikely to produce infiltration excess runoff. In these landscapes, the main mechanism of runoff generation is by saturation excess. Thus, runoff is produced in saturated areas. Soils are more likely to reach saturation in the presence of a shallow impervious layer (a clay lens or bedrock) or a shallow water table. It has been observed that in vegetated hillslopes, water infiltrates and percolates through the soil column until it reaches soil layers with low hydraulic conductivity, thereby establishing a subsurface flow parallel to the hillslope surface, known as *subsurface stormflow* (Figure 2.4). Sites with concave topography located at the valley bottom in proximity to the stream network are more likely to reach saturation because of the convergence of subsurface storm flow (Freeze, 1972). These saturated areas are sources for saturation-excess runoff, and their spatial extent varies in time both at the seasonal and at the event timescales (Dunne and Black, 1970b).

Deforestation has the effect of exposing the ground surface to rain splash compaction and overland flow erosion. The consequent loss of forest litter further exposes the soil surface to compaction with the effect of reducing the soil infiltration capacity. The watershed becomes prone to infiltration-excess runoff, particularly during intense storms, and water erosion.

2.4 Effects of Deforestation on the Hydrologic Response

The removal of forest cover affects the hydrologic response at different timescales. At the event scale, deforestation results in faster and more peaked hydrographs. At the seasonal or annual scales there is an overall increase in runoff and streamflow volumes or *water yields* (e.g., Lee, 1980; Bosch and Hewlett, 1982). The duration of these effects depends on the rates of forest regeneration, which are shorter in the warm and wet tropics and in watersheds with deep-rooted species (Chang, 2002). There is also experimental evidence that reforestation and afforestation have the opposite effects of deforestation (Figure 2.5).

2.4.1 Effects on Flood Dynamics at the Event Timescale

Deforestation results in a fundamental change in the hydrologic functioning of a watershed, leading to a shift from a system affected by saturation-excess runoff in relatively small source areas, to a landscape dominated by infiltration-excess runoff over all areas with low infiltration capacity. This transition entails a change from relatively slow pathways of rainwater delivery to the stream – mainly by subsurface storm flow – to faster transport by overland flow. Moreover, forest removal entails the loss of a variety of water storage sources (canopy, snowpack, forest floor, ground surface, pools in topographic and microtopographic depressions, and the soil column) that temporarily retain rainfall water and delay its flow (Hewlett, 1969). In other words, the concentration time of the watershed (i.e., the travel time from the most remote point in the watershed to the outlet) decreases. Classic hydrologic theories based on the so-called rational method (Mulvaney, 1851; Turazza, 1880) as well as experimental observations have shown how shorter concentration times are associated with higher peak flows and flash floods (e.g., Rodriguez-Iturbe and Rinaldo, 1997). Thus, peak flows are expected to increase after deforestation (Swank and Johnson, 1994; Jones, 2000), particularly if the forest litter layer is removed, the soil is affected by compaction, or logging requires the construction of forest roads resulting in compacted soils that turn into faster pathways of water flow during high intensity storms. Some exceptions, however, have also been reported (Chang, 2002).

2.4.2 Effects on Water Yields

Over seasonal timescales, the removal of forest vegetation increases water yields (Figure 2.5) because evaporative losses are greatly reduced (Bosch and Hewlett, 1982; Andreassian, 2004). As trees are replaced by crops or pastures, canopy interception and transpiration (see Section 2.2) decrease, while surface and subsurface runoff increase. Decades of research in forest hydrology have shown that the effect

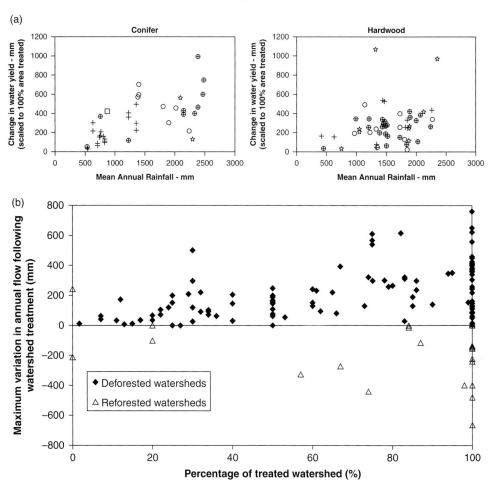

Figure 2.5. (a) Change in water yield as a function of mean annual rainfall where circles are from studies reviewed in Bosch and Hewlett (1982), plus signs from Stednick (1996), stars from Sahin and Hall (1996), and squares from additional catchments reviewed in Brown et al. (2005).
Source: Brown et al., 2005.
(b) Maximum variation in annual flow following watershed treatment as a function of percentage of basin subjected to treatment (i.e., deforested or reforested).
Source: After Andréassian, 2004.

of forest management on water yields depends on a number of factors, including treatment intensity, type of vegetation, season, mean annual precipitation, and topography. The increase in water yields increases with the reduction in forest cover, reaching a maximum in the case of clearcut. It increases with mean annual precipitation (Figure 2.5) and is hardly noticeable in semiarid climates (e.g., annual precipitation <500 mm/yr). The water yield enhancement is typically stronger in the growing season, when evaporative losses due to transpiration and interception are greater

> **Box 2.1 Plant Classification Based on Natural Habitat and Ecological Adaptation**
>
> The following terms are often used to classify plants on the basis of their natural habitat and hydroclimatic adaptation.
>
> **Phreatophytes**: plants with roots that reach the water table and rely on groundwater as a dependable source of water. While *obligate phreatophytes* are only found in areas where plants have access to groundwater by means of *tap roots* extending below the water table, *facultative phreatophytes* can also survive in upland areas without reaching the water table (e.g., Naiman et al., 2010).
>
> **Vadophytes**: plants that take up water only from the vadose zone (i.e., shallow unsaturated soil above the water table) (D'Odorico et al., 2010b). Mesophytes and xerophytes (see the following) are typically vadophytic plants.
>
> **Hydrophytes**: plants that grow in flooded areas with waterlogged soils or standing water. Hydrophytic trees are typically found in swamps and can be partly submerged. They are often adapted to flooded environments thanks to pneumatophores or other structural and morphological adaptations that favor root oxygenation (e.g., Mitsch and Gosselink, 2000).
>
> **Halophytes**: plants that grow in salt water (e.g., mangroves).
>
> **Xerophytes**: drought resistant (or *xeric*) plants adapted to arid environments.
>
> **Mesophytes**: plants adapted to mesic environments. They are found in relatively moist climates but not in flooded conditions.
>
> **Heliophytes**: plants adapted to full sunlight conditions (e.g., Agarwal, 2008).
>
> **Sciophytes**: plants that prefer shaded environments with reduced light intensity or partial sunlight (e.g., Agarwal, 2008; see Section 2.9).

(particularly with deciduous vegetation). However, in regions with "Mediterranean" climates (i.e., winter rainfall), the effect of deforestation on water yields is usually observed during the rainy season (Harr and McCorison, 1979). Moreover, this effect is stronger in evergreen conifers than in deciduous hardwood forests because of the higher leaf area indices (throughout the year) and rates of interception and transpiration (Bosch and Hewlett, 1982; Brown et al., 2005). The effect of thinning on water yields depends also on the topography and the location (within the watershed) of the areas from which trees are removed. Thinning or clearing of upland areas typically leads to smaller changes in water yields than the removal of forest vegetation from lower slopes, valley bottoms, or riparian areas. In particular, riparian trees are often phreatophytes (see Box 2.1) that sustain high transpiration rates. Thus, the removal of riparian vegetation can greatly improve water yields, increase base flow levels, and reduce the number of low flow days in forested watersheds (Chang, 2002).

Research on the impact of deforestation on water yields typically relies on measurements either in the same watershed before and after the removal (or thinning) of forest vegetation, or simultaneously in two watersheds with similar characteristics, one of which is subjected to logging, while the other (or *control*) retains the forest cover. The former approach uses exactly the same watershed and provides a more direct assessment of changes in the hydrologic response after deforestation. The main limitation, however, is that the years of observation before and after the treatment may exhibit different rainfall regimes (e.g., different mean annual rainfall, extreme events, or timing of the rainy season). While *paired watershed* experiments prevent this problem, they require access to two watersheds with similar size, geologic setting, and topography.

Unfortunately, most of these forest clearing experiments have been carried out on relatively small watersheds (mostly less than ≈10,000 ha) and in temperate regions (e.g., Bosch and Hewlett, 1982). To date, there are few studies on major river basins either because there is a lack of streamflow records prior to deforestation or because the interpretation of the results is confounded by climate fluctuations occurring concurrently with changes in land cover. As noted in Chapter 1, most deforestation is happening in the tropics; thus, the lack of experimental studies in tropical watersheds is also a major limitation to the study of the hydrologic implications of forest removal from larger tropical river basins.

2.4.3 Effect on Snowmelt

Snowmelt is an important contributor to stream flow in watersheds having their headwaters in alpine, arctic, or other snow prone regions. Forest vegetation affects the accumulation and melting of snow: Snow interception is higher in evergreen coniferous forests than deciduous hardwood forests, while the highest values of snow accumulation are observed in open areas with no canopy cover. Because of the different microclimate (lower solar irradiance, higher relative humidity, and lower diurnal temperatures), snowmelt is slower and the snowpack lasts longer in forested patches than in open canopy areas (Chang, 2002). Thus, the effect of forest cover is to reduce snow accumulation and water yields (due to rainfall and snow interception) and to slow and delay the spring snowmelt (e.g., Folliott and Thrud, 1977). Therefore, forest management approaches that reduce forest cover in snow prone watersheds are expected to enhance snow accumulation and water yields (particularly in the snowmelt season) while reducing the snow cover duration, thereby contributing to higher snowmelt peak flows in late spring (Troendle, 1983; Chang, 2002).

2.4.4 Effect of Fire

Forest fire has important impacts on infiltration, runoff, water yields, and soil erosion (Krammer and DeBano, 1965). Fires have the effect of removing at least part of the vegetation, thereby reducing water losses associated with interception and transpiration, while exposing the soil surface to rain splash compaction and erosion. Thus, some of the impacts of fires on the hydrologic response are similar to those of logging and deforestation described previously. In addition, the burning of forest biomass may dramatically reduce the soil infiltration capacity. This effect was initially attributed to soil compaction resulting from rain splash and a decrease in surface soil permeability by pore space clogging with ashy particles. Krammer and DeBano (1965) and DeBano (1966) showed that the postfire decrease in infiltration capacity is due to the development of soil water repellency (or *hydrophobicity*) by the fire. During the combustion of plant biomass some organic compounds such as fatty acids are volatilized or undergo pyrolytic reactions (e.g., Savage et al., 1972) and are transported into the soil by the strong gradients in temperature existing through the soil profile. While moving through the soil, these gases undergo cooling and eventually condense either close to the ground surface or at a shallow depth (i.e., a few centimeters below the surface), where they develop a hydrophobic coating of the soil particles (e.g., DeBano, 2000). This effect depends on fire temperature, soil texture, soil moisture, and vegetation type. Water repellency does not develop if the fire is either too "cold" (e.g., soil temperatures $T < 175$ °C) or too hot ($T > 300$–400 °C), likely because the volatilization of hydrophobic compounds is negligible at lower temperatures, while the water repellent coatings are destroyed by hot fires (e.g., DeBano, 2000; Doerr et al., 2000). Moreover, fire-induced water repellency is typically stronger on sandy soils because they have a lower specific grain surface area (i.e., total grain surface area per unit volume). Soil organic matter and soil water content also affect the formation and properties of the hydrophobic layer (Robichaud and Hungerford, 2000): Water repellency is more severe in drier soils because they reach higher temperatures (DeBano et al., 1976; Letey, 2001). The development of water repellency depends also on the type of vegetation. Doerr et al. (2000) developed a list of plants associated with the emergence of hydrophobic soils. Evergreen trees with resins, waxes, or aromatic oils (e.g., eucalyptus and pines) and shrubs from temperate, Mediterranean, and semidesert regions are typical of watersheds in which water repellency has been reported.

The organic compounds released by fire affect the physical-chemical properties of the grain surfaces by increasing the *contact angle* between the air-water interface and the grain surface, which causes a positive capillary pressure head in correspondence to the hydrophobic layer (e.g., Letey, 2001). Thus, a water drop reaching a water repellent surface is not drawn into that surface but sits on top of it. After some time, water infiltrates through the hydrophobic layer, presumably as a result of the decay of the hydrophobic organic coating by microbial biomass. This decay causes an increase in soil infiltration capacity through time (Letey, 2001). Thus, fire-induced water repellency is an overall transient property of soils, and its longevity seems to be an increasing function of fire intensity. It might last a few weeks in the case of light fires or a few years after severe fires. Overall, hydrophobic soils are very heterogeneous and patchy across a burned watershed (DeBano, 2000). Water repellent patches become runoff source areas (Figure 2.6), leading to the formation of overland flow and water erosion by surface wash and rill formation (DeBano, 2000). Thus, fire, which has been used as an effective tool for land clearing and deforestation since the onset of civilization (often known as *slash and burn*), enhances runoff generation, peakflows, and water yields.

2.5 Forest Effects on Groundwater

Trees growing on valley bottoms, on floodplains, and along riverbanks are often "phreatophytes" (Box 2.1): Their roots are deeper than the water table and tap into the shallow groundwater. These trees may affect the average depth and the diurnal fluctuations of the local water table (Naumburg et al., 2005). In fact, the removal of riparian and wetland forests causes an increase in water table elevation, a phenomenon known as *watering up* (Figure 2.7; Peck and Williamson, 1987; Borg et al., 1988; Riekerk, 1989; Dubé and Plamondon, 1995, Roy et al., 2000). Conversely, soil drainage and lower water tables are observed after planting phreatophyte vegetation in areas with shallow groundwater (Chang, 2002). This effect is generally attributed to the increase in evaporative losses (interception and transpiration) in the presence of forest vegetation and the consequent decrease in infiltration and groundwater recharge (e.g., Dubé and Plamondon, 1995), as well as to direct groundwater uptake by "tap roots" (e.g., Le Maitre et al., 1999; Naumburg et al., 2005).

In Chapter 4, we will stress how the water table rise resulting from removal of wetland or riparian forests may lead to the emergence of waterlogging with anaerobic conditions detrimental to the root system of trees that are not adapted to flooded conditions (Roy et al., 2000; Brolsma and Bierkens, 2007). These conditions may inhibit seedling establishment and forest regeneration (Ridolfi et al., 2006, 2008), enhance salt accumulation (Runyan and D'Odorico, 2010), and favor a shift in plant community composition with increased dominance of water tolerant trees or shallow-rooted invasive species (Chambers and Linnerooth, 2001; Wright and Chambers, 2002; Roy et al., 2000).

2.6 Effect of Deforestation on Wetlands

After deforestation, the reduction in evapotranspiration (ET) (as will be discussed in Section 2.7) is the primary driver leading to an increase in water availability, including the water that reaches surface water bodies and wetlands (Zhang et al., 2001). For instance, Sahin and Hall (1996) analyzed changes in water yield for 145 experimental catchments and by using fuzzy linear regression analysis (based on the average water yield changes in the first 5 years after treatment). They found that for a 10% reduction in conifer-type forest cover, annual water yield increased by 20–25 mm; in eucalypt

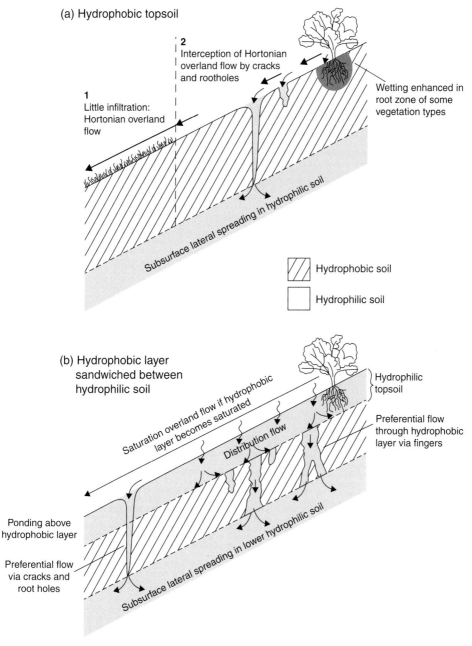

Figure 2.6. Impact of fire on soil hydrophobicity and runoff generation. if the hydrophobic layer is formed at the surface, runoff is generated by infiltration excess (hortonian overland flow), while if it forms at a shallow depth, runoff is due to saturation excess (see Section 2.3).
Source: After Doerr et al., 2000.

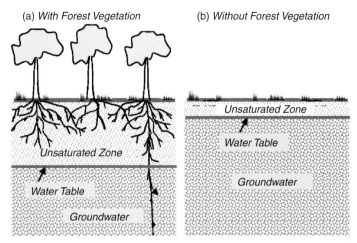

Figure 2.7. Watering up effect of forest removal (either phreatophytes or vadophytes (Box 2.1)) in areas with shallow water tables showing a system with (a) forest vegetation and (b) no forest vegetation.

forest, water yield increased by 6 mm; and in deciduous hardwoods water yield increased by 17–19 mm. The magnitude of these changes is dependent on a variety of environmental factors; water yield changes are greatest in high rainfall areas (Brown et al., 2005). Notably, these increases in water availability can convert ephemeral swamps to permanent lakes or lead to the emergence of new wetlands. Woodward et al. (2014) performed a global metaanalysis on 317 wetland sites where Holocene deforestation and changes in wetlands have been documented. They observed a significant increase in catchment water yield and change in wetland hydrologic conditions in 18% of these sites, which may underestimate this phenomenon because 35% of sites in their database did not analyze proxies that might enable hydrological changes to be detected. Moreover, using a water balance approach they observed that after complete deforestation, water available to wetlands increases by up to 15% of annual precipitation. Thus, the increase in water availability after deforestation can substantially alter surface water availability and surface water characteristics.

2.7 Evaporation and Transpiration

Water losses from the land surface to the atmosphere are due to evapotranspiration and sublimation, which entail the phase transition from liquid water and ice/snow to water vapor, respectively. Evapotranspiration comprises two different types of water vapor fluxes, namely, evaporation and transpiration; the former is a physical process taking place from wet surfaces such as plant canopies (interception), water bodies, or wet soils; the latter is a biophysical process mediated by the plants' physiology. Transpiration entails the vaporization of water and the diffusion of water vapor from the *stomata* – small cavities typically located beneath the leaf surfaces – to the

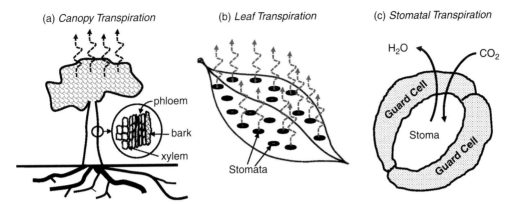

Figure 2.8. Transpiration at the canopy (a), leaf (b), and stomatal (c) scale. Notice that in panel (b) the density and size of the stomata are out of scale. Leaves may have stomata on both sides or (more commonly) only on their lower side.

atmosphere. Most terrestrial plants take up water from the soil through their root system and deliver it to the leaves through tissues whose main function is to transport water and solutes through the stem. Known as *xylem*, these tissues form a network of vessels and conducting cells (Figure 2.8). Water is then vaporized and lost to the atmosphere through the stomata as water vapor fluxes (Figure 2.8). The stomata also control the flux of atmospheric CO_2 into the plant before it is taken up by photosynthesis. In other words, important plant-atmosphere gas exchanges take place through the stomata: Vegetation loses water vapor while taking up atmospheric CO_2. Depending on leaf temperature, air humidity, light, enzyme availability, and water stress intensity, plants control the opening of the stomata by dilating or contracting the cells (*guard cells*) located around their rim (Figure 2.8c). Thus stomata physiology plays a crucial role in the coupling of the water and carbon cycles (Box 2.2).

Both evaporation and transpiration entail a phase transition (from liquid water to water vapor) that requires energy. The amount of energy used per unit mass of evapotranspired water is known as the *latent heat of vaporization* (λ_v, which is a known function of temperature). Thus, the energy flux (*LE*, energy per unit time and unit area) necessary to sustain a certain rate of evapotranspiration (*ET*, volume of water transpired per unit time and unit area) is known as the *latent heat flux*. In other words, evapotranspiration (*ET*) is related to the latent heat flux as $LE = \lambda_v \rho ET$, where ρ is the water density. This energy is provided by the Sun and results from a surface energy balance that determines the way in which the incoming net solar irradiance at the land surface (R_n) is partitioned into the latent heat flux and the fluxes of (sensible) heat associated with near-surface turbulent diffusion (*sensible heat* flux, *H*) and *conduction* through the ground (*G*). During daytime, the surface energy balance (Figure 2.9) can be expressed as $R_n = LE + H + G$. Thus, the rate of evapotranspiration [$ET = (R_n - H - G) / (\lambda_v \rho)$] depends on solar radiation as well as on factors

Box 2.2 Stomatal Regulation

Stomatal regulation reflects the plant's strategy to sustain growth through photosynthesis while preventing plant desiccation and hydraulic failure. If the soil is relatively wet, the plant will try to keep the stomata open as much as needed to sustain high productivity rates. In this case, CO_2 assimilation is limited by biochemical constraints to the photosynthesis process, mainly light and nutrient availability (Collatz et al., 1991). In conditions of limited soil water availability, plants tend to close the stomata to prevent the xylem water potential (a linear function of water pressure in the xylem) from becoming so low that it causes a loss of turgidity and the formation of air bubbles (emboli) in the xylem. Known as *cavitation*, this phenomenon interrupts the continuity of the water column in the xylem and limits the root-to-shoot water transport that is ultimately crucial for plant survival (Maherali et al., 2004). Thus, by closing the stomata during conditions of water limitation, stomatal regulation aims at maximizing carbon gain while preventing hydraulic failure. Stomatal closure reduces the risk of cavitation – the main cause of drought-induced mortality in trees (Urli et al., 2013) – but limits the rate of carbon assimilation and in the long run may lead to carbon starvation. Plants tend to adopt two main strategies of stomatal regulation: Angiosperms (i.e., hardwoods) have a better capacity to reverse embolism and repair the damage of cavitation. Therefore, because of their ability to tolerate higher cavitation risks, angiosperms operate with narrower safety margins and maintain relatively high stomatal conductances and transpiration rates even in water scarce conditions. Conversely, gymnosperms (conifers) tend to have a limited ability to reverse cavitation and therefore use wider hydraulic safety margins by closing their stomata and reducing transpiration as soon as conditions of water limitation emerge (e.g., Carnicer et al., 2013). This classification gives a good idea of the way stomatal dynamics vary across species, but it does not recognize the existence of different behaviors among the angiosperm and gymnosperm species. More generally, there is a spectrum of stomatal regulation strategies with two end members that can broadly be defined as *isohydric* and *anisohydric* behaviors (McDowell et al., 2008). Isohydric plants prevent hydraulic failure by starting stomatal closure (i.e., reducing stomatal conductance) as soon as conditions of soil water limitation ensue. These plants minimize the risk of hydraulic failure but sustain lower photosynthetic rates; during a drought, their carbon balance is often negative and in the long run, these plants are unable to meet their metabolic carbohydrate demand. Thus, isohydric plants are more likely to die from carbon starvation than hydraulic failure. Conversely, anisohydric plants maintain higher conductances and transpiration rates even in conditions of soil water limitation. In turn, they are more prone to hydraulic failure but can sustain higher photosynthetic rates.

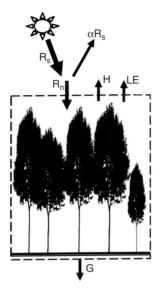

Figure 2.9. Schematic representation of the surface energy balance for a forested area (shaded in gray). The solar radiation reaching the Earth surface (R_s) is partly reflected back to the atmosphere (αR_s, where α is the forest albedo), while the remaining part (or *net radiation*, $R_n = (1 - \alpha R_s)$) is absorbed by the surface. Neglecting changes in heat storage, the net incoming solar energy, R_n, during the day is balanced by the sensible (H) and latent (LE) heat fluxes and conduction through the ground (G).

constraining the evapotranspiration process, namely, water availability (e.g., canopy wetness, soil moisture) and the transport of water vapor.

Water vapor fluxes from the evapotranspiring surface to the atmosphere are due to a diffusion process driven by concentration gradients of water vapor (or *absolute humidity*). These gradients are often expressed in terms of vapor pressure gradients or of vapor pressure deficit, which is the difference in water vapor pressure between the evaporating surface (e_s) – at saturation – and the overlying air (e_a), where e_s is the saturation vapor pressure. Thus, water vapor fluxes can be expressed as

$$ET = D(e_s - e_a) \approx De_s(1 - RH), \quad (2.1)$$

with D a diffusion coefficient (or *diffusivity*) and RH the relative humidity. The water vapor released by evapotranspiration into the near surface is removed by turbulent diffusion. Thus the diffusivity term, D, depends on wind speed and surface roughness. Moreover, in the case of transpiration, diffusivity depends on stomatal regulation, usually expressed in terms of canopy conductance, which depends on stomatal conductance and is upscaled to the whole canopy as a function of the leaf area index (LAI). In other words, the diffusivity D is proportional to the conductance of the whole canopy, which increases with the LAI. To summarize, evapotranspiration depends on (a) the net solar radiation at the land surface, (b) surface roughness, (c) leaf area index, (d) moisture

Table 2.3. *Albedo of natural surfaces*

Surface	Albedo
Fresh snow	0.75–0.95
Old snow	0.40–0.70
Desert	0.20–0.45
Glacier	0.20–0.40
Sandy soil	0.25–0.45
Clayey soil	0.20–0.35
Cropland	0.18–0.25
Grass	0.15–0.26
Deciduous forest	0.10–0.20
Coniferous forest	0.05–0.15
Mixed forest	0.10–0.15
Water	0.03–0.10

Source: Values compiled from Campbell and Norman, 1998; Bonan, 2008a; and Chang, 2002.

availability (i.e., soil moisture and canopy wetness), (e) vapor pressure deficit, (f) stomatal regulation (Box 2.2), and (g) wind speed. The presence/absence of a forest canopy may affect evapotranspiration both directly and indirectly. Directly, depending on canopy brightness or darkness, vegetation determines the surface *albedo*, which is the fraction of solar radiation reflected by the land surface (Table 2.3). Forest canopies are typically darker (i.e., have lower albedo) than pastures or bare soil (e.g., Bonan, 2008a) and therefore absorb more solar radiation (i.e., higher R_n). Thus, more energy is available for evapotranspiration in a forest than in a pasture. Moreover, the presence of a canopy greatly enhances evapotranspiration by increasing surface roughness and leaf area index. Trees often have deeper roots and can take up water from greater soil depths than many crops and grasses; thus, they have better access to soil moisture. There are also some indirect effects associated with the impact of tree canopies on microclimate, which in turn affects water pressure deficit, canopy conductance (e.g., Collatz, et al., 1991), and soil moisture (D'Odorico et al., 2007).

In sum, the removal of forest vegetation leads to an increase in albedo (hence a decrease in net solar radiation) and a decrease in surface roughness, leaf area index, and root depth. Therefore, the overall effect of deforestation is a decrease in both evaporation and transpiration.

2.7.1 The Effect of Climate Change on Evapotranspiration from Landmasses

Climate change trends both from the recent past and from future projections indicate a global scale increase in air temperature (IPCC, 2014). The effect on global scale evapotranspiration can be investigated through equation (2.1), where D is a term for global scale diffusivity. Assuming that both D and RH are not greatly affected by climate warming (see Section 2.2.2 for an explanation), we should expect global evapotranspiration to increase because e_s is an increasing function of temperature (T)

$$e_s = a \exp\left(\frac{bT}{T+c}\right) \quad (2.2)$$

where $a = 0.611$ kPa, $b = 17.5°$C, and $c = 249.93°$C (Campbell and Norman, 1998). Differentiating equation (2.2) with respect to temperature, T, it has been found that there should be a 6.8% increase in global ET per Celsius degree of climate warming (e.g., Katul and Novick, 2009). This result, however, is not consistent with global records of precipitation (P; over long timescales P should be equal to global evapotranspiration, i.e, $P ≈ ET$), which show only a 2–3% increase per Celsius degree of warming (Katul et al., 2012).

The analysis of the terrestrial portion of the water cycle (i.e., over landmasses only) highlights an unresolved conundrum (Katul et al., 2012). Over long timescales we expect average terrestrial evapotranspiration (ET) to be related to average terrestrial precipitation (P) and runoff (RO) as $ET = P - RO$. Interestingly, stream gauge records show a positive trend in the past 50–100 years. Because such a trend cannot be solely explained by the positive trend in P, a negative trend must exist in terrestrial evapotranspiration. Possible explanations include (a) solar "dimming" due to increased cloudiness and aerosols; (b) altered plant responses to increased CO_2 induced by a decrease in transpiration due to more rapid uptake of CO_2, though this hypothesis has not been conclusively supported by experiments in enriched CO_2 conditions; and (c) land use change associated with deforestation, which – as indicated in previous sections of this Chapter – should increase runoff and decrease evapotranspiration (Katul et al., 2012).

2.8 Effect of Forest Vegetation on Precipitation

Vegetation affects the rainfall regime by modifying the exchange of energy and water vapor with the atmosphere (Section 2.7). The major mechanisms determining the effect of deforestation on regional precipitation are based on 1) the impact of the reduction in evapotranspiration and precipitation recycling (see Section 2.1), 2) changes in surface energy balance and their effects on the stability of the atmospheric boundary layer, 3) the impact of vegetation on the atmospheric concentration of organic aerosols and cloud microphysics, and 4) mesoscale circulations resulting from the heterogeneous forest cover and the opening of forest gaps.

Overall, decades of research on this topic have shown that large-scale deforestation is expected to lead to a decrease in regional precipitation (e.g., Henderson-Sellers and Gornitz, 1984; Lean and Warrilow, 1989; Shukla et al., 1990; Hasler et al., 2009).

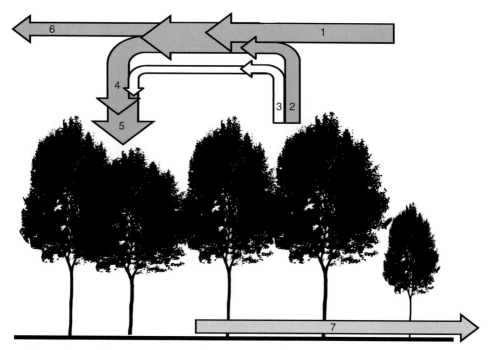

Figure 2.10. Schematic representation of moisture sources contributing to regional precipitation: 1, Advection of atmospheric moisture from external sources (e.g., evaporation from the ocean); 2, fraction of atmospheric moisture from regional evapotranspiration that is exported from the region; 3, fraction of atmospheric moisture from regional evapotranspiration that contributes to regional precipitation, also known as precipitation recycling, P_r; 4, fraction of regional precipitation contributed by external moisture sources P_e; 5, regional precipitation, $P = P_r + P_e$; 6, atmospheric moisture exported from the region by advection; 7, net outgoing surface runoff. Arrows in white represent the phenomenon of precipitation recycling, often quantified in terms of the recycling ratio $r = P_r/P$.

2.8.1 Effect of Deforestation on Precipitation Recycling

By reducing evapotranspiration and the supply of water vapor to the atmosphere, the removal of forest vegetation leads to a reduction of atmospheric humidity and precipitation (e.g., Eltahir and Bras, 1996). In fact, in some regions, a relatively important fraction of precipitation is contributed by water evapotranspired within the same region (Salati et al., 1979). Known as *precipitation recycling* (Figure 2.10), this phenomenon is stronger in continental than coastal regions, and during the growing season, when evapotranspiration is stronger (Eltahir and Bras, 1996). The notion of precipitation recycling allows us to identify the components of the regional water cycle that are more sensitive to changes in vegetation cover (Figure 2.10).

These components, however, are not easy to quantify (Table 2.4) and scientists have often relied on methods based on isotope geochemistry, trajectory analyses, and model simulations. Geochemical methods (e.g., Salati et al., 1979; Kendall

Table 2.4. *Fraction of precipitation contributed by regional evapotranspiration, or recycling ratio (r, see Figure 2.10); as a general rule, the recycling ratio increases with the size of the study region and with the distance from the ocean*

Continent	r
North America	0.27
South America	0.36
Africa	0.45
Europe	0.22
Asia	0.24
Oceania	0.18

Source: The values reported here refer to recycling ratios determined by van der Ent et al., 2010, for entire continents.

and McDonnell, 1998) are based on the different isotopic signatures of water vapor from transpiration or evaporation. This approach allows us to determine the relative importance of terrestrial versus marine sources of moisture and evaluate the fraction of precipitation contributed by regional transpiration. Trajectory analyses rely on the use of atmospheric circulation models and Lagrangian trajectory algorithms for the reconstruction of pathways of water vapor transport in the atmosphere. These methods allow for the identification of the sources of moisture to a given study region (Dirmeyer and Brubaker, 2007; van der Ent et al., 2010). Recently, a more comprehensive approach (Spracklen et al., 2012) has used trajectory analyses in conjunction with remote sensing observations of leaf area index (from MODIS, Moderate Resolution Imaging Spectroradiometer) and precipitation (from TRMM, Tropical Rainfall Measuring Mission, and other satellites) and found that in 60% of the tropical land surface, air that moves through extensively vegetated areas (i.e., potentially high ET) yields at least twice as much rain as air passing over areas with modest vegetation cover (i.e., potentially low ET). These results provide quantitative evidence of the significance of precipitation recycling to regional precipitation (Spracklen et al., 2012).

2.8.2 Impact on Surface Energy Balance and Boundary Layer Dynamics

By modifying albedo, surface roughness, root depth, and latent heat fluxes (Section 2.7), changes in vegetation cover modify the surface energy balance and the stability of the atmospheric boundary layer. The overall effect of large-scale deforestation is to reduce evapotranspiration and near-surface turbulence, and to increase the stability of the atmospheric boundary layer. These processes are reviewed in greater detail in Chapter 4.

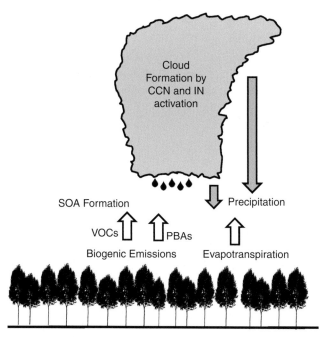

Figure 2.11. Forest-cloud interactions associated with biogenic emissions of primary biological aerosols (PBAs), plant fragments, and fungal spores and volatile organic compounds (VOCs) that undergo photochemical oxidation and gas-to-particle transformations conducive to the formation of secondary organic aerosols (SOAs). SOAs serve as major CCNs and INs. Biogenic emissions may play a crucial role in cloud microphysical processes and precipitation formation in large forested regions removed from other aerosol sources (e.g., urban pollution or desert dust). *Source:* Redrawn and modified from Pöschl et al., 2010.

2.8.3 Effect of Forest Vegetation on Cloud Microphysics

Forest vegetation may emit some trace gases (e.g., volatile organic compounds, VOCs) that subsequently undergo photochemical reactions conducive to the formation of organic aerosols. These aerosols may serve as cloud condensation nuclei (CCNs) and ice nuclei (INs) with possible effects on the rainfall regime (Pöschl et al., 2010). For instance, in the case of the Amazon it has been found that most of the cloud condensation nuclei are secondary organic aerosols (SOA) formed by photochemical oxidation and gas-to-particle conversion of volatile organic compounds emitted by vegetation. In addition, forests emit primary biological aerosol (PBA) particles (plant detritus and spores) that can also serve as CCNs. Thus, a positive feedback seems to exist between biogenic emissions of VOCs and PBAs and cloud microphysical processes conducive to the formation of precipitation that sustains these ecosystems (Figure 2.11). Likewise, forests may also affect cloud microphysics by means of the smoke particles released by fires. These organic aerosols can impact rainfall formation: Clouds forming within the smoke plume of a forest fire tend to exhibit high concentrations of cloud condensation nuclei; this tendency leads to an inefficient process of condensation whereby too many tiny cloud droplets are formed without reaching the size required for precipitation (Rosenfeld, 1999). This effect, however, can constitute just a short-term impact of slash and burn deforestation on precipitation.

2.8.4 The Effect of Mesoscale Circulations Induced by "Small-Scale" Canopy Gaps

Relatively small forest gaps (10–100 km) can favor the emergence of mesoscale circulations often known as *canopy breezes*. These breezes are induced by heterogeneities in vegetation cover that are associated with temperature gradients between the open gap and the forest (higher temperature in the gap because less energy is used for evapotranspiration) and convergence of moist air toward the gap. These conditions are favorable for the uplift of moist air parcels above the gap, which leads to cooling, condensation, and precipitation (Avissar et al., 2002). Observational studies have provided evidence in support of a local increase in precipitation (Negri et al., 2004; Chagnon and Bras, 2005; Butt et al., 2011). Thus, the early stages of deforestation, when only small gaps are open in the forest, may be associated with an increase instead of a decrease in precipitation (Lawrence and Vandecar, 2015), and land clearing needs to exceed a critical size for the change in vegetation cover to have a negative effect on precipitation (Saad et al., 2010).

2.9 Effect of Forest Vegetation on Microclimate

Forest vegetation modifies the microclimate within the canopy (e.g., Geiger, 1965; Raynor, 1971; Lee, 1978; Germino and Smith, 1999; Davies-Colley et al., 2000; Newmark, 2001; Bonan, 2008a); it maintains lower maximum temperatures and higher minimum temperatures than those observed in adjacent areas with no forest cover (e.g., Chen et al., 1993; Renaud and Rebetez, 2009; Villegas et al., 2010; Royer et al., 2011), while the opposite effect is observed at night (e.g., D'Odorico et al., 2013; Li et al., 2015). This pattern of diurnal cooling and nocturnal warming has been reported in boreal, temperate, alpine, and tropical forests (Young and Mitchell, 1994; Newmark, 2001; Renaud and Rebetez, 2009, Maher et al., 2005; Bader et al., 2007).

In these ecosystems, the nocturnal warming induced by the presence of a forest canopy is generally due to a reduction in radiative cooling (Figure 2.12) that may limit the exposure to frost stress and improve woody plant survival and growth, as discussed in Chapter 4 (Örlander, 1993; Groot and Carlson, 1996; Langvall and Örlander, 2001; Voicu and Comeau, 2006; Langvall and Ottosson Löfvenius, 2002). In the absence of forest cover, nocturnal radiative cooling is more intense because the longwave radiation is more effectively lost to the atmosphere. Tree canopies slow nocturnal cooling by absorbing part of the radiation emitted by the ground and

Figure 2.12. Difference in radiative cooling between forest patches and open canopy areas. In forested areas, part of the long-wave radiation (L_{Sfc}) emitted by the ground surface is not lost to the free atmosphere but is absorbed by vegetation or reflected back to the ground surface (rL_{Sfc}). The vegetation itself emits long wave radiation (L_{Veg}), which can be absorbed by the ground surface.
Source: After D'Odorico et al., 2013.

reradiating it back to the ground (Figure 2.12) (Chen et al., 1993; Grimmond et al., 2000). This warming effect depends on vegetation density, wind, and cloud cover (Geiger, 1965).

Forests may also modify the microclimate by altering the surface albedo both in summer and in wintertime (Betts and Ball, 1997). As noted in Section 2.7, tree canopies are darker than grasses and can therefore absorb more solar irradiance (Beringer et al., 2005). In regions affected by winter snow, the wintertime effect of forest vegetation on albedo is even stronger because tree canopies emerge from the snow mantle and render the soil surface darker, thereby decreasing the albedo of the otherwise highly reflective snow cover (Betts and Ball, 1997; Eugster et al., 2000; Chapin et al., 2005; Sturm et al., 2005). The lower albedo of forest vegetation allows for a greater absorption of solar irradiance and an overall warmer microclimate than in the absence of forest cover (Bonan et al., 1992; Bonan, 2008a).

Thus, forest vegetation has a consistent effect of increasing nocturnal temperatures both with and without snow cover. This warming effect improves seedling survival, particularly in areas close to the altitudinal and latitudinal limits of woody plants (i.e., tree lines; see Chapter 1). Seedlings growing within forest patches close to a tree line typically experience higher frost survival rates than those in open areas (Maher et al., 2005; Maher and Germino, 2006). Similar nurse plant effects have been reported in proximity to *inverse tree lines* that form in correspondence to treeless topographic depressions also known as *frost hollows* (Dy and Payette, 2007).

These depressions typically exhibit lower temperatures than the surrounding forest, particularly during calm and cloudless nights when strong radiative cooling occurs. In these conditions, the near-surface air is cooler than the ambient air, leading to downslope (katabatic) flows and accumulation of cold air in the hollows. The establishment and survival of tree seedlings close to this inverse tree line can be improved by the presence of adult trees because of their ability to provide a warmer nocturnal microclimate by reducing the radiative surface cooling (Dy and Payette, 2007; Figure 2.12).

The nurse plant effects presented in this section are associated with the ability of adult trees to reduce the exposure of cold sensitive plants (Box 2.3), particularly juveniles, to low temperature extremes (D'Odorico et al., 2013). There are also nurse plant effects associated with soil cryoturbation. In arctic and subarctic soils, the warming effect of forest canopies may reduce seedling exposure to damage from *frost heaving*, a phenomenon resulting from the formation of belowground ice crystals and their growth toward the surface. This phenomenon is observed in cold climates with no snow cover, when the near-surface temperature is below freezing and there is a constant supply of water to the depth where ice is formed (de Chantal et al., 2007). Frost heaving can uproot tree seedlings or damage their roots. By reducing radiative cooling, adult trees may limit the risk of frost heaving, thereby improving the conditions for seedling establishment and survival (de Chantal et al., 2007, 2009).

High elevation areas experience both low temperatures at night and high sunlight during the day. This combination of environmental conditions may inhibit photosynthetic carbon gain and reduce seedling survival (Germino and Smith, 1999, 2000; Egerton et al., 2000) because of a phenomenon known as *low-temperature photoinhibition* (LTP). LTP is due to an excess of sunlight absorption. Some metabolic processes such as carboxylation can be slowed by low temperatures experienced by the plant during the previous night, making the plant unable to utilize all the light it absorbs. The excess of available light can inhibit assimilation by damaging enzymes that are crucial to photosynthesis (e.g., Krause, 1994). LTP occurs even when exposure to high light and low temperatures is not simultaneous but frost occurs in the night before exposure to excess sunlight (Lundmark and Hällgren, 1987; Germino and Smith, 1999; Figure 2.13).

This phenomenon has been observed in several alpine species such as *Picea engelmannii* (Ronco, 1970) and *Eucalyptus pauciflora* trees (Ball et al., 1991; Egerton et al., 2000). Germino and Smith (2000) compared the LTP sensitivity in two coexisting coniferous evergreen species typical of some regions of the Rocky Mountain tree lines, namely, *Abies lasiocarpa* and *Picea engelmannii*. *P. engelmannii* has both a higher LTP tolerance (i.e., light saturated photosynthetic rates under frost and sunlight exposure) and a higher structural avoidance of LTP (i.e., lower ratios of silhouette to total leaf area). Conversely, *A. lasiocarpa* is a sciophytic (Box 2.1) species

Box 2.3 Plant Sensitivity to Low Temperatures

Cold sensitivity in woody plants can be due to a variety of physiological mechanisms, including cold induced reduction of photosynthesis and primary production, growth limitation, reduced regeneration capacity, frost damage, and winter *frost desiccation* (Tranquillini, 1979; Körner, 1998; Grace et al., 1989). Moreover, cold sensitivity may limit the access to soil resources as a result of inadequate root development due to exposure to low temperatures (Stevens and Fox, 1991).

In cold areas (e.g., close to tree lines) the carbon balance becomes unfavorable for trees. Low temperatures and short growing seasons limit carbon assimilation rates and plants' ability to maintain a positive carbon budget. In years with short growing seasons new plant tissues do not mature completely and remain more sensitive to frost desiccation (Tranquillini, 1979). CO_2 uptake is further inhibited by frost occurrence in the previous night and low temperature photoinhibition (Section 2.9). Independently of the existence of an unfavorable carbon balance, growth limitations can result from the fact that in some woody plants new cell production and tissue differentiation require a minimum temperature (Körner, 1998). Cold stress can also limit seed production and germination, thereby constraining *reproduction* of woody plants and ability to compete with grasses (e.g., Körner, 1998). In cold climates, plants may also experience frost desiccation when they are unable to take up water from frozen soil, and plant tissues have not developed adequate cuticular resistance to limit cuticular transpiration. In extratropical tree lines, frost desiccation is a more severe disturbance of woody vegetation than direct frost damage (Tranquillini, 1979), while in tropical tree lines the soils are not seasonally frozen and direct *frost damage* plays a more important role in determining tree dominance along altitudinal gradients (Bader et al., 2007). This phenomenon is mostly associated with mechanical and structural impacts of ice formation on plant tissues. To prevent frost damage, plants use a variety of strategies, including microhabitat selection and physiological adaptation. In perennial plants, physiological tolerance to freezing results from a variety of gene regulation mechanisms in response to relatively gradual changes in seasonal temperatures that stimulate important metabolic processes (Ouellet, 2007). Such processes often prevent ice crystal formation in living cells through compounds that alter the osmotic potential and lower the freezing point during the cold period (Hughes and Dunn, 1996).

In woody species, sudden freezing due to rapid and extreme cold events can induce another form of frost drought associated with *freezing-induced xylem embolisms* (Medeiros and Pockman, 2011; Buchner and Neuner, 2011; Cavender-Bares et al., 2005). As ice crystals are formed, dissolved gases are forced out of solution in the xylem, thereby causing enlargement of gas bubbles and embolism. Even though in this case the mechanisms leading to embolism are different from those underlying drought-induced cavitation (Box 2.2), the impact on the plant's hydraulic failure is similar.

The impact of cold environmental conditions on the carbon budget and regeneration of woody plants is overall controlled by diurnal temperatures, whereas direct frost damage is caused by extreme minimum temperatures generally occurring at night

(D'Odorico et al., 2013). Frost damage is a major disturbance for woody plants growing in tropical tree lines, in desert shrub-grass ecotones, and at the limits of the latitudinal range of mangroves (Loik and Nobel, 1993; Pockman and Sperry, 1997; Bader et al., 2007; Krauss et al., 2008; Stuart et al., 2007; D'Odorico et al., 2013). Therefore, the geographic location of these tree lines is determined by minimum nocturnal temperatures. Conversely, arctic and extratropical alpine tree lines are more sensitive to diurnal temperatures during the growing season and to soil temperatures (Körner, 1998; Körner and Paulsen, 2004).

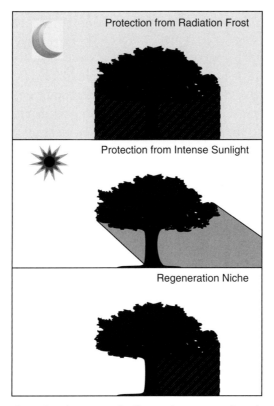

Figure 2.13. The diurnal and nocturnal effects of tree canopies: At night the canopy reduces the radiative cooling while during the day it reduces exposure to intense radiation. The zone shown as the "regeneration niche" provides conditions favorable for seedling establishment in species sensitive to LTP.
Source: Redrawn from Ball et al., 1991.

with high LTP sensitivity; its seedlings establish only under the canopy of adult trees, which provide a warmer nocturnal microclimate while limiting light exposure (Germino and Smith, 2000; Germino et al., 2002; Maher and Germino, 2006). This double advantage offered by understory microsites (Figure 2.13) justifies the use of shelterwood logging (i.e., thinning without greatly altering the habitat provided by forest vegetation) as a forestry technique that favors regrowth (Ball et al., 1991).

2.10 Effects of Deforestation on Large-Scale Climate

The effect of land use and land cover change on regional climate varies with the geographic setting (Bonan et al., 1992, Bonan, 2008a; Pielke et al., 2011). Research on the effect of deforestation on the regional climate has often concentrated on the Amazon Basin and the Boreal forest (Bonan et al., 1992; Foley et al., 1994; Chapin et al., 2000a,b). In the Amazon, the replacement of forest with pastures causes an increase in albedo roughly from 0.13 to 0.18 and consequently, a decrease in net radiation that has been estimated around 11% (Gash and Nobre, 1997). Nevertheless, dry season surface temperatures are found to be about 1.5°C higher in pastures than in forest (Loarie et al., 2011) because of the higher evapotranspiration rates afforded by deep-rooted trees with respect to grasses (Pielke et al., 2011). The lower latent and greater sensible heat fluxes experienced by pastures cause an increase in temperature that outweighs the effect of the increase in albedo. Pastures also exhibit enhanced planetary boundary layer (PBL) development, and – depending on the size of the clearing (see Section 2.8) – canopy breezes and near-surface convergence of relatively moist air from the surrounding forest (Souza et al., 2000). Deforestation also affects cloud climatology and the onset of the rainy season (Wang et al., 2009; Butt et al., 2011). As noted in Section 2.8, regional deforestation is expected to cause an overall decrease in precipitation (particularly in continental regions) and increase in the occurrence of droughts (Lee et al., 2011).

While most of past research has concentrated on the effect of forest removal on mean annual or seasonal precipitation, recent work is now pointing to important effects of deforestation on seasonal and interannual rainfall variability (Figure 2.14). Deforested regions seem to exhibit a stronger seasonal variability of precipitation and fewer rainfall occurrences than those that have forest cover (Webb et al., 2005; Lee et al., 2011). The possible impact on forest ecosystems of such an increase in seasonal and interannual variability still needs to be assessed with coupled forest-climate models.

Land use and land cover change are recognized as important forcings to the regional climate. Their effect on the global climate system, however, remains poorly understood. Most research has concentrated on the impact of large-scale forest removal on the climatology of the deforested region, or on the combined effect of land atmosphere feedbacks and large-scale forcings on regional climatology (e.g., Wang and Eltahir, 2000a, 2000b; Zeng and Neelin, 2000). Teleconnections between land use/land cover change in one area (or region) and climatic changes in another have only recently started to be addressed (Ray et al., 2006; Pielke et al., 2011). The impact of regional deforestation on large-scale climatic patterns (e.g., the El Nino Southern Oscillation, the North Atlantic Oscillation, or large-scale atmospheric circulation) is also a new frontier for future research (Pielke et al., 2011).

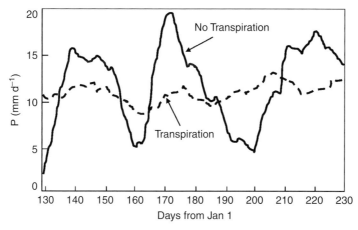

Figure 2.14. Seasonal precipitation fluctuations in Borneo (latitude 1.4°S; longitude 113°E) from model simulations accounting for (dashed line) or excluding (solid line) the effect of forest transpiration (Lee et al., 2012). These lines show the 10-day running averages of precipitation. Forest removal is expected to enhance the amplitude of seasonal fluctuations.
Source: Partially redrawn from Lee et al., 2012.

2.11 Summary

Forest removal affects the dynamics and rate of hydrologic processes in deforested watersheds. Decades of research in forest hydrology have provided overwhelming evidence of how the removal of forest vegetation reduces evapotranspiration and consequently increases water yields. The loss of forest cover enhances soil compaction and reduces soil bioturbation, thereby favoring runoff generation by infiltration-excess overland flow. As the litter layer is lost, the water storage capacity of the ground surface is also reduced; the hydrologic response becomes faster, with shorter concentration times and higher flow peaks. These effects are further enhanced when forest removal is contributed by forest fires that are capable of developing soil hydrophobicity, thereby reducing the soil infiltration capacity and increasing runoff. This phenomenon is spatially heterogeneous and typically lasts a few weeks to a few years.

In watersheds receiving part of the precipitation as winter snow, deforestation causes an increase in snow accumulation as a result of the reduced losses from canopy interception. The snowpack melts faster and earlier because of the more direct exposure to solar irradiance; as a result, the snowmelt peak flow is higher.

In cloud forests or areas frequently immersed in fog, the effect of deforestation on water yields and runoff can contrast with what has been reported for most other forests: In this case the presence of a forest enhances the input of water to the watershed because the canopy provides surfaces (i.e., leaves) upon which fog and cloud water can be deposited. Therefore, forest removal is expected to reduce occult precipitation and water yields. Similarly, microclimatic changes due to climate warming or land use shifts in nearby areas could affect the frequency of the cloud/fog immersion conditions that characterize these forests, thereby leading to a possible shift in plant community composition.

In regions with a relatively shallow water table, forest removal can cause a "watering up" effect, whereby the water table rises closer to the ground surface, a phenomenon that can lead to salt accumulation as a result of the evaporation of groundwater moved up by capillary rise (*exfiltration*). "Watering up" often causes the emergence of waterlogging in the root zone, thereby preventing forest regrowth.

Deforestation also greatly affects the interactions between the land surface and the atmosphere. It reduces evapotranspiration by increasing land surface albedo, decreasing roughness, leaf area index, and access to deeper soil moisture by plant roots. Because of the reduction in evapotranspiration, large-scale deforestation can modify the microclimate and rainfall regime. Model simulations have shown that large-scale forest removal reduces precipitation recycling and alters the surface energy balance, thereby reducing regional precipitation. Conversely, small-scale deforestation (at scales >10 km) can induce mesoscale circulations that enhance local convergence and precipitation. Ongoing research is also investigating the effect of deforestation on biogenic trace gas emissions and the associated formation of organic aerosols that could serve as cloud condensation nuclei (CCNs) and ice nuclei (INs); thus, forest removal could alter cloud microphysics and the rainfall regime by reducing the supply of CCNs and INs. Forest removal also affects the temperature regime: At the global scale, forests maintain a warmer planetary temperature because of their lower albedo; however, at smaller scales (ranging from forest gaps to entire regions) they can cause diurnal warming and nocturnal cooling, thereby enhancing the exposure to extreme cold events.

3
Biogeochemical Impacts

Terrestrial vegetation is often limited by nutrients, particularly nitrogen and/or phosphorus, and in some cases potassium, calcium, sulfur, magnesium, silica, and other micronutrients or trace minerals (Vitousek and Howarth, 1991). On a global scale, Fisher et al. (2012) found the average reduction in terrestrial plant productivity due to nutrient limitation to be between 16% and 28% (Figure 3.1). One major factor altering patterns of nutrient cycling is land use change due to deforestation. Deforestation alters nitrogen and phosphorus cycling, both of which can feed back to affect the carbon cycle and atmospheric CO_2 concentrations. In this chapter, we examine carbon, nitrogen, and phosphorus cycling beneath forests and the effect of deforestation on these cycles.

3.1 Carbon Cycle

In this section, we briefly describe the carbon cycle in undisturbed forests to provide the basis for understanding how deforestation alters this cycle. Next, we examine global pools of carbon stored in forests and the soils beneath these forests as well as the rate at which forests across different latitudinal belts take up carbon. Finally, we consider how deforestation might alter the carbon balance of forests and forest soils.

3.1.1 Carbon Cycle in Undisturbed Forests

Gross primary production (GPP) of an ecosystem represents the gross uptake of atmospheric CO_2 that is used for photosynthesis. Plants use energy in the synthesis of new plant tissue and the maintenance of living tissues (Luyssaert et al., 2007). Because of the costs associated with growth and maintenance of leaves, wood, and roots, some photoassimilated compounds are lost from the ecosystem as autotrophic respiration (R_a) (Figure 3.2; Luyssaert et al., 2007). The fraction that is used for maintenance respiration can vary widely (i.e., 0.23–0.83 for different forest types as determined from a literature review of 60 different studies), yet in many forest studies it is generally

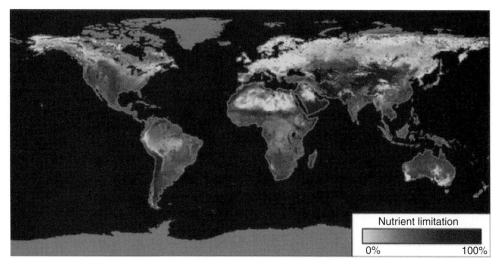

Figure 3.1. Global map depicting the percentage that vegetation growth is limited by available soil nutrients, with 0 representing no nutrient limitation, and 100 being completely nutrient limited.
Source: Fisher et al., 2012.

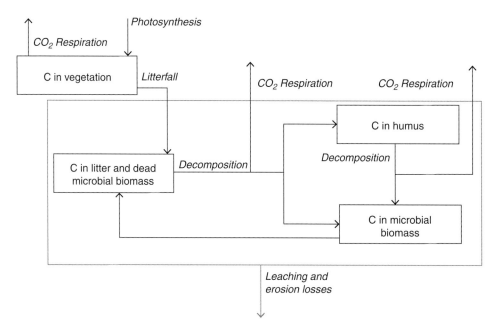

Figure 3.2. Schematic representation of the C cycle described in Section 3.1.1. The quantity of carbon loss in CO_2 respiration during litter decomposition is a fraction, R_h, of the litter carbon that is decomposed (R_h ranges between 0.6 and 0.8). Similarly, the carbon loss as CO_2 emissions (respiration) during the decomposition of humus is R_h multiplied by the quantity of carbon as humus that is being decomposed. The portion of litter reaching the humus pool is equal to the isohumic fraction times the quantity of litter that is decomposed.

assumed to be a constant value at 0.5 (DeLucia et al., 2007). The energy that is not used for respiration is the net primary production (NPP) and is equal to GPP − R_a. A large fraction of NPP is used in the production of leaves, wood, and roots and a portion of the standing biomass is transferred annually to litter (Luyssaert et al., 2007). This carbon enters the soil when C contained in leaf and woody litter, dead roots, mycorrhizal turnover, and carbon exudates from roots are transferred to the forest soil (Figure 3.2). These carbon pools are subjected to decomposition by microbial activity, a process defined as heterotrophic respiration (R_h). The sum of R_h and R_a represents the total ecosystem respiration (R_e). Net ecosystem production (*NEP*) is determined from the difference between GPP and R_e.

The amount of leaf litter, dead roots, and other plant residues that are present on the forest floor and in the root zone varies depending on plant phenology, the amount of leaf biomass that is present, and the rate of litter decomposition (Chapin et al., 2002). Litter decomposition rates (*k*) are controlled by factors such as temperature, precipitation intensity, litter and vegetation type, and litter quality characteristics such as the concentration of N and P as well as the C:N ratio and C:P ratio of the litter (Figure 3.3; Aerts, 1997). Zhang et al. (2008) compiled a large data set of litter decomposition rates (hereafter indicated also as "*k* values") from 110 research sites from across the globe. They found that the combination of total litter nutrient concentrations (i.e., N, P, K, Ca, and Mg) and the C:N ratio of that litter accounted for 70% of the variation in the litter decomposition rates, suggesting that litter quality is one of the most important regulators of litter decomposition at the global scale. Litter decomposition occurs over timescales ranging from days to years and depends on (i) the environmental conditions of the soil, such as soil moisture and temperature; and (ii) the quality of added residues as a food source for soil organisms. Rapid decomposition occurs when there is sufficient soil moisture, good aeration (≈60% of soil pore space is filled with water), warm soil conditions (20–30°C), and close to a neutral pH (Skopp et al., 1990; Linn and Doran, 1984). Thus, turnover of soil C tends to be more rapid in tropical than temperate forest soils (Feller and Beare, 1997). During decomposition, some quantity of C is oxidized and lost as CO_2 to respiration (i.e., heterotrophic respiration; Figure 3.2). This fraction ranges between 0.6 and 0.8 (Brady and Weil, 2008). Of the remaining decomposing litter, a portion of this C is transformed into stable organic complexes or humus (i.e., the isohumic fraction, which ranges from 0.15 to 0.35; Aber and Melillo, 1991), while the remainder is incorporated into the microbial biomass (Brady and Weil, 2008; Figure 3.2). Microbes can also decompose the more recalcitrant organic C fraction. When microbes die, the C stored in their biomass becomes stored in the soil carbon pool. C losses from the soil also result from methane (CH_4) efflux, hydrologic leaching of dissolved and particulate carbon compounds, and erosion of particulate carbon compounds (Figure 3.2). Over longer timescales (more than a year), all or part of the carbon that has been built up by the accumulation of NEP can leave the ecosystem and eventually return to the atmosphere by forest fires,

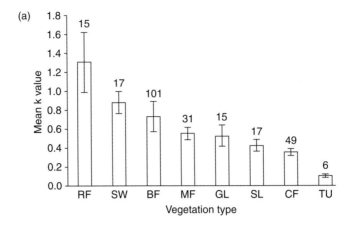
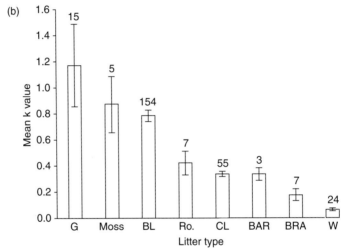

Figure 3.3. Variation of k value (g g^{-1} yr^{-1}) with different vegetation types (a) and litter types (b) at the global scale. Data at the top of each bar are the number of sample k values for each group. Vegetation types include rain forest (RF), swamp (SW), broadleaved forest (BF), mixed forest (MF), grassland (GL), shrub land (SL), coniferous forest (CF), and tundra (TU). Litter types include grass leaf (G), moss (M), broadleaved litter (BL), roots (R), conifer needles (CL), bark (BAR), branch (BRA), and woody litter (W).
Source: Zhang et al., 2008.

harvests, and/or erosion (Randerson et al., 2002; Amiro et al., 2006). Thus, the net ecosystem carbon balance (NECB) is the term applied to the total rate of organic carbon accumulation (or loss) from ecosystems (Chapin et al., 2006).

GPP is strongly dependent on climatic conditions (Table 3.1). From their review of 513 forest sites, Luyssaert et al. (2007) found that climatic conditions (i.e., temperature and precipitation) explain 71% of the variability in global GPP patterns. GPP increases with increasing temperatures and precipitation once ecosystems are

Table 3.1. *Mean carbon fluxes and NPP components as well as climatic characteristics including their standard deviation for different forest biomes in g C m^{-2} yr^{-1}*

	Boreal humid		Boreal semiarid		Temperate humid		Temperate semiarid	Med. warm	Tropical humid
	Ev.	Ev.	De.	Ev.	De.	Ev.	Ev.	Ev.	Ev.
GPP	973 ± 83	773 ± 35	1201 ± 23	1762 ± 56	1375 ± 56	1228 ± 286	1478 ± 136	3551 ± 160
NPP	271 ± 17	334 ± 55	539 ± 73	783 ± 45	738 ± 55	354 ± 33	801 ± NA	864 ± 96
NEP	131 ± 79	40 ± 30	178 ± NA	398 ± 42	311 ± 38	133 ± 47	380 ± 73	403 ± 102
R_e	824 ± 112	734 ± 37	1029 ± NA	1336 ± 57	1048 ± 64	1104 ± 260	1112 ± 100	3061 ± 162
R_a	489 ± 83	541 ± 35	755 ± 31	951 ± 114	673 ± 87	498 ± 58	615 ± NA	2323 ± 144
R_h	381 ± 40	247 ± 26	275 ± 31	420 ± 31	387 ± 26	298 ± 16	574 ± 98	877 ± 96
Climatic characteristics									
Mean winter temp. (°C)	−9 ± 7	−18 ± 6	−20 ± 8	4 ± 5	2 ± 9	0 ± 5	10 ± 3	23 ± 4
Mean summer temp. (°C)	13 ± 4	13 ± 4	13 ± 4	17 ± 4	20 ± 5	14 ± 3	23 ± 3	24 ± 3
Winter precip. (mm)	205 ± 110	52 ± 33	47 ± 31	449 ± 337	183 ± 164	356 ± 182	239 ± 212	685 ± 664
Summer precip. (mm)	144 ± 88	183 ± 105	156 ± 86	194 ± 234	356 ± 259	81 ± 99	106 ± 127	469 ± 395

Note: Evergreen is abbreviated as Ev., deciduous as De., and Mediterranean as Med.
Source: Luyssaert et al., 2007.

Table 3.2. *Estimated C pools in Pg C and area-weighted C densities for forest vegetation, litter, and forest soils across different terrestrial ecosystems*

Biome	Total forest area, 2007 (Mha)	Total living biomass	Dead wood	Litter	Soil	Total C stock	Carbon density (Mg C ha^{-1})
Boreal	1135.2	53.9	16.1	27	174.5	271.5	239.2
Temperate	766.7	46.6	3.3	12.1	56.7	118.6	154.7
Tropical intact	1392.2	228.2	44.4	4	116.6	393.3	282.5
Tropical regrowth	557.2	33.9	9.1	0	34.7	77.7	139.4
All tropics	1949.4	262.1	53.6	4	151.3	471	241.6
Global total	3851.3	362.6	72.9	43.1	382.5	861.1	223.6

Source: Pan et al., 2011.

not limited by either low precipitation (<800 mm) or low mean annual temperatures (<5° C). However, these effects are not observed above 1,500 mm. Climatic conditions are less likely to explain the variability in NPP and NEP. Thus, on the global scale, GPP is mainly climate driven while the global pattern in NEP is mainly determined by nonclimatic conditions such as successional stage, management, site history, and site disturbance.

3.1.2 Global Estimates of C Pools, Emissions, and Uptake in Terrestrial Ecosystems

Globally, forest vegetation and soils contain roughly 1,146 petagrams (Pg) of carbon, with approximately 37% of this carbon stored in low-latitude forests, 14% in midlatitude forests, and 49% in high-latitude forests (Dixon et al., 1994; Table 3.2). Deforestation has largely altered the quantity of C stored in forests and their soils. This trend is apparent when examining temporal changes in average aboveground biomass since the mid-19th century (Figure 3.4; see also Chapter 1). Currently, the global rate of deforestation has decreased since the 1990s from 16 Mha yr^{-1} to 13 Mha yr^{-1}, yet this rate remains particularly high in forests located in Central and South America and Southeast Asia (FAO, 2010). Notably, however, the FAO definition of forest loss does not include trees in plantations and agriculture, and some estimates suggest that the rate at which forest cover in the tropics was lost rose between 2000 and 2012 (Hansen et al., 2013).

A compilation of global forest inventory data from Malhi et al. (1999) suggests that there is a net global biomass accumulation of 1.4 Pg C yr^{-1} in boreal, temperate, and tropical forests. Approximately 50% of this accumulation occurs in boreal and temperate forests. For instance, Dixon et al. (1994) estimated that northern temperate and boreal forests accumulate 0.7 ± 0.2 Pg C yr^{-1}, while Houghton (1996) estimated a net

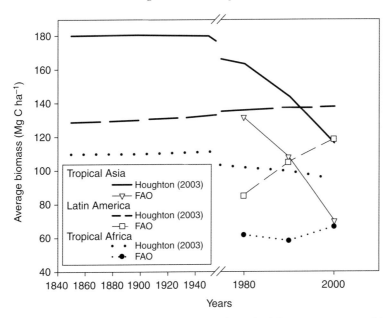

Figure 3.4. Estimates of the average biomass of tropical forests, as reported by the FAO (1980, 1990, and 2000 assessments) (light lines) and as modeled from changes in land use (1850–2000) (Houghton, 2003) (heavy lines) (from Houghton, 2005). Note the change in temporal scale between 1940 and 1980.

uptake of 0.8 Pg C yr^{-1} for the same area. This difference may be due to Houghton's estimate also including net uptake of living biomass (0.5 Pg C yr^{-1}), wood products (0.14 Pg C yr^{-1}), slash (0.08 Pg C yr^{-1}), soil (0.08 Pg C yr^{-1}), and peatlands (0.02 Pg C yr^{-1}). Phillips et al. (1998) compiled data from forest inventories across 68 sites in undisturbed tropical forests and found that despite large variability among plots, there is a net accumulation of 0.6 ± 0.3 Pg C yr^{-1} in neotropical (i.e., American) forests. Further, they found no apparent net increase in biomass in African and Asian forests, yet limited data were available to definitively support this conclusion.

Whether a given area is a net source or sink of C depends on the balance between C uptake and emissions. Global emissions from land cover change vary widely; however, they have been estimated to range from 0.9–1.6 Pg C yr^{-1} (McGuire et al., 2001) to 2.0 Pg C yr^{-1} during the 1980s and 2.2 Pg C yr^{-1} during the 1990s (Houghton, 2003). Houghton (2003) estimated that the long-term (1850–2000) flux of carbon from changes in land use released 156 Pg C to the atmosphere, about 60% of it from the tropics (Figure 3.5). Outside the tropics, the average net flux of carbon attributable to *land use* change decreased from a source of 0.06 Pg C yr^{-1} during the 1980s to a sink of 0.02 Pg C yr^{-1} during the 1990s (Houghton, 2003). During the 1990s, carbon emissions from deforestation accounted for approximately 2.11 Pg C yr^{-1} (rate of deforestation = 14.8 10^6 ha yr^{-1}), while temperate and boreal zones accounted for 0.13 Pg C yr^{-1} (rate of deforestation of natural forest cover in Canada, United States,

Table 3.3. *Carbon sinks and sources (Pg C year^{-1}) in the world's forests*

Region of the world	Net forest carbon flux (1990–1999)	Net forest carbon flux (2000–2007)
Tropical	0.13	0.07
Temperate	−0.62	−0.72
Boreal	−0.51	−0.50

Note: Negative values represent C sinks, positive values represent C sources.
Source: From Pan et al., 2011.

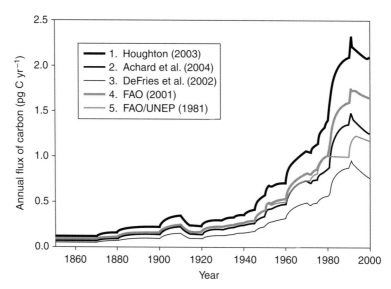

Figure 3.5. Annual emissions of carbon from land use change in the tropics according to alternative rates of tropical deforestation and alternative estimates of average forest biomass.
Source: Houghton, 2005.

Russia, Norway, Finland, Sweden, and Japan = 0.9 10^6 ha yr^{-1}) (Houghton, 2003). Pan et al. (2011) estimate a source of 1.3 ± 0.7 Pg C yr^{-1} from tropical land use change over the period 1990 to 2007, consisting of a gross tropical deforestation emission of 2.9 ± 0.5 Pg C yr^{-1} partially compensated by a carbon sink in tropical forest regrowth of 1.6 ± 0.5 Pg C yr^{-1} (Table 3.3).

3.1.3 Changes in the Carbon Cycle as a Result of Deforestation

3.1.3.1 Direct C Losses Due to Logging

In forests that are logged, there is a direct loss of C contained in the forest due to the transport of timber out of the forest. The quantity of C contained in aboveground timber that is exported out of the system varies, depending on the location of the forest being logged (see Table 3.2 for a range of values of the C contained per hectare of forest area), logging frequency, and percentage of forest removed. Over the past century, most logging has occurred in temperate and boreal zones (Malhi et al., 1999). Rates of logging in boreal forests have increased from approximately 1 Mha yr^{-1} in 1850 to 3.5 Mha yr^{-1} in 1980, and in temperate forests from 3 Mha yr^{-1} in 1850 to 6 Mha yr^{-1} in 1980 (Houghton, 1996). In contrast, logging in tropical forests was less than 0.5 Mha yr^{-1} in 1850, and less than 2 Mha yr^{-1} in 1950, but accelerated to 8 Mha yr^{-1} in 1980 (Houghton, 1996). The flux of C to the atmosphere due to deforestation activities during the 1990s is estimated at 2.1 Pg C yr^{-1} from the tropics and 0.13 Pg C yr^{-1} from boreal and temperate zones (Houghton, 2003).

3.1.3.2 Pyrogenic C Losses in Forests during "Slash and Burn"

One source of disturbance in tropical forests is deforestation due to shifting cultivation (or "slash and burn"), which affects approximately 400 million ha, or 27% of the world's arable land (Kleinman et al., 1996; Seubert et al., 1977; Giardina et al., 2000). Shifting cultivation consists of clearing forests by slash and burn, using the land for cropping, then leaving the land fallow, during which time the forest regenerates. During and after biomass burning, losses of C result from the emission of CO_2, CH_4, and CO during the burn as well as the erosion of C contained in ashes after the burn. Van der Werf et al. (2006) examined global C emissions over an eight year study period and determined that biomass burning resulted in emissions of 2.5 Pg C yr^{-1}, consisting of 8.9 Pg CO_2 yr^{-1}, 0.4 Pg CO yr^{-1}, and 0.02 Pg CH_4 yr^{-1} (Figure 3.6). High levels of emissions were observed in boreal forests of North America and Eurasia, tropical Central and South America, Africa, Southeast Asia, and Australia. These emissions accounted for 4.4% of the total carbon loss from terrestrial ecosystems.

C losses also result from the erosion of ashes that contain C. For instance, during three controlled burn experiments, Kauffman et al. (1993) found that after the burn, 1.5 ± 0.2 Mg C ha^{-1} (prefire aboveground C in biomass was 33.6 ± 4.2 Mg C ha^{-1}), 1.2 ± 0.2 Mg C ha^{-1} (prefire aboveground biomass was 33.8 ± 2.1 Mg C ha^{-1}), and 0.9 ± 0.2 Mg C ha^{-1} (prefire aboveground biomass was 33.7 ± 1.9 Mg C ha^{-1}) remained as ash. Thus, ash accounted for ~4% of the prefire aboveground biomass. Of this, 57% of the ash was lost to erosion 17 days after the burn.

3.1.3.3 Changes in Soil Carbon Following Deforestation

The top one meter of the world's soils contains approximately 1,500 Pg C (Johnson and Henderson, 1995; Bruce et al., 1999; Figure 3.7), and there is evidence that deforestation is depleting soil carbon and subsequently increasing atmospheric CO_2 (Bruce et al., 1999; Houghton, 1999; Murty et al., 2002). After deforestation, there can be an initial increase in decomposition rates due to warmer and wetter soils (Kirschbaum, 1995). Small increases in soil temperature can substantially increase the rate of soil

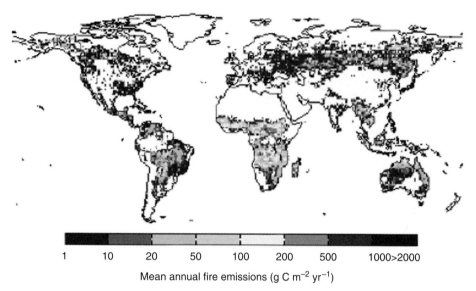

Figure 3.6. Mean annual fire emissions (g C m^{-2} yr^{-1}) averaged over the period 1997–2004. See also color figures at the end of the book.
Source: van der Werf et al., 2006.

respiration in the tropics, leading to rapid decomposition and the drawdown of carbon pools (Townsend et al., 1992). For instance, tropical soils are estimated to emit 0.2 Pg C yr^{-1} as a result of land use changes, accounting for 10%–30% of the total C emission from deforestation (Houghton, 1999; Achard et al., 2004). However, these trends are not necessarily observed for extended periods because of the decline in available (and nonrecalcitrant) substrate that leads to a reduction in decomposition rates of soil organic matter.

A decline in decomposition rates may also result when deforestation alters the soil microbial population (e.g., Henrot and Robertson, 1994; Borneman and Triplett, 1997; Caldwell et al., 1999; Nusslein and Tiedje, 1999; Waldrop et al., 2000; Bossio et al., 2005; Chaer et al., 2009). For instance, Borneman and Triplett (1997) found that significant microbial population differences exist between a mature forest soil and an adjacent pasture soil after deforestation. Similarly, Henrot and Robertson (1994) found that six months after deforestation, the microbial population decreased by 50% and within 15 months, the microbial biomass had decreased to 35% of its original population. Cerri et al. (1985) found that biomass burning eliminated all of the microbial biomass from the top 10 cm of soil, and two-thirds of the total microbial biomass in the soil profile. Although these factors drive declines in decomposition rates after deforestation, land use change can lead to a substantial loss of soil carbon and a decline in the rate at which carbon is added to the soil.

Previous reviews have shown that conversion of forest to cultivated agriculture leads to considerable decreases in soil carbon; however, no consistent changes occur

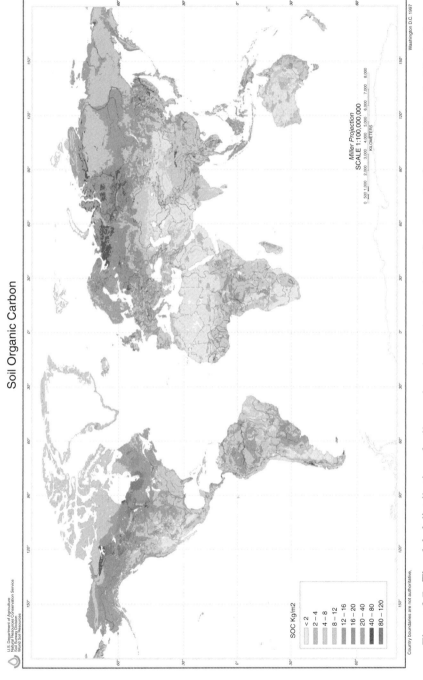

Figure 3.7. The global distribution of soil organic carbon to 1 meter depth. See also color figures at the end of the book. *Source:* Reproduction is attributed to the U.S. Department of Agriculture and obtained from FAO-UNESCO, 2006.

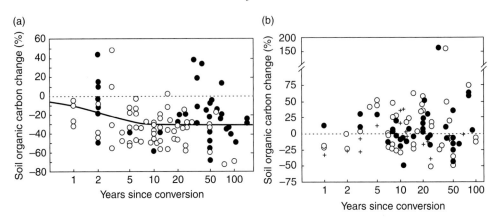

Figure 3.8. Changes to soil carbon following (a) conversion from forest to cultivated land use and (b) conversion from forest to uncultivated pasture. Closed circles (●) show data where bulk-density effects have been considered by the authors, whereas open circles (○) show data where it is uncertain whether bulk density effects were accounted for. Accounting for changes in bulk density following land use change is necessary as agricultural soils have on average 13% higher bulk density.
Source: Murty et al., 2002.

when forests are converted to uncultivated pasture (e.g., Murphy and Lugo, 1986; Murty et al., 2002; Figure 3.8). For instance, Garcia-Oliva et al. (1994) found no change in soil organic carbon (SOC) seven years after the conversion from forest to pasture in a Mexican forest. Murty et al. (2002) conducted a global review of 75 studies and found that when forest is converted to cultivated land, the mean percentage change in soil carbon 10 or more years after conversion is approximately –30% (Figure 3.8). However, in contrast to the global review by Murty et al. (2002), Don et al. (2011) found that forest conversion into grassland reduced SOC stocks by 12% in tropical systems. In their recent metaanalysis of 385 studies on land use change in the tropics, Don et al. (2011) found that the highest SOC losses were caused by conversion of primary forest into cropland (–5%) and perennial crops (–30%) (Table 3.4). The decline in SOC following deforestation has also been observed in temperate forests. Poeplau et al. (2011) conducted a metaanalysis of 29 studies in the temperate zone and found that 20 years after the conversion of forest to cropland, there was a 31% decline in SOM. For instance, Balesdent et al. (1987) showed a decrease in SOC 23 years after the conversion from a French pine forest to cornfield from 24 g C/kG of soil to 14.5 g C/kg soil.

Management practices can significantly influence the SOC content after deforestation. For instance, Beare et al. (1994) show that after 11 years, the SOC of no till surface soils was 18% higher than that of conventional tillage practices for similar levels of plant productivity. Under very well-managed pasture near Manaus, Brazil, C in the top 20 cm of soil returned to approximately the same level present under original forest eight years after clearing (Cerri et al., 1991; Choné et al., 1991).

Table 3.4. *Mean absolute and relative SOC stock changes for different land-use change (LUC) types from a meta-analysis of studies in the tropics*

Land-use change (LUC) type	Absolute SOC change (Mg ha^{-1})	Relative SOC change (%)	SOC prior LUC (Mg ha^{-1})	Sampling depth (cm)	Time after LUC (years)	Number of studies
Primary forest to grassland	−12.6 ± 3.0	−12.1 ± 2.3	73 ± 7	36 ± 3	25 ± 3	93
Primary forest to cropland	−20.1 ± 5.2	−25.2 ± 3.3	83 ± 9	36 ± 4	28 ± 4	56
Primary forest to perennial crops	−32.0 ± 3.5	−30.3 ± 2.7	105 ± 20	48 ± 8	49 ± 12	20
Primary forest to secondary forest	−12.6 ± 2.4	−8.6 ± 2.0	91 ± 9	39 ± 4	28 ± 3	71

Source: Don et al., 2011.

Organic carbon is also removed from the system as a result of leaching losses from the upper soil layer (Dosskey and Bertsch, 1997). After deforestation, leaching beneath the soil profile can increase as a result of a reduction in transpiration and subsequent increase in soil moisture. Cerri et al. (1991) estimated that after deforestation in a central Amazon rain forest, 10^3 t C/ha are leached annually from the top 0.2 m of the soil. Similarly, Williams et al. (1997) found after slash and burn in an Amazonian forest that the volume-weighted mean concentration of DOC in leachate water was ~40% greater in the water collected beneath the deforested site (i.e., 194 µM at the deforested site versus 121 µM in the forested site). They also measured significantly greater vertical flows of water in the burned versus forested plot.

The increase in both the quantity of water leached beneath the rooting zone and DOC concentration in this water enhances C losses from the system and can lead to an increase in streamwater DOC concentrations. For instance, Williams et al. (1997) found that streamwater DOC concentrations were more than 45% higher after deforestation (i.e., 73 versus 158 µM in forested versus deforested streamwater concentrations, respectively). Sollins and McCorison (1981) found that DOC concentrations were higher in a clear-cut watershed in comparison to a forested watershed.

After deforestation, there can also be an increase in erosion of C containing particles. Forest vegetation can stabilize the soil and, thus, decrease the erosion of C containing particles as a result of (i) the presence of shallow roots that bind soil aggregates, thereby enhancing soil cohesion (Angers and Caron, 1998); and ii) the deep roots of woody species that bind the upper soil layer to saprolite substrates and bedrock, as explained in Chapter 4 (e.g., O'Loughlin and Pearce, 1976; Wu et al., 1979; Wu and Swanston, 1980). Erosion may be a major pathway of SOC loss (at the plot scale) for insufficiently aggregated soils that are typical of tropical regions (van Noordwijk et al., 1997; Berhe et al., 2007). Whether deforested areas are used for pasture or cultivation will also affect erosion losses. Land plowing increases soil erodibility and different soil conservation techniques (i.e., rilling, contouring, terracing) can be used to limit soil losses. During cultivation, the particulate organic matter fraction is highly susceptible to erosion. This may be important in tropical soils, where 20% to 40% of the soil organic carbon is associated with the sand-size particulate organic matter fraction as compared to 2% to 14% for many temperate soils (Feller and Beare, 1997).

3.1.3.4 Carbon Emissions from Deforestation and Land Use Change

While global emissions from land cover change vary widely – for example, they have been estimated to range from 0.9 (McGuire et al., 2001) to 2.2 Pg C yr^{-1} during the 1990s (Houghton, 2003) – Van der Werf et al. (2009) used updated estimates on carbon emissions from both deforestation and fossil fuel combustion and found that carbon emissions from deforestation accounted for 1.2 Pg C yr^{-1} over the period 1996–2007, or an estimated 12% of global anthropogenic CO_2 emission (Van der Werf et al., 2009). Moreover, for about 30 developing countries, including Brazil and Indonesia, deforestation is the largest source of CO_2 (Van der Werf et al., 2009). Deforestation contributes to atmospheric CO_2 emissions largely through combustion of forest biomass and decomposition of remaining plant material and soil carbon (Van der Werf et al., 2009).

The method of land clearing strongly affects the temporal dynamics of atmospheric C emissions. If deforestation occurs through slash and burn, most of the carbon (a large fraction of which is biomass) is released to the atmosphere within a year during the burning process. In contrast, if deforestation occurs through clear-cut or selective logging, most of the useful wood is removed off-site and is used to make paper, furniture, and other wood products that oxidize over much longer timescales (Ramankutty et al., 2007).

Whether pasture and agricultural soils are a net sink or source of C after deforestation depends on their management. In well-managed pastures in formerly forested areas, the rooting system of pasture grass can redistribute carbon to deeper layers where it is less susceptible to decomposition (Nepstad et al., 1991). Under more commonly occurring conditions in the Amazon, Serrão and Falesi (1977) found a decline of ~50% in the C concentration in pasture soil after 11 years of use in Mato Grosso, while Eden et al. (1991) found a 15% decline in the C concentration in pasture after 12 years of use in Roraima, Brazil. Well-managed pastures are not typical of the Amazon (Trumbore et al., 1995), and Fearnside and Imbrozio Barbosa (1998) found that pasture soils in the Brazilian Amazon are a net carbon source, with the upper 8 m releasing an average of 12.0 t C ha^{-1} in land maintained as pasture in the decades following deforestation.

In the tropics, increasing C inputs to the soil are obtained by improving the fertility and productivity of cropland and pastures. In extensive systems with vegetated fallow periods (e.g., shifting cultivation), planted fallows and cover crops can increase C levels over the cropping cycle. Use of no-till, green manures and agroforestry are other beneficial practices (Paustian et al., 1997). In temperate regions, key management strategies to reduce emissions involve increasing cropping frequency and reducing bare fallow, increasing the use of perennial forages (including N-fixing species) in crop rotations, retaining crop residues, and reducing or eliminating tillage (i.e., no-till; Paustian et al., 1997). Thus, carbon emissions can be reduced after deforestation by adopting management practices that increase the soil C content.

3.2 Nitrogen Cycle

In this section, we begin by describing the nitrogen cycle in undisturbed forests to provide the basis for understanding how deforestation alters nitrogen cycling. Next, we examine how nitrogen limited forests are distributed globally, including factors that lead to nitrogen limitation in terrestrial ecosystems. Finally, we describe how deforestation might alter the nitrogen balance of forests and their soils.

3.2.1 Nitrogen Cycle in Undisturbed Forests

Nitrogen (N) is essential for plants because it is an integral part of enzymes that mediate biochemical reactions in which C is reduced (i.e., photosynthesis) or oxidized (i.e., respiration). Nitrogen is present in the atmosphere, organic matter, mineral soil, and dissolved in water. Most plants almost exclusively take up mineral nitrogen in the form of ammonium (NH_4) and nitrate (NO_3) that become available during N-fixation and decomposition of soil organic matter (SOM; Figure 3.9). During decomposition, SOM is oxidized by microbes, which produce ammonium and nitrate via the processes of ammonification and nitrification, respectively (Figure 3.9). There are two main processes by which atmospheric N becomes available for plant uptake: fixation by lightning and by biological processes. N-fixation converts atmospheric N_2 to a form that can be used by biota; it primarily occurs via biological fixation. In fact, rates of N-fixation by lightning are <10 Tg N yr^{-1}, whereas rates of terrestrial biological N-fixation are estimated to range between 90 and 140 Tg N yr^{-1} (Vitousek et al., 1997a). Biological N-fixing species include symbiotic and free-living microbes. Symbiotic fixation by plants can add substantial amounts of N to ecosystems (Cleveland et al., 1999), whereas fixation by soil heterotrophic bacteria contributes less substantial amounts (i.e., 1–4 kg N ha^{-1} yr^{-1}) to most terrestrial ecosystems (Crews et al., 2000; Vitousek and Hobbie, 2000). It is estimated that asymbiotic fixation contributes ~31% to the total biological N-fixation rate, while the remainder is from symbiotic fixation in higher plants (Schlesinger, 1997). Biological nitrogen

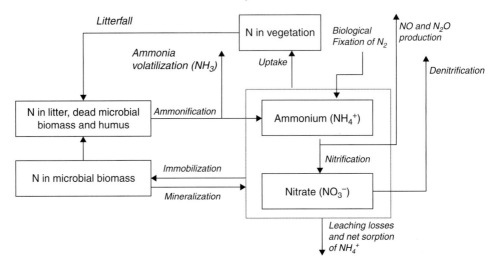

Figure 3.9. Schematic representation of the nitrogen cycle as described in Section 3.2.1.

fixation is highly variable temporally and spatially (Gerber et al., 2010). N fixing plant species in temperate and boreal regions (e.g., alders) are few and limited to recently disturbed communities, whereas tropical N fixing species (e.g., legumes) are abundant even in mature plant communities (Crews, 1999). Individual N fixing plants in tropical forests down-regulate their N-fixation if sufficient N is available in local environments (Barron, 2007).

Plant demand for mineral N is met by two different mechanisms of root uptake: either passively through the soil solution during transpiration or actively by a diffusive flux that is driven by concentration gradients produced by the plant (Engels and Marschner, 1955; Larcher, 1995). Active uptake only occurs when the nitrogen demand by plants is higher than the passive supply by transpiration. If both mechanisms are insufficient to meet the plant demand, the plants are subject to nitrogen limitation.

In N-limited environments, the rate of decomposition is strongly controlled by the SOM C/N ratio because microbial growth requires a fixed proportion of C and N (Tables 3.5 and 3.6; Figure 3.10). As a consequence, if the nitrogen content of the organic matter being decomposed is high (i.e., C/N < 24), mineralization proceeds unrestricted and excess mineral N is released into the soil (Brady and Weil, 2008). In contrast, when the litter is nitrogen poor (i.e., C/N > 24), microbes use the mineral nitrogen through the process of immobilization (Brady and Weil, 2008; Figure 3.10). Thus, soils beneath vegetation with a high litterfall C/N ratio tend to have low rates of mineralization (Gosz, 1981), whereas the opposite is true for soils with a low litterfall C/N ratio.

During transformations of N in the soil, a variety of N gases are emitted as products and by-products of microbial activity (Schlesinger, 1997). In soils, ammonium may be converted to ammonia gas (NH_3) and subsequently emitted to the atmosphere (Langford et al., 1992; Sutton et al., 1993). During decomposition of organic matter,

Table 3.5. *Observed C/N ratios of decomposer and substrate biomass with standard deviation in parenthesis*

Substrate type	Climatic region	Decomposer C:N	Substrate C:N	References
Litter	BR	12.4 (5.7)	52.8 (39.3)	Baath and Soderstrom (1979); Wagener and Schimel (1998); Schimel et al. (1999); Ross et al. (1999); Hannam et al. (2007)
	TM	11.0 (2.3)	44.1 (14.8)	Stark (1972); Hobbie et al. (2001); van Meeteren et al. (2008)
	TR	25.8 (5.7)	19.9 (1.8)	Stark (1972)
O Layer	BR	10.7	22.1	Wagener and Schimel (1998)
Soil organic matter	BR, TM, TR	7.37	12.3	Cleveland and Liptzin (2007)
Wood	TM	15.1 (7.1)	343.6 (320.2)	Edmonds and Lebo (1998); Hart (1999); Hobbie et al. (2001)

Note: Climatic regions are: BR, polar, subpolar, and boreal; TM, temperate; and TR, tropical and subtropical.
Source: Manzoni et al., 2010.

Table 3.6. *Mean foliar and litter nutrient ratios +/− 1 standard error*

Forest type	Foliar C:N	Foliar C:P	Senesced litter C:N	Senesced litter C:P
Overall	43.6 ± 3.5	1334.1 ± 137.6	66.2 ± 6.3	3144.1 ± 341.9
Temperate broadleaf	35.1 ± 3.7	922.3 ± 77.3	58.4 ± 3.7	1702.4 ± 170.3
Coniferous	59.5 ± 7.0	1231.8 ± 140.3	87.8 ± 6.9	2352.9 ± 350.3
Tropical	35.5 ± 4.1	2456.9 ± 503.7	60.3 ± 13.2	4116.0 ± 577.4

Source: McGroddy et al., 2004.

many soil organisms excrete NH_x (i.e., NH_3 and NH_4^+) either directly or indirectly and, thus, NH_3 fluxes are closely related to biological activity in the soil (Schlesinger and Hartley, 1992). Approximately 1% of the NH_4^+ present in the topsoil escapes as NH_3 gas (Bouwman et al., 1997). However, a large fraction of the NH_3 emitted from the soil surface may be absorbed by plant leaves. On a global scale, it is estimated that about 50% of NH_3 soil emissions are captured by canopies (Bouwman et al., 1997). Losses of NH_3 are generally low in natural forest and grassland soils (i.e., the global flux is $\sim 10 \times 10^{12}$ g N yr^{-1} from these sources; Schlesinger and Hartley, 1992) with losses of NH_3 being greatest from fertilized soils (Terman, 1980). Both nitric oxide (NO) and nitrous

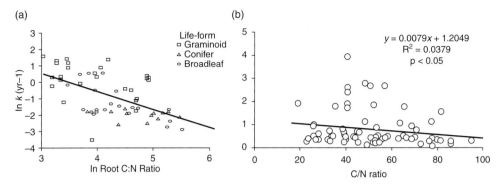

Figure 3.10. Decomposition constant (or k value (yr^{-1})) as a function of (a) the root C/N ratio.
Source: A metaanalysis of 175 studies by Silver and Miya, 2001.
(b) The litter C/N ratio.
Source: A metaanalysis of 110 global studies compiled by Zhang et al., 2008.

oxide (N_2O) are generated as microbial by-products of nitrification (Williams et al., 1992; Figure 3.9). These two gases play an important role in the chemical and radiative properties of the atmosphere (Schlesinger, 1997). Tropospheric N_2O is a major greenhouse gas, while NO leads to the formation of nitric acid and acid rain. Approximately 1%–3% of the N passing through the nitrification pathway is volatilized as NO with the global flux from natural soils contributing ~10–12 × 10^{12} g N yr^{-1} (Prather et al., 1994; Potter et al., 1996). Nitrate is also converted to NO, N_2O, and N_2 in the process of denitrification (e.g., Knowles, 1982). Annual denitrification rates in forest soils range from < 0.1 to 40 kg N ha^{-1} yr^{-1} (e.g., < 0.1–2.4 kg N ha^{-1} yr^{-1} for undisturbed coniferous and 0.3–2.8 kg ha^{-1} yr^{-1} for undisturbed deciduous; Barton et al., 1999). The relative importance of denitrification as a source of these gases depends on environmental conditions, with moist soils where the diffusion of oxygen is slower contributing to higher rates of denitrification (Schlesinger, 1997).

Other processes that reduce mineral N in the soil are leaching and net sorption of ammonium. Nitrate is readily soluble in water, a property that facilitates its uptake by plants, yet also makes it prone to leaching losses at high soil moisture levels. In contrast, the positive charge of ammonium ions attracts them to the negatively charged surfaces of clays and humus, thus partially protecting them from leaching, particularly in clayey soils. Despite its being in an exchangeable form under these conditions, fixed ammonium has a rate of release that is often too slow to fulfill plant demand (Porporato et al., 2003).

3.2.2 Location of N-Limited Forests

Prior to anthropogenic disturbance, N-limited forests were largely distributed throughout temperate and boreal ecosystems (Hietz et al., 2011). However, acid

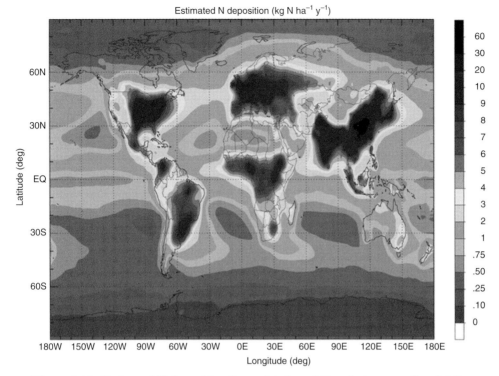

Figure 3.11. Estimated N deposition from global total N emissions, totaling 0.1 Pg N yr^{-1}. See also color figures at the end of the book.
Source: Galloway et al., 2008.

deposition associated with the burning of coal, oil, and gas; the use of N fertilizer; and the increase in N fixing crops has altered these patterns (Galloway et al., 1995; Figure 3.11), leading to increased deposition of N containing compounds in temperate forests that were once N-limited (Aber et al., 1989). In the 1990s, nitrogen fertilizer used in food production and nitrogen emitted to the atmosphere during fossil-fuel combustion amounted to more than 160 Pg N per year, which is more than that supplied by natural biological nitrogen fixation on land (Gruber and Galloway, 2008). Notably, the extent of N-limited forests is widespread across biomes (LeBauer and Treseder, 2008) and the pattern of N deposition is likely to shift increasingly to the tropics (Galloway and Cowling, 2002; Lamarque et al., 2005). Deforestation in the tropics has also altered patterns of N-limitation such that unlike primary forests growing in the same areas, some secondary tropical forests in the early stages of old-field succession have been documented as N-limited (Davidson et al., 2004). Secondary forests account for approximately 40% of tropical forests and thus, N-limitation may be widespread throughout the tropics (Brown and Lugo, 1990; FAO, 1995). For example, the growth of tropical forests located on young volcanic soils in Hawaii was highly responsive to N additions, suggesting that the relatively high availability of P in these young soils leads to conditions where productivity is more likely to be constrained by nitrogen (LeBauer and Treseder, 2008).

N deposition to ecosystems in the absence of anthropogenic influence is ~0.5 kg N ha^{-1} yr^{-1} (Dentener et al., 2006); however, there are now large regions of the world where average N deposition rates exceed 10 kg N ha^{-1} yr^{-1} (Galloway et al., 2004). By 2050, this estimate has been projected to double, with some regions reaching 50 kg N ha^{-1} yr^{-1} (Galloway et al., 2004). Elevated N deposition may stimulate plant growth (as well as shifts in forest composition) in N-limited regions and cause substantial CO_2 uptake in Northern Hemisphere forests (Galloway et al., 2008). It is projected that tropical regions will receive the most substantial increases in N deposition over the next few decades (Zhu et al., 2005). However, the response of P-limited tropical forests and N-saturated temperate forests to these increased N inputs could differ substantially from those of N-limited temperate ecosystems, possibly resulting in elevated concentrations of nitrate, aluminum, and hydrogen in streams, increased cation leaching from soils, and nitrate losses. These changes would lead to reduced soil fertility and increased acidity and possibly increasing emissions of trace gases such as nitrous oxide (Matson et al., 1999; Aber et al., 1989).

3.2.3 Change in the N Cycle Following Deforestation

3.2.3.1 Increase in N Mineralization and Losses to Leaching

After deforestation, there is an increase in leaching and runoff losses of nitrate to groundwater and streamwater (e.g., Sollins and McCorison, 1981). These losses increase when the harvesting method is intensified (Bengtsson and Wikstrom, 1993) by repeated cuttings (Bengtsson and Wikstrom, 1993; Liski et al., 1998) or by shortening the rotation length (Aber et al., 1978). After the removal of forest vegetation, Likens et al. (1970) observed that nitrate concentrations in streams draining the deforested catchment were 41- and 56-fold higher the first and second year after deforestation in comparison to undisturbed conditions. Nitrate export to streamwater increases after deforestation because of reduced plant uptake, increased rates of N mineralization and nitrification (Likens et al., 1970, Aber et al., 2002; Bormann and Likens, 1979; Matson and Vitousek, 1981; Vitousek and Matson, 1985), and increased leaching (see Section 3.3.2). Moreover, in-stream uptake of nitrate by aquatic biota may substantially underestimate measurements from streamwater samples of these losses from the soil profile (Bernhardt et al., 2003).

Deforestation can increase soil temperature and moisture, thereby enhancing conditions for mineralization (i.e., the conversion of organic to inorganic N; Attiwill and Adams, 1993; Figure 3.12). Subsequently, nitrification may be so rapid that uptake by regrowing vegetation and immobilization by microbes are insufficient to prevent large losses of NO_3^- to streamwater and groundwater (Schlesinger, 1997). An

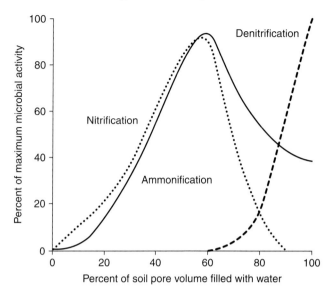

Figure 3.12. Rate of microbial activity related to the various phases of nitrogen transformation as a function of the soil water content.
Source: Porporato et al., 2003.

increase in nitrification rates results because in the absence of vegetation, the soil microflora rapidly oxidize ammonium to nitrate despite potential decreases in soil pH (Likens et al., 1969). The increase in soil temperature and soil moisture can also lead to rapid ammonification and, thus, increase the availability of NH_4^+ (Schlesinger, 1997; Figure 3.12).

3.2.3.2 Decrease in Organic Nitrogen

Substantial quantities of N can be lost during deforestation through the removal of organic N contained in logged wood. To estimate the quantity of N removed from the forest during clear-cut, the percentage of aboveground biomass removed multiplied by the C/N ratios of the aboveground biomass could be used. Finer et al. (2003) estimated the change in N pools after clear-cutting in an old-growth Norway spruce mixed boreal forest in eastern Finland and found that although most of the C pool was contained in the living trees (i.e., 60%), only 15% of the ecosystem N pool (i.e., biomass plus soil) was in living trees. During partial clear-cut, 3% of the ecosystem N pool was reduced, and if all the trees had been removed, 11% of the ecosystem N pool would have been exported from the system. Similar results pertaining to the distribution of N in vegetation versus soils have been found for other ecosystems (i.e., subtropical; Yang et al., 2005, where 4% of the ecosystem N pool was removed during clear-cut).

Sollins and McCorison (1981) estimated that 400 kg N ha^{-1} was removed in tree boles after deforestation of a Douglas fir dominated watershed in the Oregon Cascade

Mountains. This was a relatively large amount in comparison to the streamwater losses of total dissolved N after cutting (<2.0 kg ha^{-1} yr^{-1}). At the control (i.e., forested) site in their study, ammonium concentrations accounted for 18%–33% of the total dissolved N in solution, organic N accounted for virtually all of the remainder, and nitrate concentrations were very low. In contrast, nitrate accounted for a large fraction of solute N concentrations and organic N accounted for a relatively small fraction at the clear-cut site. Fredriksen (1975) found that after forest clear-cut at two experimental watersheds in western Oregon, sharp increases in stream N concentrations were attributed to decreased plant N uptake and increased detritus N that was subject to mineralization into ammonium.

In contrast to forests that are logged, fire differently alters patterns of N cycling. During burning there is a loss of both N stored in vegetation (via smoke and airborne ash) and soil N, which is lost to wind and water erosion (Khanna et al., 1994; Raison et al., 1985; Kauffman et al., 1993). For instance, Kauffman et al (1993) conducted three experimental burns in a tropical dry forest, where 78%, 88%, and 95% of the aboveground biomass were consumed by fire. During these burns, as much as 96% of the prefire aboveground N pool was lost during the combustion process. After fire, up to 47% of the residual aboveground biomass was in the form of ash, which was rapidly lost to wind erosion. Despite these losses of organic N, fire can lead to a short-term increase in N mineralization rates. For instance, postfire studies of N pools in a variety of ecosystems have shown 2- to 26-fold increases in soil ammonium that are relatively short lived (<2 yr) and 2- to 5-fold increases in soil nitrate concentrations (Covington and Sackett, 1992; Wan et al., 2001).

3.2.3.3 Loss to the Atmosphere

Atmospheric losses of N result both during and after deforestation. The quantity of N lost is dependent on the mechanism of forest removal. Fires associated with land clearing volatilize 300–700 kg ha^{-1} of N from vegetation and surface litter (Nye and Greenland, 1964; Ewel et al., 1981). Fire volatilizes N from vegetation and litter, reducing N stored in the burned ecosystem but often increasing mineralization of the remaining organic matter (e.g., Turner et al., 2007a). During biomass burning, significant amounts of NH_3 are injected into the atmosphere (Warneck, 1988). Approximately 11% of the NH_3 emitted into the atmosphere is due to biomass burning as a result of deforestation (Bouwman et al., 1997). The loss of the forest canopy also leads to a decrease in the amount of NH_3 emitted by the soil that is captured by plant leaves.

As noted in Section 3.2.1, denitrification is the reduction of nitrogen oxides to di-nitrogen gas by microorganisms. Studies comparing undisturbed and disturbed coniferous forests have shown up to a 10-fold increase in denitrification in disturbed sites, which has been attributed to increased nitrate availability (Hulm and Killham, 1988; Myrold et al., 1989; Dutch and Ineson, 1990). However, increased denitrification rates after clear felling may only occur for a relatively short period. For example, denitrification has been shown to return to prefelling rates after four years (Dutch and Ineson, 1990). High soil moisture levels after deforestation may also increase the rate of denitrification and the release of nitrogen gases.

3.3 Phosphorus Cycle

We begin this section by describing the phosphorus cycle in undisturbed forests to provide the basis for understanding how deforestation alters phosphorus cycling. Next, we examine how phosphorus limited forests are distributed globally, including factors that lead to phosphorus limitation. Last, we describe how deforestation might alter the phosphorus balance of forests and their soils.

3.3.1 Phosphorus Cycling in Undisturbed Forests

Soil phosphorus (P) availability to an ecosystem is supplied almost entirely from the weathering of underlying mineral soil and parent rock material or from atmospheric inputs that are external to the ecosystem because P does not have a significant gaseous component (Toy, 1973). P inputs via atmospheric deposition are important in areas where the availability of P is low and there is a relatively high flux of P containing dust being transported and deposited to that ecosystem (Reed et al., 2011; Figure 3.13). P is found in two major pools, organic and inorganic; however, different factors control the rate at which P becomes available from these pools. During early stages of soil development, P becomes available for plant uptake via weathering of calcium phosphate minerals, such as apatite (Figure 3.14). The total inorganic P pool includes the available fraction ($P_{nonoccluded}$) and the less available, occluded fraction ($P_{occluded}$). The available fraction can account for a relatively small percentage of the total P pool (i.e., <1% in highly weathered soils; Resende et al., 2010), whereas the occluded pool can account for 30%–70% of P found in soils, and the average organic fraction ranges between 30% and 65% of total P (Sample, 1980; Harrison, 1987; Jones and Oburger, 2011). P sorption to more occluded forms removes both inorganic and organic P from soil solution into a relatively unavailable form for plant uptake (Berg and Joern, 2006). Soils with a high capacity for P sorption include those with a high percentage of secondary clays such as kaolinite, where P is adsorbed, and highly weathered soils, where P is bound with Fe or Al (Reed et al., 2011).

Plant roots and soil microbes take up inorganic soil P and use this P to synthesize biomass growth. Phosphorus is an essential nutrient for plants and is the primary component of adenosine triphosphate (ATP), which drives the uptake of nutrients and their transport within the plant. Moreover, P is used to synthesize fundamental plant processes such as photosynthesis and root growth. Phosphorus is necessary for

94 *Global Deforestation*

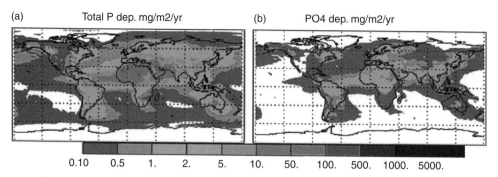

Figure 3.13. Model estimated deposition fluxes of (a) total phosphorus (TP) and (b) available P (PO_4). See also color figures at the end of the book.
Source: Mahowald et al., 2008.

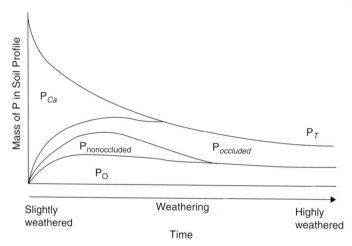

Figure 3.14. Conceptual model of Walker and Syers (1976) showing variations in soil P pools during pedogenesis. Total soil P (P_T) includes all P pools during a given time: calcium phosphates (P_{Ca}); organic P (P_O); sorbed P that is relatively unavailable to organisms ($P_{occluded}$); and P that is relatively available for uptake by plants, oftentimes called "labile" P ($P_{nonoccluded}$).
Source: Walker and Syers, 1976.

photosynthesis because ATP is required during photosynthesis to convert light energy to chemical energy. Synthesis of vegetation biomass requires inorganic P to form phospholipids and genetic material; without P the synthesis of metabolic active tissue is not possible (Westheimer, 1987).

In regions with sufficient P availability, passive uptake of P is sufficient to satisfy vegetation demand of P. In regions with low P availability, vegetation relies on active P uptake mechanisms. P deficiency can be minimized because plant roots, microorganisms, and fungi can produce extracellular phosphatases that directly mineralize (i.e., "mine") unavailable forms of P under low levels of labile P (McGill and Cole,

1981; Dakora and Phillips, 2002; Chapin et al., 2002; Vance et al., 2003; Barroso and Nahas, 2005). This is advantageous to vegetation because the ester linkages that bind phosphorus to organic matter must be cleaved to produce a form of P that is water-soluble and can be taken up by plants (e.g., Chapin et al., 2002). The strongest control on enzyme activity is P availability. Under low levels of available P, phosphatase enzymes are produced to mineralize P (McGill and Cole, 1981; Nannipieri et al., 2011), while phosphatase production is repressed in the presence of an adequate supply of available P (Jones and Oburger, 2011; McGill and Cole, 1981). P availability is also enhanced via cluster roots that combine specialized structure and physiology to maximize P acquisition from soils with low P availability (Lambers et al., 2008a). Cluster roots produce large amounts of carboxylates, which release P from strongly sorbed forms by either replacement of P bound to Al or Fe in acid soils or to Ca in alkaline soils or by local reduction of pH in highly alkaline soils (Lambers et al., 2008a). Nutrient acquisition from nutrient-impoverished soils also involves symbiotic structures (e.g., mycorrhizae, root nodules) that increase access to nutrients by scavenging large volumes of soil (Lambers et al., 2008a). Mycorrhizae can increase the volume of soil that is explored by roots and can acquire P at distances of several millimeters up to 15 cm away from the root (Chapin et al., 2002). Experiments with radioactive P have found that 0%–30% of P is taken up via mycorrhizal fungi (Smith et al., 2004; Jakobsen et al., 2005). Scavenging is achieved by rapid root growth, localized root proliferation in enriched zones, development of root hairs, or establishment of symbioses with mycorrhizal fungi (Lambers et al., 2008a). The mining strategy dominates in highly weathered soils because most P is sorbed to soil particles, and little P is in solution; the scavenging strategy is common in younger soils, where the P concentration in the soil solution is sufficient to support mycorrhizal symbioses (Lambers et al., 2008a). It is estimated that approximately 70% of angiosperms are mycorrhizal; coniferous species always are mycorrhizal (Trappe, 1987).

P is cycled back to the forest floor in organic forms via litterfall, microbial biomass, and decayed roots (Figure 3.15). In P-limited areas, species are often more efficient in retranslocating P from leaves prior to leaf senescence (McGroddy et al., 2004; Reed et al., 2011). For instance, McGroddy et al. (2004) found that live leaves lost during the growing season (e.g., as a result of a storm) have nearly twice as much P per unit biomass as leaves that drop after senescence. Wood C:P ratios tend to be two to three orders of magnitude higher than the C:P ratios of canopy leaves (e.g., McGroddy et al., 2004; Table 3.7). The amount of P stored in leaf litter and other plant residues varies, depending on decomposition rates and seasonality, yet is generally low relative to the amount of organic P (P_O) stored in other compartments, which is in part the result of the rapid turnover time of this pool (less than one year). For instance, Resende et al. (2010) found that for a Brazilian Cerrado, only 2% of total P_O was stored in leaf litter and dead root biomass. The residence time of leaf litter is dependent on the quality of that litter such that litter with a high C:P ratio will take

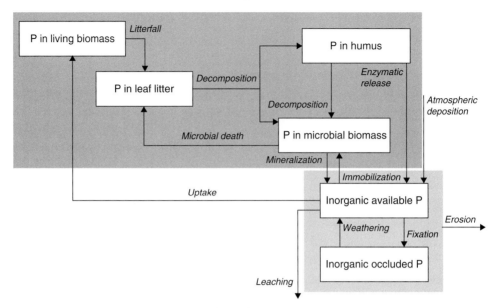

Figure 3.15. Schematic representation of the P cycle as discussed in Section 3.3.1. In this schematic, the dark gray box represents organic pools while the light gray box represents inorganic P pools.
Source: After Runyan and D'Odorico, 2012.

a longer time to decompose and could therefore contribute to P limitations (Aerts, 1997; Vitousek et al., 1994; Hobbie, 1992). Wieder et al. (2009) found that P availability strongly controlled rates of litter decomposition in a lowland tropical forest in Costa Rica, a finding that corroborates results from other studies (e.g., Aerts, 1997; Hobbie and Vitousek, 2000; Santiago, 2007; Table 3.7).

The fraction of organic P (P_O) stored in the microbial biomass typically accounts for 20%–30% of organic P in soils (Oberson and Joner, 2005; Chapin et al., 2002). Storage within this compartment is important because the turnover time of microbial P is only a few months, and this fraction is protected from reactions with the mineral phase of the soil, which makes it more available over the long term than available inorganic P (Chapin et al., 2002). Similarly, although P in microbial biomass may be temporarily immobilized, it remains in an available form that can be readily mineralized (i.e., within 24 hours) after microbial death (Jones and Oburger, 2011).

3.3.2 Location of P-Limited Forests

There are six mechanisms that can cause P limitation to terrestrial ecosystems: P depletion, soil barriers, transactional constraints, low-P parent materials, P sinks, and anthropogenic forcings (Vitousek et al., 2010). P depletion occurs when there is a progressive loss of P during long-term soil and ecosystem development (Walker and

Table 3.7. *Observed C:P ratios of decomposer and substrate biomass*

Substrate type	Climatic region	Decomposer C:P	Substrate C:P	References
Litter	BR	71.4		Baath and Soderstrom (1979)
	TM	82.7 (82.5)	417.1 (448.4)	Stark (1972); van Meeteren et al. (2008)
	TR	186.5 (20.4)	502.3 (294.1)	Stark (1972)
O Layer	TM	204.9 (27.8)	370 (99.0)	Cromack et al. (1975)
Soil organic matter	BR, TM, TR	23	72	Cleveland and Liptzin (2007)
Wood	TM	158.5 (147.8)	2857.5 (3206.7)	Cromack et al. (1975); Edmonds and Lebo (1998)

Note: Climatic regions are: BR, polar, subpolar, and boreal; TM, temperate; and TR, tropical and subtropical. The standard deviation is in parentheses.
Source: Manzoni et al., 2010.

Syers, 1976; Figure 3.14). This mechanism of P limitation tends to occur in tropical forests of Hawaii, South America, and Africa that are low in nonoccluded, inorganic P because, unlike in cold-temperate or high-latitude forests, their soils reach ages of 40,000 years, without having their parent material P replenished by glacial or periglacial processes (Vitousek et al., 2010). Soil barrier driven P limitation occurs when portions of the soil are inaccessible to roots and physically separate biota from P and P-bearing minerals. This occurs in areas with clay-rich soil horizons or where iron pans form in high rainfall regions (Vitousek et al., 2010). Transactional P limitation occurs when N fixers are abundant early in ecosystem development relative to the availability of P. Parent material based P limitation occurs because the P content of varying parent materials can differ by more than two orders of magnitude (McBirney, 1993). As an example, forests found on sandy Spodosol soils with low soil P stocks are limited by this mechanism (Davidson et al., 2004). Sink driven P limitations occur when there is strong adsorption of P on colloids and P precipitates with iron, aluminum, manganese, or calcium. This limitation exists in dry tropical forests of the southern Yucatan Peninsula found on highly calcareous soils where P is chemically bound to calcium and clay constituents that have a high capacity for P-fixation (Vitousek, 1984; Silver, 1994). Anthropogenic P limitations occur when human activity induces P limitation by affecting the supply of other resources, most often N (Vitousek et al., 2010). Human enhanced N deposition is now concentrated in many formerly N-limited areas of Europe, central and eastern North America, and East Asia (Galloway and Cowling, 2002). For example, because of high levels of atmospheric N deposition, temperate forests in northwestern Europe are now limited

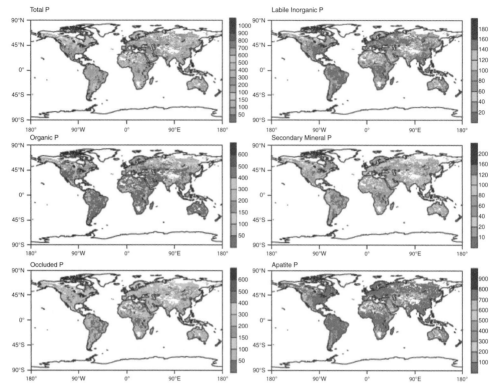

Figure 3.16. Schematic representation of the P cycle as discussed in Section 3.3.1. In this schematic, the dark gray box represents organic pools while the light gray box represents inorganic P pools. See also color figures at the end of the book.
Source: Yang et al., 2013.

primarily by P (Verhoeven and Schmitz, 1991; Aerts et al., 1992), and similar results have been obtained in controlled experimental settings in North America (Tilman, 1987; Huenneke et al., 1990; Suding et al., 2005).

In general, total P is lowest in tropical regions, where soils have gone through millions of years of soil development and have lost much of their P through leaching or erosion (Walker and Syers, 1976; Figure 3.16). This trend results because weathering of parent material and leaching of P from soils are enhanced in warm and humid climates of tropical regions and parent P material is not replenished via glacial processes. The weathering stage of the soils is reflected by the global distribution of apatite P (Figure 3.16). As discussed previously, total P is composed of organic and inorganic P. Organic soil P tends to be lower in tropical regions and higher in temperate regions (Figure 3.16; Yang et al., 2013). The low organic P in tropical regions is the result of several processes: loss of P through leaching, adsorption of P on secondary aluminum and iron oxide minerals, and increasing occlusion of P by Al and Fe minerals under low soil pH conditions, all of which increase with soil development (Figure 3.16; Yang et al., 2013). In addition, the warm and humid climate in tropical regions enhances soil organic matter decomposition and mineralization of organic P. In contrast, most soils in temperate regions are of intermediate stages. In these soils, there are a continuing supply of inorganic P from weathering of parent material and moderate losses of P due to leaching and erosion; these characteristics lead to an adequate supply of P for plants and microbes and the highest P_o in soils (Yang et al., 2013). The relatively slow organic matter decomposition in temperate regions also contributes to the accumulation of P_o. The proportion of occluded P in total P generally increases with soil development (Yang et al., 2013).

3.3.3 P Losses after Deforestation

3.3.3.1 P Exported from the System during Deforestation

P losses resulting from deforestation will vary, depending on the method of forest removal. For forests that are logged, large-scale P losses will result from the P contained in exported timber that is removed from the system. Another type of deforestation that is widespread in P-limited areas is shifting cultivation, where forests are cleared by slash and burn. During the burning phase, P stored in vegetation is lost as smoke and airborne ash, while soil P is lost to wind and water erosion (Khanna et al., 1994; Raison et al., 1985; Kauffman et al., 1993). The quantity of P stored in vegetation lost during the burning phase is highly dependent on the fraction of total biomass consumed by fire. Kauffman et al. (1993) found for a controlled burn in a neotropical dry forest where 78% of total aboveground biomass was consumed by fire that only 3.5% of aboveground plant P was lost. In contrast, during a burn where 95% of the total aboveground biomass was consumed by fire, there was a net loss of 56% of the aboveground biomass P pool. After biomass burning, there is an immediate, large P input to the soil via P transported in ash and other deposited organic material as well as transformations in the soil due to pyromineralization (Lawrence and Schlesinger, 2001). Pyromineralization can increase the amount of available P from more resistant forms of P in the top 10 cm of the soil (Saa et al., 1993; Kauffman et al., 1993; Giardina et al., 2000; Lawrence and Schlesinger, 2001). This increase in P content due to pyromineralization can be substantial (e.g., Giardina et al., 2000). For instance, Garcia-Montiel et al. (2000) studied a pasture chronosequence in the Amazon and found an initial increase in available P pools during the first three to five years after pasture establishment, followed by a gradual decline thereafter. The process of gradual reacidification of the soil can lead to a decrease in available P because cations from ash are lost to erosion or accumulate in biomass (Markewitz et al., 2001; de Moraes et al., 1996).

Large losses of P immediately after a burn are in part due to the high percentage of P that is contained in ash, which is very susceptible to entrainment and removal by air and water flow during and after fire. Kauffman et al. (1993) found that after fire, up to 84% of the residual aboveground P was in the form of ash, which was quickly lost through erosion. The export of harvested crops following cultivation and increase in erosion and soil P leaching when the vegetation cover is removed (during harvesting) further contribute to P losses from the system (Stoorvogel et al., 1993; Runyan et al., 2012b).

3.3.3.2 Loss of Microbes and the Alteration of Microbially Mediated P Fluxes

Soil microbial biomass aids in reducing net losses of P by temporarily storing P in a relatively available form that can readily be mineralized after microbial death (Runyan and D'Odorico, 2013). Microbial P can be mineralized readily as a result of the low microbial C to P ratio (e.g., Manzoni et al., 2010). This is important because P cannot be biologically replenished. Thus, when P is stored in the microbial biomass, losses to less available forms via chemical fixation by iron and aluminum oxides and from the system via leaching and erosion are reduced (Resende et al., 2010). After deforestation there can be a considerable reduction in microbial biomass (Henrot and Robertson, 1994; Borneman and Triplett, 1997; Caldwell et al., 1999; Nusslein and Tiedje, 1999; Waldrop et al., 2000; Bossio et al., 2005; Chaer et al., 2009). The loss of microbial biomass can contribute to enhanced P losses from leaching and to less available forms (e.g., Runyan and D'Odorico, 2013).

In addition to the role that the soil microbial biomass plays in mitigating nutrient losses, forest vegetation is often dependent on arbuscular mycorrhizal fungi for increasing nutrient availability (Bever et al., 2001; Lambers et al., 2008a). Mycorrhizae cannot establish without a higher plant host, and removal of the forest can cause a loss of the mycorrhizal fungus that in turn makes it difficult for trees and mycorrhizae to reestablish after deforestation (Perry et al., 1989). In the absence of a host, populations of mycorrhizal fungi can decline rapidly because they become vulnerable to consumption by saprophytes and to loss by soil erosion (Perry et al., 1987; Janos, 1988; Corman et al., 1987; Perry et al., 1989). The loss of mycorrhizae and plant roots leads to a reduction in available P because mycorrhizae i) exude phosphatases that cleave C-P ester bonds (Colpaert et al., 1997; Tarafdar and Marschner, 1994); ii) influence solubility (Easterwood and Sartain, 1990) through exudation of charged organic compounds that compete with PO_4^{3-} for binding surfaces on other elements; iii) alter soil pH (i.e., through acidification or alkalinization), which influences P binding with elements such as Fe, Al, and Ca (Chapin et al., 2002); and iv) increase the scavenging volume of the soil (Lambers et al., 2008a).

3.3.3.3 Increase in Physical Losses of P

Although there is no gaseous form of P that can be lost in significant amounts during fire or from the soil, after deforestation, nutrient losses via leaching (i.e., percolation beneath the rooting zone) increase (e.g., Lawrence et al., 2007). This is in part due to the resulting changes in soil moisture and the lack of uptake by tree roots. While removal of forest vegetation leads to an increase in soil evaporation resulting from the increase in solar radiation reaching the ground surface and soil temperatures (Sun et al., 2001), in humid and mesic (i.e., moderately moist) climates, this increase in evaporation only partially offsets the reduction in transpiration. The large reduction in transpiration is due to the loss of trees and elimination of intercepted water (Bosch and Hewlett, 1982), both of which increase soil moisture (Best et al., 2007; National Research Council, 2008; see Chapter 2). The increase in soil moisture as a result of deforestation leads to higher nutrient losses in runoff via overland flow and leaching losses via increases in the soil hydraulic conductivity. Nutrient losses are further exacerbated by the reduction in roughness elements that can trap nutrients either adsorbed to eroded soil particles or transported within runoff water. Soil erosion also preferentially removes the clay and organic matter fractions (both rich in P), leaving behind the coarser P-poor particles. The increase in runoff water containing P and erosion of soil rich P particles have been documented via an increase in stream P concentrations in deforested versus adjacent forested watersheds. For instance, Biggs et al. (2004) found that total in-stream concentrations of dissolved and particulate P in heavily deforested watersheds were 2.3 ± 1.5 and 1.9 ± 0.8 times those of forested watersheds, respectively.

Removal of the forest canopy can also lead to a considerable reduction in atmospheric inputs of P (Lawrence et al., 2007; Das et al., 2011; Runyan et al., 2013). Results from prior studies have documented the possible existence of a positive feedback between P deposition and the forest canopy whereby the presence of the canopy augments the amount of P deposited by throughfall to the forest floor. Changes in the size, leaf area, or complexity of the canopy enhance deposition as a forest matures. The canopy plays an active role in enhancing nutrient deposition from the air masses that move through it, both by dry deposition and by canopy condensation or fog deposition. This quantity can be considerable. For instance, Lawrence et al. (2007) found that the lost input following deforestation represents >9% of the available phosphorus pool in a mature seasonal dry forest.

3.4 Conclusion

At the global scale, the availability of nutrients and the alteration of the N and P cycles as a result of deforestation affect the production and storage of biomass in terrestrial ecosystems. In turn, these processes influence atmospheric C concentrations (Zhang et al., 2011; Goll et al., 2012). If vegetation growth is constrained by nutrient availability, less carbon will be fixed to structural biomass, and therefore it will be readily respired and released to the atmosphere (Körner, 2006; Canadell et al., 2007). Moreover, increasing concentrations of atmospheric CO_2 can lead to reduced transpiration (Niklaus et al., 1998; Del Pozo et al., 2007), because more rapid uptake of CO_2 is possible with higher stomatal closure (Samarakoon and Gifford, 1995). This can limit the capacity of vegetation to take up nutrients (Pritchard and Rogers,

2000; BassiriRad et al., 2001) and lead to a progressive limitation of nutrients that can quickly limit the initial increase in plant production under elevated atmospheric CO_2 concentrations (Murray et al., 2000; Zak et al., 2003; Luo et al., 2004). Hence, nutrient dynamics in soils and vegetation are crucial for understanding the terrestrial vegetation feedback on the global C cycle (Sardans and Penuelas, 2012; Cernusak et al., 2013) and how these dynamics are modified after deforestation. Increased atmospheric deposition of N in P limited areas (e.g., Galloway et al., 2008) creates an N/P imbalance that may induce significant alterations in Earth's ecosystems and a P limitation of natural ecosystems growth and future agricultural production (Penuelas et al., 2013).

4
Irreversibility and Ecosystem Impacts

4.1 Background on Irreversibility and Bistability in Deforested Ecosystems

In this chapter, we examine the potential situations in which deforestation induces a change in the physical and/or the chemical environment that leads to a loss of environmental conditions necessary to sustain forest vegetation. The reversibility or irreversibility of deforestation is often determined by the absence or presence of positive feedbacks of adequate strength.

Sudden and often irreversible changes in the structure and functioning of ecosystems are typically associated with the existence of multiple stable ecosystem states (e.g., May, 1977; Holling, 1973). We focus on the case of a system that can be stable both with and without forest vegetation (e.g., Box 4.1), although bistable forest dynamics can also emerge in systems with two forested states but with different species compositions (e.g., Pastor and Post, 1988; Ridolfi et al., 2008). The presence of alternative states or "attractors" is commonly associated with positive feedbacks (i.e., a sustained sequence of processes) between forest vegetation and its physical environment, though bistability may emerge in nonlinear dynamics even in the absence of such feedbacks (e.g., Ridolfi et al., 2011; Petraitis, 2013). A change among attractors may be an effect of changes in environmental conditions or disturbance regime that are sustained by changes in forest vegetation (e.g., Wilson and Agnew, 1992). The magnitude of the perturbation required to push the system into the basin of attraction of the stable "deforested state" depends on the resilience of the "forest state," a property defined as the ability of the system to recover that state after a disturbance (Holling, 1973). The occurrence of shifts between ecosystem states depends both on the magnitude of the external disturbance and on the resilience of the initial state of the system (e.g., Folke et al., 2004). As the resilience of an ecosystem's state declines, it becomes increasingly vulnerable to state shifts such that progressively smaller external events can cause regime shifts (Holling, 1973). When a disturbance imposed on a system causes some critical bifurcation point to be passed, this can produce a shift to an alternative stable state (e.g., a state of low vegetation) (Kuznetsov, 1995; Figures 4.1 and 4.2).

Box 4.1 Large-Scale Reforestation Projects

As discussed in Chapter 1, human history has seen several cases of large-scale deforestation in which tree cover loss has occurred within relatively short periods (e.g., a few decades). Reforestation is the opposite process, which has often occurred as small-scale restoration projects or as the result of spontaneous forest regeneration after the cleared land has been abandoned and left fallow. A major ecological restoration project is under way in the *Brazilian Atlantic forest*, an important biodiversity hotspot. This forest was affected by anthropogenic disturbance prior to European settlement, and deforestation intensified during the colonial period (Galindo-Leal and de Gusmao Camara, 2003), first for wood extraction and then for clearing of land for cattle ranching (16th–18th century), sugarcane production (18th century), and coffee plantations (19th–20th century). Presently, the forest is reduced to less than 14% of its cover prior to the European colonization. The landscape is very fragmented and less than 20% of the remaining forested area is in fragments exceeding 50 ha (Melo et al., 2013). In 2009, a large group of stakeholders, including environmental groups, NGOs, private companies, research institutions, and government agencies, launched a large-scale project of forest restoration known as the *Atlantic Forest Restoration Pact*. This project has the ambitious goal of restoring native forest species in 15 million ha by 2050, thereby extending forest cover from 14% to 30% of the pre-European forest area (Melo et al., 2013). This is without a doubt a unique effort that is moving the technology, science, policy, and economics of forest restoration forward. Scientists are addressing questions related to the impact of restoration on biodiversity, water supply, flood regulation, and the development of effective protocols for restoration technology based on assisted natural regeneration and plantation (Rodrigues et al., 2011). At the same time, the success of this initiative strongly depends on its social and economic viability through a system of incentives, payment for ecosystem services and forest products, and other opportunities associated with the need of agribusiness companies to be environmentally certified in order to gain access to foreign markets (Rodrigues et al., 2011). The reforestation process partly capitalizes on current trends of land abandonment in the region (Melo et al., 2013). Interestingly, a recent evaluation (Rodrigues et al., 2011) of 32 restoration projects covering more than 5,000 km^2 of sugarcane and mixed farmland has shown that these landscapes have a relatively low resilience: Despite its diversity (Box 4.2) the remnant forest cover is generally too small and fragmented and the land too degraded by agricultural practices to allow for successful autogenic restoration. Thus, planting high-diversity native species over the entire area appears to be the restoration technique needed in this region (Rodrigues et al., 2011). This suggests that, at least in the area evaluated by Rodrigues et al. (2011), ecosystem dynamics are bistable: After losing its forest cover for a prolonged period, the landscape is unable to recover spontaneously and remains locked in the degraded state. A massive reforestation is then needed to restore the habitat for other plant species and pollinators and enhance forest connectivity, microbial biomass, and other ecosystem functions that are crucial to the stable persistence of the Atlantic forest.

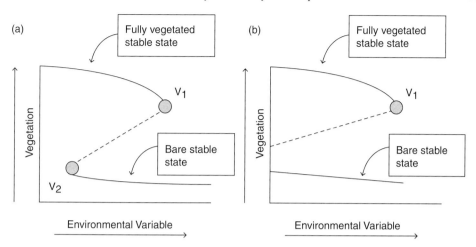

Figure 4.1. Conceptual illustration of hysteresis. (a) When some disturbance is imposed on the system that causes the environmental variable to exceed the bifurcation point, V_1 the system shifts to the lower state. Once this state change occurs, it is not sufficient to restore conditions present before the state change (i.e., with the environmental variable between V_1 and V_2) in order to return to the fully vegetated stable state. Rather, the lower state needs to become unstable; that means that conditions as determined by the quantity of vegetation present and the environmental variable must be restored to V_2. In this figure, stable states of the system are represented by solid lines and these two stable states are separated by an unstable state as represented by the dashed line. (b) Irreversible threshold changes where a change in the controlling variable can shift the system into an alternate stable state from which it cannot recover. *Source*: Runyan et al., 2012a.

Once this shift occurs, the system's dynamics remain locked in the other stable state (Box 4.1) either permanently or until a change in the environmental conditions controlling the vegetation dynamics restores the initial forested state. When the system is in the basin of attraction for the low or bare vegetation state (Figure 4.1a), restoring the environmental conditions present before the collapse is not sufficient to switch back to the fully vegetated state (point V_1 in Figure 4.1). This concept, which is known as *hysteresis*, indicates that in order for the system to return to the other stable state (i.e., the fully vegetated state), the environmental conditions must be restored beyond another bifurcation point (point V_2 in Figure 4.1). Thus, understanding environmental factors that affect a system's resilience is important as an incremental change in environmental conditions has the potential to trigger an irreversible state shift to an unproductive state (Scheffer et al., 2001).

There are several different ways in which deforestation feedbacks can be classified. In this book, we classify them depending on whether they affect vegetation dynamics by modifying the disturbance regime (e.g., fire, exposure to freezing, soil erosion, or waterlogging) or resource availability (e.g., water, nutrients, light). Positive feedbacks that modify resource availability involve interactions of forest vegetation with

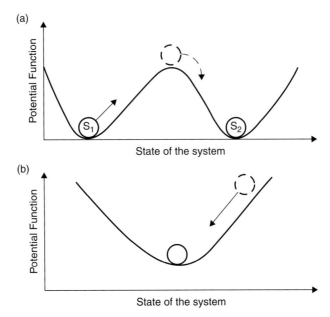

Figure 4.2. Potential function associated with (a) bistable and (b) monostable dynamics. Plots of the potential function are a way to visualize the difference in dynamics between a bistable and a monostable system. The interpretation of these plots is analogous to a landscape. For the bistable system (i.e., (a), the two stable states or attractors are represented as valleys (S_1 and S_2) that are separated by an unstable state located on the hill ridge (indicated by the dashed ball in (a). When disturbed, the ball will roll down toward either of the valley bottoms (i.e., stable states) depending on where the ball is initially located with respect to the unstable state.

precipitation, atmospheric deposition, soil moisture dynamics, water table fluctuations, nutrient cycling, and permafrost melting. Feedbacks that affect the disturbance regime involve interactions of forest vegetation with landslides, fire, waterlogging, salt accumulation in the rooting zone, and exposure to freezing events. For each of these feedbacks, we examine the operating principle, geographical extent, and observational evidence. We also evaluate management practices that enhance forest resilience. Notably, the presence of certain species within an ecosystem or a large array of functional groups can enhance resilience, while their loss can alter the disturbance regime or resource availability (Box 4.2).

4.2 Feedbacks That Modify Resource Availability

4.2.1 Precipitation-Vegetation

Operating Principle. Deforestation alters both the water and the energy budget (Chapter 2). The reduction in surface roughness, leaf area, root depth, and incoming radiation resulting from deforestation (Culf et al., 1996; von Randow et al., 2004;

Box 4.2 Relationship between Diversity and Resilience

Diversity in species and functional groups (i.e., the spectrum of ecosystem functions fulfilled by different species) may be important for helping to maintain ecosystem services (Luck et al., 2003). The *diversity–stability hypothesis* suggests that diversity minimizes the chance of large ecosystem changes in response to environmental change (McNaughton, 1977). This concept was originally proposed by Elton (1958) and became less popular in the 1970s after the work of May (1973). It has since been rigorously tested experimentally and via models (Tilman et al., 2014). The primary arguments of the diversity–stability hypothesis at the community level have been classified into two, not mutually exclusive theories called the *averaging effect* (Doak et al., 1998) and the *negative covariance effect* (Tilman et al., 1998; McCann, 2000). These theories argue that diversity (species richness) increases stability at the community level because diverse plant communities respond differentially to variable background processes. Over time, the different responses of populations lead to more stable community dynamics (McCann, 2000).

Community-level variance tends to decrease with increased diversity (McGrady-Steed and Morin, 2000). The *insurance hypothesis*, which has been used to explain this observation, consists of two components: i) Increasing diversity increases the odds that at least some species will respond differentially to variable conditions and perturbations; and ii) greater diversity increases the odds that an ecosystem has functional redundancy by containing species that are capable of functionally replacing important species (Naeem and Li, 1997; Naeem, 1998; Lawton and Brown, 1993). Yachi and Loreau (1999) tested the insurance hypothesis using a general stochastic dynamic model to assess the effects of species richness on ecosystem processes such as productivity. Results showed two major effects of species richness: i) a *buffering effect* whereby species richness reduced the temporal variance of productivity resulting from asynchronicity of the species responses to environmental fluctuations; and ii) a *performance enhancing effect* whereby species richness increased diversity. Thus, the larger the number of functionally similar species in a community, the greater the probability that at least some of these species will survive environmental changes and maintain the current properties of the ecosystem (Chapin and Shaver, 1985). If different species respond differently to environmental changes, the contribution of some species to ecosystem processes may decrease while that of others may increase when the environment changes. *Response diversity* is observed when the range of reactions to environmental change varies among species contributing to the same ecosystem function. Systems where whole functional groups become extinct as a result of environmental change are characterized by low response diversity and, thus, low resilience (Elmqvist et al., 2003).

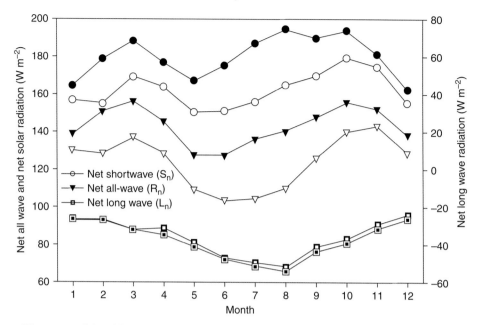

Figure 4.3. Monthly averages of radiation balances from von Randow et al. (2004) of shortwave radiation (S_n), long wave radiation (L_n), and net all-wave radiation (R_n) over forest (closed symbols) and pasture (open symbols). Measurements of energy fluxes were obtained from a forest and pasture site in southwest Amazonia in 1999–2002.

Bonan, 2008b) limits evapotranspiration (ET) and the amount of water vapor that can be recycled into the atmosphere locally through ET (Eltahir and Bras, 1994; Costa and Foley, 1999). Thus, removal of forest vegetation may lead to a decrease in precipitation recycling, which is the fraction of regional precipitation contributed by ET from the same region (see Section 2.8; Eltahir and Bras, 1996). Deforestation alters the surface energy balance (Section 2.7) by (i) increasing the albedo; (ii) increasing the amount of diurnal surface long wave radiation (pastures tend to be warmer than forests during the day, but colder at night); and (iii) reducing the latent heat fluxes through the decrease in ET resulting from the replacement of forests with croplands or pastures (Leopoldo et al., 1993; Nepstad et al., 1994; Jipp et al., 1998; von Randow et al., 2004; Figure 4.3). An increase in albedo results in a reduction in absorbed solar radiation that leaves less energy available to sustain conditions favorable for precipitation (e.g., Eltahir and Bras, 1996; Shukla et al., 1990).

The reduction in ET after deforestation increases the height of the convective boundary layer because of the stronger sensible heat flux over pastures (Gash and Nobre, 1997). When the net radiative energy at the surface is converted primarily into sensible heat, the boundary layer tends to be deep and well-mixed. On the other hand, when a greater proportion of the net radiative energy at the surface is converted into latent heat, a shallow, moist boundary with low cloud base height tends to result

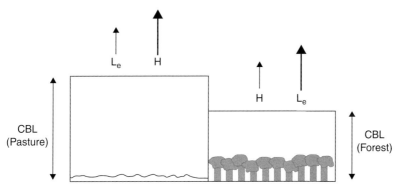

Figure 4.4. The difference in boundary layer dynamics between a forest and a pasture. Over the pasture there is an increase in the sensible heat flux (H) and a reduction in the latent heat flux (L_e), which contribute to a relatively higher convective boundary layer. In turn, this leads to conditions less favorable to the formation of precipitation.

(e.g., Bonan, 2008b). Fisch et al. (2004) found from measurements in southwestern Amazonia that the height of the convective boundary layer for pasture was 1,650 m compared to 1,100 m for forest. The larger latent heat flux over forests also increases the instability of the boundary layer. Collectively, the conditions found over a forest are more favorable to rainfall formation because they support (a) the crossing of the boundary layer height and lifting condensation level (i.e., the height at which a parcel of air reaches saturation) and (b) convection once it is initiated (Juang et al., 2007). Overall, deforestation may reduce precipitation either through its effect on the regional water cycle (decrease in precipitation recycling) or by altering the surface energy balance and the depth, moisture content, and stability of the boundary layer (Costa and Foley, 1999; Figure 4.4).

These effects of vegetation on precipitation imply a positive feedback because the presence of vegetation enhances precipitation, while the removal of vegetation leads to a reduction in precipitation that inhibits the reestablishment of vegetation.

Geographic Extent. As noted, the effect of vegetation on regional rainfall is often investigated in terms of precipitation recycling (Section 2.8), which is the part of regional precipitation contributed by water evapotranspired from the same region (e.g., Salati et al., 1979; Eltahir and Bras, 1996). Van der Ent et al. (2010) found that geographic areas where regional precipitation recycling ratios (i.e., ratios of recycled to total regional precipitation) are the highest include very wet areas, such as tropical forests of South America, Africa, and Southeast Asia, as well as mountainous regions (Figure 4.5). Other regions of the world where this feedback is strong are areas that have relatively low advective moisture fluxes such as subtropical high-pressure zones and convergence zones (Trenberth, 1999). Low advective fluxes are typical of continental areas away from marine sources of atmospheric moisture (e.g., Eltahir and Bras, 1996). The effect of vegetation on precipitation is also important in areas where

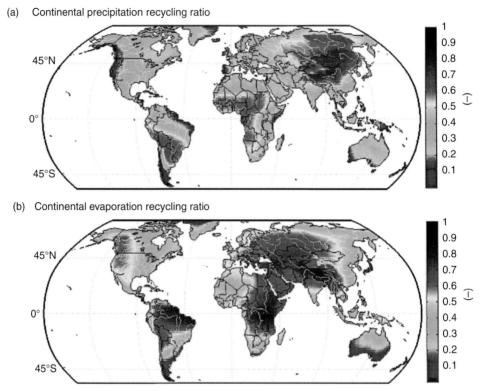

Figure 4.5. (a) Average global precipitation recycling (1999–2008), which is the part of precipitation falling in a region that originates from evaporation within that same region. This ratio describes the region's dependence on evaporation from within the region to sustain precipitation. (b) Average global evaporation recycling (1999–2008), which is the amount of evaporated water that returns as precipitation in the same region. High values of ε_c indicate locations from which the evaporated moisture will fall again as precipitation over continents. See also color figures at the end of the book. *Source*: After van der Ent et al., 2010.

dry season rainfall, which is largely derived from localized convection, tends to dominate (Malhi et al., 2008).

Observational Evidence. A reduction in precipitation in tropical forests has been shown to reduce ecosystem resilience and increase tree mortality. Results from two recent manipulation experiments support these observations. By excluding a fraction of the incoming precipitation to one-hectare rain forest plots in eastern Amazonia, these experiments shifted a rainfall regime typical of evergreen forest into one near to or within the savanna bioclimatic zone (Nepstad et al., 2007; Fisher et al., 2007; Malhi et al., 2009). This shift induced an increase in tree mortality three years after the start of the experiment (Nepstad et al., 2007). The increase in tree mortality under the altered rainfall regime supports the sensitivity of this system to a change in rainfall regime. In addition to the adverse effect on vegetation by the decrease in precipitation, the loss of forest vegetation may also act to sustain the altered rainfall regime. Data from southern and eastern portions of the Amazon have shown that such a transition may already be under way. Evidence for changes in temporal and spatial patterns of precipitation, such as an extended dry season length, has emerged at local and regional scales where deforestation is widespread (Davidson et al., 2012). Historical evidence also suggests that this feedback led to altered climatic and landscape conditions. For instance, this feedback has been used to explain the strong greening of the Sahara in the early and mid-Holocene, some 6,000–9,000 years ago (Claussen et al., 2003).

Modeling Techniques. Minimalist modeling studies have showed that the enhancement of rainfall by vegetation can lead to a positive feedback (Dickinson and Henderson-Sellers, 1988; Foley et al., 1996; Brovkin et al., 1998; Zeng et al., 1999; Da Silveira and Sternberg, 2001). Zeng et al. (1999) used this type of modeling framework to examine the effect of this feedback on ecosystem dynamics and demonstrated that if the general climatic state tends to wet conditions, only the fully vegetated equilibrium exists, whereas at the driest overall climatic conditions, only the unvegetated equilibrium exists. As an effect of the feedback, over a range of intermediate climatic situations, two stable states exist, one with vegetation and the other without.

Complex multidimensional models have been used extensively to investigate the role of deforestation in affecting the rainfall regime (e.g., Claussen, 1997; Xue et al., 1991; Dickinson and Henderson-Sellers, 1988; Foley et al., 1996; Avissar and Liu, 1996). Results from global climate model (i.e., GCM) simulations – particularly for the case of the Amazon – support the existence of a positive feedback with lower precipitation following deforestation at larger spatial scales of deforestation (i.e., 10^5 km^2 and greater) (Henderson-Sellers and Gornitz, 1984; Lean and Warrilow, 1989; Shukla et al., 1990; Nobre et al., 1991; McGuffie et al., 1995; Lean and Rowntree, 1999; Kleidon and Heimann, 1999; Costa and Foley, 2000; Kleidon et al., 2000; Wang and Elathir, 2000c; Werth and Avissar, 2002; Medvigy et al., 2011). For the Amazon, results from these types of models suggest that removal of 30%–40% of the forest could push much of the Amazon into a permanently drier climate regime (Oyama and Nobre, 2003).

Modeling studies have also shown, however, that smaller-scale deforestation associated with the opening of gaps on the order of a few hundred square kilometers may induce pressure gradients that sustain "canopy breezes" toward the cleared area (Section 2.8.4). In turn, convergence resulting from such "breezes" may enhance precipitation (Avissar and Liu, 1996; Figure 4.6).

Management Practices. The spatial scale at which deforestation occurs can influence the strength and direction of this feedback (Baidya Roy and Avissar, 2002; Avissar et al., 2002). Observational data from the Amazon suggest that small-scale deforestation can lead to an increase in precipitation (Chu et al., 1994; Easterling

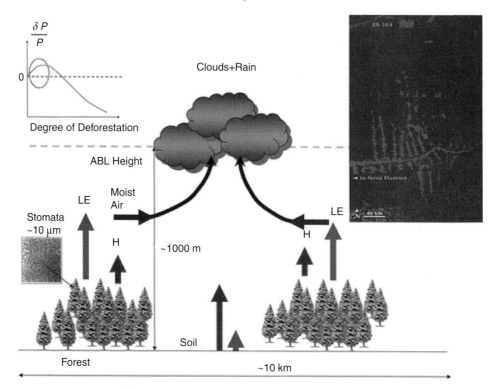

Figure 4.6. The effect of deforestation on water cycling and precipitation. The air temperature and pressure contrast between the forest (high-pressure zone) and the bare soil (low-pressure zone) induces a "canopy breeze." If the sensible heat flux is sufficiently large over an area of similar length or width to the boundary layer height, then the convective updrafts originating from the soil surface vertically lift the moist air originating and advecting from the forested area (and into the deforested area) to the top of the atmospheric boundary layer (ABL). This canopy breeze and its vertical lifting by convective eddies can enhance the predisposition of the deforested area to rainfall because moist air has a lower lifting condensation level (LCL). However, if deforestation progresses further, the amount of water vapor in the atmosphere is reduced by reduced regional ET. Although this reduction results in a higher ABL height, the even higher LCL of the drier air reduces the predisposition to rainfall. *Source*: Katul et al., 2012.

et al., 1996, 2000; Chen et al., 2001; Folland et al., 2001; Negri et al., 2004; Chagnon et al., 2004; Chagnon and Bras, 2005; Baidya Roy and Avissar, 2002). Heterogeneous small-scale reductions in forest cover that occur over tens of kilometers may increase local vertical instability of atmospheric temperature and humidity, thereby leading to rising air motion and enhancing precipitation (e.g., Eltahir and Bras, 1996; Avissar et al., 2002). Baidya Roy (2009) showed by using a regional atmospheric model that a specific type of deforestation pattern (i.e., fishbone) common in the Amazon can lead to higher temperatures over deforested areas that create horizontal pressure gradients, forcing cool, moist air from adjacent forests to converge over the clearings.

This increased convergence generates shallow cumulus clouds preferentially located over the bare patches, leading to increased precipitation over the bare patches (Baidya Roy, 2009).

Once deforestation occurs in larger areas, there can be a reduction in precipitation, which can limit economic activities that were dependent on this source of moisture. For example, once the land is being used to grow crops and the system shifts into the basin of attraction of the deforested stable state, crops may now be limited by available water. The presence of water-limited conditions would lead to a reduction in crop yield. Moreover, in order to sustain the crop yield that would have been supported prior to the state shift, the use of irrigation systems, which may be cost-prohibitive, could be required.

4.2.2 Canopy Deposition

Operating Principle. Plants can trap nutrients and fog water, thereby enhancing atmospheric deposition (see Section 2.2.2). Atmospheric deposition occurs in three separate processes, (i) wet deposition, which is the deposition of precipitation and the nutrients contained in that water; (ii) dry deposition, which is the direct deposition of aerosol particles and gases to vegetation, soil, or surface water; and (iii) occult deposition, which is the deposition of water (and the nutrients contained in that water) from low lying clouds or fog (Lovett, 1994). Plant canopies influence both dry and occult deposition because the larger surface area provided by the foliage enhances the removal of airborne particulates and fog/cloud droplets from the air moving through the canopy (Weathers et al., 2001; Das et al., 2011; DeLonge et al., 2008). Characteristics of both the individual plant as well as the forest influence the trapping strength of the forest canopy: (a) The structure of the canopy affects the intensity of turbulent mixing, which determines the capture efficiency of the particle (i.e., the likelihood that the particle will stay deposited within the local forest area); (b) the shape, orientation, and spatial arrangement of leaves and stems determine the efficiency of particle and droplet transfer across the boundary layers of these surfaces (i.e., leaves and stems); (c) the overall density of the canopy determines the surface area available for deposition (Merriam, 1973; Shuttleworth, 1977; Lovett and Reiners, 1986; Lovett, 1994); and (d) characteristics of the leaf such as stickiness, wettability, morphology, and the presence of hairs on the leaf surfaces, which determine the capture efficiency of a particle once it lands on the plant (Chamberlain and Little, 1981). In turn, the dependency of atmospheric deposition on characteristics of the forest canopy can lead to a reduction in either nutrients or water when deforestation occurs.

Geographic Extent. In areas where the predominant water input is via fog or clouds, the forest canopy can enhance available water because water either condenses or is deposited by low clouds and fog on the surfaces of leaves and then drips down

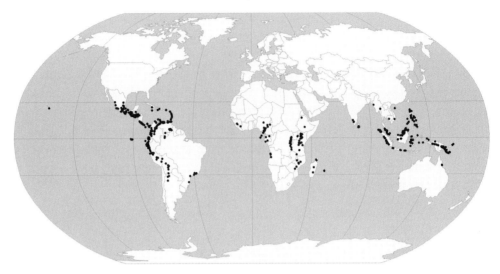

Figure 4.7. Tropical montane cloud forest sites.
Source: Compiled by UNEP-WCMC and published in Aldrich et al., 1997.

to the forest floor (Wilson and Agnew, 1992). This type of trapping can occur in locations where mountaintops protrude into low-lying cloud layers (i.e., cloud forests, which account for 12% of tropical forests; Bubb et al., 2004), when orographic clouds are formed by the rising of air over a mountain range, and when sea or ocean fogs are advected into coastal areas (Lovett and Kinsman, 1990; Lovett, 1994; Dawson, 1998; Figure 4.7). Other areas susceptible to this feedback include phosphorus (P) limited forests such as Florida and the Amazon that receive P inputs from nutrient enhanced atmospheric sources such as dust. Likewise, this feedback can be important in nitrogen limited ecosystems. For instance, N-limited temperate forests in southern Chile (Weathers and Likens, 1997) and Hawaiian montane forests (Heath and Huebert, 1999) are sustained by high inputs of nitrogen received via cloud deposition.

Observational Evidence. The ability of the canopy to increase nutrient or water availability in the area underneath the canopy has been documented for a wide range of ecosystems (Charley and West, 1975; Kellman, 1979; Schlesinger et al., 1990; Matson, 1990; Schlesinger and Pilmanis, 1998; Das et al., 2011; D'Odorico et al., 2011). Ecosystems where this feedback has been documented include semiarid forests in Chile (del-Val et al., 2006; Gutierrez et al., 2008), cloud forests in Central America and the Andes (Holder, 2006; Cavelier et al., 1996; Cavelier and Goldstein, 1989), tropical seasonal forests in Southwest China (Liu et al., 2004), coastal forests in California (Dawson, 1998), high altitude tropical forests in Australia (McJannet et al., 2007), nitrogen limited temperate ecosystems in southern Chile (Weathers and Likens, 1997), and Hawaiian montane forests (Heath and Huebert, 1999), as well as P limited ecosystems such as the Florida Everglades (Redfield, 2002; Wetzel et al., 2005) and dry tropical forests (Das et al., 2011; Runyan et al., 2013).

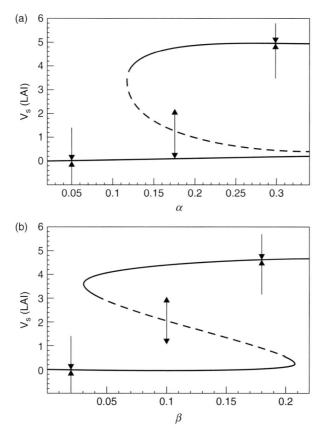

Figure 4.8. Hysteresis in vegetation dynamics due to P deposition. (a) Stable vegetation state, V_s as a function of changes in α; α represents the strength of the deposition–canopy relationship. The solid and dashed lines show stable and unstable solutions, respectively. Arrows show the movement of a change of V_s toward stable and away from unstable states. (b) V_s as a function of β, where β represents the quantity of P inputs that are not associated with canopy cover.
Source: DeLonge et al., 2008.

Modeling Techniques. Minimalist process based models have been used to investigate the canopy trapping feedback (DeLonge et al., 2008). For a nutrient or water limited system, this positive feedback could lead to the presence of two alternate stable states (i.e., one state having high vegetation cover and high canopy trapping potential, and the other having low vegetation cover and low canopy trapping potential). The existence of these two states was shown by DeLonge et al. (2008) for the case of a P limited dry tropical forest (Figure 4.8; Box 4.3). In the model, the dynamics of soil P (P_s) and forest vegetation (V_s) were considered. P_s was dependent on vegetation (using the leaf area index, LAI, as a proxy of tree biomass and canopy density). The steady states of this system were obtained by setting the differential equations expressing the dynamics of P_s and vegetation (V_s) equal to zero and solving

Box 4.3 Minimalist Models of Bistable Ecosystem Dynamics

Minimalist models such as that in DeLonge et al. (2008) express the temporal dynamics of the system through a set of first-order differential equations where equilibrium conditions are determined by setting the temporal derivatives of the state variables equal to zero and solving the associated set of algebraic equations. This simplified representation of the system dynamics comprises two major components: (1) the response of vegetation to changes in the limiting resource (soil P, in this case); and (2) the response of the limiting resource to changes in vegetation cover (Scheffer et al., 2005).

Using this type of modeling framework, changes in vegetation biomass, V, can be modeled through the logistic equation (e.g., Noy-Meir, 1975),

$$\frac{dV}{dT} = aV\left(1 - \frac{V}{V_c}\right) \qquad (4.1)$$

where V_c is the ecosystem carrying capacity [dimensionless], and a regulates the temporal response of the system [time^{-1}]. Ecosystem carrying capacity represents the maximum amount of vegetation sustainable given the available resources (e.g., temperature, light, and nutrients) and the existing disturbance regime. The sensitivity of vegetation to a limiting resource (E) such as soil P can be accounted for using a piecewise function,

$$V_c = \begin{cases} 0, & E < E_{lim} \\ \left(1 - e^{[-\omega(E-E_{lim})]}\right), & E \geq E_{lim} \end{cases} \qquad (4.2)$$

where E_{lim} represents the minimum amount of resources required for vegetation growth/survival and ω [L^{-1}] represents the sensitivity of vegetation to the limiting resource. Given that vegetation controls the limiting resource, a linear relationship (as in DeLonge et al., 2008) has been used to describe the control of vegetation on the availability of this resource,

$$E = \alpha V + \beta \qquad (4.3)$$

where α represents the sensitivity of the limiting resource to vegetation cover and β represents the value of the limiting resource in the absence of vegetation. DeLonge et al. (2008) showed for the case of atmospheric P deposition how both the background deposition as well as the strength of the deposition–canopy relationship affect the number and type of steady-state solutions (Figure 4.8). These relationships show a clear hysteresis such that changes in vegetation cover could drive a catastrophic shift to an alternative and undesirable stable state.

for V_s. This led to two relationships for V_s as a function of P_s (see Box 4.3). Plots of these equations in the (V_s, P_s) coordinates are known as isoclines. Solutions of this set of equations are steady states of the system and correspond to intersections of the isoclines in the (V_s, P_s) plane. Depending on the isoclines, there will be one to three (or more) equilibrium states (i.e., intersection points).

Management Practices. Deforestation that leads to complete forest removal (i.e., clear-cutting) would lead to the largest reduction in atmospheric deposition of the limiting resource and, thus, the greatest reduction in system resilience. On the other hand, selective logging would contribute to the smallest relative change in canopy characteristics and would have the smallest effect on system resilience. For nutrient limited systems that are dependent on the slow buildup of that nutrient within the soil, frequent removal of the forest canopy would further contribute to a decline in resilience (Runyan et al., 2012b). This is less relevant to water limited systems dependent on a canopy enhanced atmospheric input of water given the shorter residence time of soil water that is necessary to support plant growth.

For economies that practice shifting cultivation, a loss of either nutrients or water (whichever is limiting to crop growth) would lead to a reduction in yield if fertilizers or irrigation were unavailable to supplement the nutrients/water that are no longer being deposited as a result of the reduction in forest canopy. Moreover, using fertilizers (e.g., Lawrence et al., 2007) or irrigation systems (e.g., Tilman et al., 2002) may not be economically viable (Naylor, 1996). A loss of land that no longer supports crop growth could lead to the expansion of deforestation in order to maintain the same level of agricultural productivity as well as the loss of ecosystem services that follow deforestation (Chapter 5).

4.2.3 Soil Moisture

Operating Principle. Shading from the plant canopy can reduce evaporation from the soil surface, thereby maintaining higher soil water content in the shallow soil (Vetaas, 1992; Scholes and Archer, 1997; Zeng et al., 2004; D'Odorico et al., 2007). The ability of plant roots to increase soil infiltration capacity (Greene, 1992; Joffre and Rambal, 1988; Greene et al., 1994, 2001) can also enhance soil moisture. This effect is further favored by the presence of a forest or shrub canopy that decreases rain splash compaction and maintains a higher infiltration capacity (Boeken and Orenstein, 2001). Soil moisture conditions are also enhanced by stem flow at the base of trees and shrubs that funnels rainwater down to the ground surface (Pressland, 1976; Martinez-Meze and Whitford, 1996; Whitford et al., 1997). Collectively, these mechanisms explain the increase in soil moisture often found beneath the plant canopy in comparison to the intercanopy area (Breman and Kessler, 1995). In dryland regions, a loss of woody vegetation can lead to conditions that are too dry to sustain the regrowth of vegetation. A similar feedback may exist in the case of herbaceous vegetation (Zeng et al., 2004). Among the primary drivers causing a shift to the bare state in systems affected by this feedback are grazing and deforestation.

Geographic Extent. In arid and semiarid ecosystems where water is the most limiting resource, the presence of woody vegetation can be crucial to enhancing soil moisture that supports the establishment of plant seedlings (e.g., Scholes and Archer, 1997). The enhancement of infiltration is crucial in non-sandy soils, which in the absence of roots have only a limited soil infiltration capacity. The effect of canopy shading and reduced soil evaporation can be more important in sandy soils, though it only affects the shallow soil. Moreover, this effect needs to be strong enough to offset the fact that under the canopy, throughfall is smaller than precipitation because of interception (i.e., the evaporation of intercepted water) (Carlyle-Moses, 2004; Návar and Bryan, 1990). Greater soil moisture reserves can also be maintained via hydraulic redistribution, where plants transport water from the upper soil to deeper soil layers (Burgess et al., 1998). Hydraulic redistribution tends to occur after a period of rain when the shallow soil is wetter than deeper soil layers. This allows plants to store water for later use when the upper soil layers become dry (Burgess et al., 1998). Moreover, water stored in the deep soil layers is protected from solar radiation and is not affected by soil evaporation (D'Odorico et al., 2007).

Observational Evidence. Spatial patterns characterized by intercanopy soils that are too dry for woody vegetation growth and survival, contrasting with moister subcanopy soils that are capable of sustaining seedling establishment and growth, are indicative of a vegetation–soil moisture feedback (Greene, 1992; Greene et al., 1994, 2001; Breman and Kessler, 1995) and suggest the presence of two stable states (e.g., Wilson and Agnew, 1992; Rietkerk and van de Koppel, 1997; D'Odorico et al., 2007). Once the existing vegetation is removed, the system is no longer capable of enhancing soil moisture to sustain plant growth and it remains in the dry, bare soil state until there is a change in soil moisture conditions that allows the vegetation to reestablish.

Modeling Techniques. This feedback has been examined for dryland ecosystems by a number of minimalist models (Walker et al., 1981; Rietkerk and van de Koppel, 1997; van de Koppel et al., 1997; Zeng et al., 2004; D'Odorico et al., 2007). Results from these models show that high values of mean soil moisture (or precipitation) are associated with stable vegetated conditions, while low values correspond to a system that is stable with no canopy cover. Over an intermediate range of values, bistable conditions emerge as a result of the feedback. As mean soil moisture decreases, the resilience of dryland woody vegetation decreases, indicating that smaller and smaller disturbances are needed to cause a shift to a state with no canopy cover.

4.2.4 Water Table

Operating Principle. The presence of forest vegetation exerts control on the amount of water that percolates past the rooting zone through rainfall interception and relatively high transpiration rates of forest vegetation in comparison to a surface with low or no vegetation cover. These factors control the amount of water that recharges the groundwater and, thus, the depth to groundwater table (e.g., Le Maitre et al., 1999). In turn, the removal of forest vegetation in areas with relatively shallow groundwater causes a rise in the water table (Section 2.5). When the water table rises to the surface, saturated conditions in the rooting zone ensue. Waterlogging in the root zone causes anaerobic conditions that can be detrimental to plant growth, particularly during periods when soil oxygen movement to the plant roots is critical to maintain adequate respiration (i.e., during the growing season) (Rengasamy et al., 2003; Davison and Tay, 1985). Thus, oxygen is the limiting resource to forest vegetation for this feedback. While plants vary in their sensitivity to waterlogged conditions, continued exposure to anaerobic conditions will adversely affect many woody plant types and either result in their death or hinder the ability of seedlings to reestablish (e.g., Jones et al., 2006).

Geographic Extent. In areas that have a shallow depth to groundwater table or are susceptible to rising water tables, deforestation can cause a rise in the water table toward the surface (e.g., Peck and Williamson, 1987). Areas susceptible to this feedback are found throughout the world. These environments range from riparian ecosystems (Wilde et al., 1953; Peck and Williamson, 1987; Borg et al., 1988; Riekerk, 1989; Chambers and Linnerooth, 2001) to forested wetlands (Dubé and Plamondon, 1995; Roy et al., 2000).

Observational Evidence. Broad experimental evidence supports this feedback between vegetation and groundwater. For instance, significant increases in water table elevation were observed after the removal of riparian vegetation (Wilde et al., 1953; Peck and Williamson, 1987; Borg et al., 1988; Riekerk, 1989), while the opposite effect was observed after planting vegetation in areas with shallow water tables (Wilde et al., 1953; Chang, 2002). Chen et al. (2002) found that after 50 days of flooding in an area of the western United States with shallow water tables, there was 80% less total biomass than in areas that were not flooded. This feedback has also been documented following deforestation in the Murray-Darling Basin, Australia. Over the last 200 years, 63% of forests in the Murray-Darling Basin have been cleared for agricultural use (Walker et al., 1993). Deforestation led to an increase in downward water fluxes below the root zone from 0.1 mm yr^{-1} to 50 mm yr^{-1} (Allison et al., 1990; Petheram et al., 2000; Gordon et al., 2003). These changes led to a rapid rise in the groundwater table at a rate of ~1 m per year (Ruprecht and Schofield, 1989; Ghassemi et al., 1995). For instance, Peck and Williamson (1987) found that immediately after clear-cutting of eucalypts, the water table depth was 3.5 m below the surface; however, three years after clearing, the water table intersected the surface.

Modeling Techniques. Mechanistic models relating plant–water table feedbacks have been used to examine the presence of multiple stable states in vegetation biomass (Ridolfi et al., 2006, 2007). These models show that when the (undisturbed)

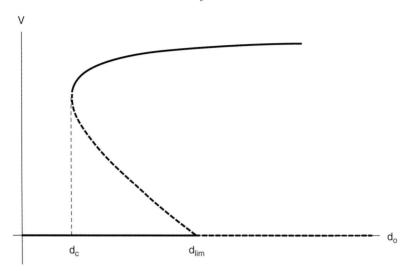

Figure 4.9. Stable (solid line) and unstable (dashed line) states of a system exhibiting the water table feedback as a function of the undisturbed water table depth (d_0) from Ridolfi et al. (2006). d_{lim} represents the threshold of vegetation tolerance to shallow water tables. When d_0 decreases below a critical value (d_c), the stable vegetated state disappears.

water table depth (d_0, i.e., the water depth that would exist in the absence of vegetation) is greater than a certain value, the ecosystem has only one stable state (fully vegetated); however, when d_0 is less than this depth, a bistability emerges, with one state associated with the existence of vegetation and the other corresponding to bare soil conditions (Figure 4.9). When the undisturbed water table depth decreases below a critical level, bistability disappears and the system again has only one stable state, the bare stable state.

Management Practices. Land use practices that reduce the resilience of systems affected by this feedback include large-scale forest clearing, especially of those forests with a high transpirative demand. When there are relatively small changes in the depth to water table because a small area is being deforested, the hydraulic gradient would induce flow away from the open area because of the lower hydraulic head beneath the forested area relative to the open area. This gradient would aid in lowering the water table over relatively small cleared areas where a rise in the water table occurred. However, when a decrease in the water table occurs over a much larger area, the edge effects previously described would no longer contribute to a substantial reduction in the water table.

Waterlogged soils can decrease the productivity of an area when forests are being used for timber production. Anaerobic conditions would prevent the regrowth of forests and render the land economically less valuable. In addition, when forests are cleared for the purpose of agricultural production and a rise in the water tables ensues, options to reduce the water table such as pumping could be cost prohibitive. Once a state shift to the bare state occurs, there could be a relatively permanent loss of the ability of this land to grow a forest when means to reduce the water table are not feasible. Therefore, selective logging, thinning, and other silvicultural techniques are often recommended (instead of clear-cutting) to limit waterlogging and favor postharvest forest recovery (e.g., Dubé and Plamondon, 1995; Roy et al., 2000).

4.2.5 Permafrost

Operating Principle. Permafrost–water table–vegetation feedbacks strongly affect the resilience of boreal forests in regions underlain by permafrost (Runyan and D'Odorico, 2013). Permafrost occurs in grounds where the earth remains below $0°C$ for more than two years in a row (Muller, 1945). The presence of forest vegetation helps to preserve the permafrost layer by maintaining lower soil temperatures via the accretion of an organic layer at the surface (Dyrness, 1982; Vitt et al., 1994). The organic layer has a low bulk density and low thermal conductivity, which insulates the mineral soil by lowering soil temperature (Bonan and Shugart, 1989). Forests also reduce soil thermal conductivity by reducing the soil water content during summertime as an effect of plant uptake (Vitt et al., 1999). Collectively, these processes lead to lower temperatures in soils beneath forest vegetation compared to adjacent bare soils.

A deeper permafrost table enables tree growth, and the presence of trees further feeds back to increase the depth to permafrost table because tree uptake and interception reduce the amount of water that percolates beneath the rooting zone (Runyan and D'Odorico, 2013). A deeper active layer (i.e., the surficial layer above permafrost, which thaws during summer and freezes again in winter; Muller, 1947) after deforestation can be a result of a number of potentially overlapping factors (Figure 4.10): (i) the presence of a deeper, impermeable permafrost layer (Sharratt, 1998; Magnusson, 1994) or layer of reduced permeability (Lloyd et al., 2003); (ii) an increase in the amount of water reaching this layer as a result of reduced interception, transpiration, and sublimation rates from forest canopies (Dubé and Plamondon, 1995; Molotch et al., 2011); and (iii) the loss of soil elevation caused by volume contraction from the melting ice (Schuur et al., 2008) and increased decomposition of the organic soil layer (Schuur et al., 2008). Abundant meltwater from thawing permafrost can create ponded areas that cause trees to die as a result of the reduced aeration of the rooting zone (Osterkamp et al., 2000; Jorgenson et al., 2001). Waterlogging in the root zone causes anaerobic conditions that are detrimental to plant growth (Wilde et al., 1953; Davison and Tay, 1985). None of the northern conifer species have particularly strong mechanisms to withstand anaerobic conditions (Conlin and Lieffers, 1992), and thus saturated conditions result in slow growth rates or mortality of these species (Lieffers and Rothwell, 1986).

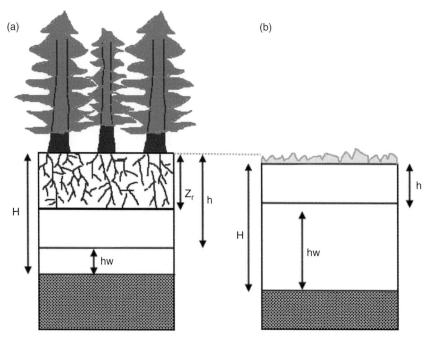

Figure 4.10. (a) Under a fully vegetated state, the depth to active layer does not constrain vegetation biomass growth because the low soil temperature promotes the buildup of the organic soil layer and high evapotranspiration rates reduce the amount of water reaching the active layer and the root zone (Z_r) is not waterlogged. (b) The active layer depth (H) and the water table depth (h) under low vegetation or a bare state after deforestation. After forest removal, the increase in drainage beneath the rooting zone due to the reduction in transpiration and interception increases the amount of water that reaches the active layer. The increase in the soil temperature melts the permafrost layer, thereby causing subsidence (as shown by the light gray dashed line) and leading to an increase in the thickness of the active layer.
Source: Runyan and D'Odorico, 2013.

Geographic Extent. Permafrost regions occupy approximately 22.8 million km², or 24% of the exposed land surface in the Northern Hemisphere (Zhang et al., 2003), much of which is overlain by evergreen boreal forest (Armstrong et al., 2001; Figure 4.11). Notably, the emergence of saturated conditions following forest removal in areas underlain by permafrost is not likely to result at well-drained upland sites where the permafrost is not ice-rich (e.g., Lloyd et al., 2003).

Observational Evidence. Simultaneous measurements of soil temperature at an adjacent clear-cut versus forested site showed that after deforestation, soil temperatures increased during the summer months by 5°C (Machimura et al., 2005). Similarly, Kallio and Rieger (1969) found for the first three years after forest clearing that soil temperatures at 0.2 m depth in a forest and adjacent grassland were, respectively, 5.2°C and 14.1°C over the course of the growing season and −6.0°C and −4.4°C during winter. Studies have also shown that thawing of ice-rich permafrost

Irreversibility and Ecosystem Impacts 123

(a)

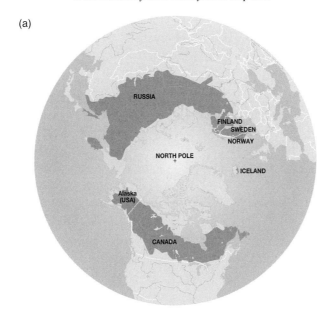

(b)

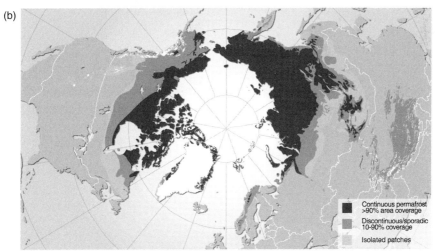

Figure 4.11. (a) Extent of boreal forest.
Source: Available at: http://maps.grida.no/go/graphic/boreal-forest-extent.
(b) Northern Hemisphere permafrost extent.
Source: Available at: http://maps.grida.no/go/graphic/permafrost-extent-in-the-northern-hemisphere.

can lead to the wholesale conversion into wetland systems (Drury, 1956; Vitt et al., 1994; Osterkamp et al., 2000; Jorgenson et al., 2001).

Modeling Techniques. Minimalist models have been used to investigate the conditions under which bistable conditions emerge for systems exhibiting this feedback. For instance, Runyan and D'Odorico (2013) coupled a model of vegetation biomass

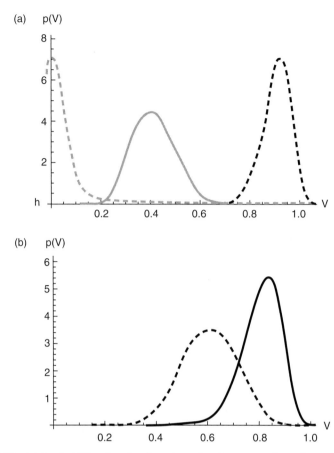

Figure 4.12. (a) Probability density functions for the vegetation given a high interannual variation in precipitation (black dashed line), low interannual variation in precipitation (solid grey line), and no interannual variation in precipitation (dashed grey line). (b) Low autocorrelation (black dashed line), and strong autocorrelation (solid black line).
Source: Runyan and D'Odorico, 2013.

growth (where growth was limited by saturated conditions) with a model of the depth to active layer to investigate this feedback. Both a steady state and a dynamic model were developed. Results from the steady state analysis showed that when vegetation exerted a strong control on soil temperature, the system was more resilient to disturbance and better able to absorb increases in soil temperature. Bistable conditions resulted under steady state conditions when the water budget was in balance; however, under variable (stochastic) hydrologic conditions the system exhibited only one attractor. This preferential state of the stochastic dynamics was found to depend on rainfall characteristics, specifically interannual rainfall variability and autocorrelation of precipitation between years. The forest was most resilient with higher interannual rainfall variability and autocorrelation in annual precipitation (Figure 4.12).

Management Practices. Forests susceptible to this feedback are those located on poorly drained ice-rich forests. Thus, widespread clearing of forests in these areas could lead to the emergence of saturated conditions and a low vegetation biomass. The size of the area being deforested affects the resilience of the system, with selective logging followed by small-scale clearing being practices that would most enhance forest resilience. This is the case because these practices would lead to the smallest increase in water percolating beneath the rooting zone.

4.2.6 Nutrient Cycling

Operating Principle. Deforestation can alter nutrient cycling in ways that can lead to state changes in the vegetation dynamics. A feedback between vegetation and nutrient cycling can result from physical and/or biogeochemical processes.

Physical Processes. Physical nutrient losses associated with deforestation are due to erosion of soil, ash, and vegetation matter; volatilization of nutrients during biomass burning; and/or export of nutrients contained in aboveground biomass for forests that are logged (Runyan et al., 2013). After deforestation, nutrient losses via leaching (i.e., percolation beneath the rooting zone) are exacerbated (e.g., Lawrence et al., 2007) because of the resulting changes in soil moisture and the lack of uptake by tree roots. Although removal of forest vegetation leads to an increase in soil evaporation due to the increase in shortwave radiation reaching the soil surface (Sun et al., 2001), in humid and mesic (i.e., moderately moist) climates, this increase in evaporation only partially offsets the reduction in transpiration. This is due to the loss of trees and elimination of intercepted water (Bosch and Hewlett, 1982), both of which tend to increase soil moisture (Best et al., 2007; National Research Council, 2008). The increase in soil moisture as a result of deforestation leads to higher nutrient losses in runoff via overland flow and leaching losses via the increase in hydraulic conductivity. Nutrient losses are further exacerbated by the reduction in roughness elements that can trap nutrients either adsorbed to eroded soil particles or transported dissolved in runoff water. When deforestation is caused by fire, nutrients are mineralized during the fire and (in the case of nitrogen) partly lost through volatilization. Immediately after biomass burning, nutrients are further lost via erosion of ash containing nutrients as well as via leaching and runoff due to the increase in nutrient solubility (Kauffman et al., 1994).

Biogeochemical Processes. The soil microbial biomass influences the release of immobilized nutrients and, in turn, the availability of these nutrients for mineralization and vegetation uptake (Robertson et al., 1993; Bauhus and Barthel, 1995; Diaz-Ravina et al., 1995; Wardle, 1998). The microbial biomass reduces net losses of nutrients by temporarily storing them in a relatively available form as a result of the low carbon to nutrient ratio of microbes in comparison to the carbon to nutrient ratio of other soil organic matter (i.e., litter; Resende et al., 2010). For P, this is important as it can also reduce losses to unavailable P by chemical fixation and leaching

(Resende et al., 2010). Deforestation can lead to a reduction in the microbial biomass (Henrot and Robertson, 1994; Borneman and Triplett, 1997; Caldwell et al., 1999; Bossio et al., 2005; Chaer et al., 2009), and changes in nutrient availability can limit the reestablishment of microbes (e.g., Cleveland et al., 2002). Thus, removal of vegetation can decrease nutrient availability and lead to conditions less favorable for the growth of vegetation, thereby inducing state changes in the vegetation dynamics (Runyan and D'Odorico, 2013).

In addition to the role that the soil microbial biomass plays in mitigating nutrient losses, forest vegetation is often dependent on arbuscular mycorrhizal fungi (i.e., a type of mycorrhiza in which the fungus penetrates the cortical cells of the roots of a vascular plant) for increasing nutrient availability (Bever et al., 2001). Plants with the potential to form mycorrhizae occur in most ecosystems (Lambers et al., 2008b). Mycorrhizae can affect P availability by i) exuding phosphatases that cleave C-P ester bonds (Colpaert et al., 1997; Tarafdar and Marschner, 1994); ii) influencing solubility (Easterwood and Sartain, 1990) through exudation of charged organic compounds that compete with PO_4^{3-} for binding surfaces on other elements; and iii) altering soil pH (i.e., acidification or alkalinization), which influences P binding with elements. Mycorrhizae cannot establish without a higher host plant, and removal of the forest can cause a loss of mycorrhizae that in turn makes it difficult for trees to reestablish (Perry et al., 1989).

Geographic Extent. These feedbacks persist in nutrient limited areas where the nutrient, most often P, is not biologically replenished over timescales necessary for the regrowth of vegetation. Thus, P-limited areas are most vulnerable to state changes brought about by this feedback. These areas tend to occur within 15° north or south of the equator (see Figure 3.16 in Chapter 3 as well as the discussion of mechanisms causing P limitations). Feedbacks between nutrient limitations and forest growth, however, can also be found outside this latitudinal band.

Observational Evidence. There are several studies that support a state shift following deforestation in areas where these processes dominate. In the Siskiyou Mountains of southwestern Oregon and northern California, the loss of mycorrhizae following deforestation led to a failure of forest vegetation to reestablish (Perry et al., 1989). Despite three attempts at reforestation, the area remained under a state of low vegetation for 15–20 years after clear-cutting. Similar results have been observed after deforestation associated with intensified shifting cultivation in P-limited areas. For instance, Blaike and Brookfield (1987) observed a loss of mycorrhizae following the intensification of shifting cultivation, which led to the conversion of large areas of moist tropical forest to scrublands. The loss of microbial biomass has also been documented after deforestation in areas affected by this feedback. Henrot and Robertson (1994) found that six months after deforestation, the microbial population decreased by 50% and within 15 months, the microbial biomass had decreased to 35% of its original population.

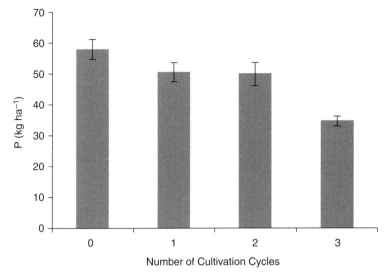

Figure 4.13. Available P as a function of the number of cultivation cycles in the top 15 cm of the soil for a P limited ecosystem in southern Mexico. Error bars represent the standard error. The diamonds with numerical values beneath are results obtained by running the model from Runyan et al. (2012b) for this system. Running the model for four and five cycles resulted in 33 kg P ha^{-1} and 31 kg P ha^{-1}, respectively, in the top 15 cm.
Source: Data from Lawrence et al., 2007, and Runyan et al., 2012b.

Modeling Techniques. Process based models have shown how repeated deforestation can lead to high nutrient losses and ultimately an irreversible phase shift in vegetation dynamics for nutrients that are not biologically replenished over timescales necessary for the regrowth of vegetation. Runyan et al. (2012b) assessed how shifting cultivation affects vegetation and P dynamics over multiple cycles of shifting cultivation. Shifting cultivation consists of clearing forests (typically by slash and burn), using the land for cropping, and then leaving the land fallow, during which time the forest regenerates. For P-limited ecosystems undergoing multiple cycles of shifting cultivation, available P can quickly be depleted from the soil (Figure 4.13), and this depletion can lead to a phase shift in the vegetation dynamics. Process based models have also been used to examine how physical and biogeochemical processes can drive state changes. For example, Runyan and D'Odorico (2013) developed a process based model of the P cycle to examine whether deforestation can lead to a shift to a stable treeless state. A large reduction in the recalcitrant organic P pool resulting from a decline in the microbial pool, a reduction in immobilization, and an increase in P losses led to a state change following deforestation from which the vegetation and microbial biomass did not recover.

Management Practices. Physical losses of P can be exacerbated in areas that are repeatedly deforested. Runyan et al. (2012b) found that reducing the length of

Figure 4.14. Stages of vegetation regrowth during shifting cultivation. A reduction in the duration of the fallow period (red arrow) reduces the recovery of the labile soil P pool and eventually leads to its depletion, thereby preventing forest regrowth. See also color figures at the end of the book.
Source: Photos were taken by the authors at field sites in southern Mexico.

the cultivation period from one to three years and extending the fallow period from 12 to 25 years resulted in a 57% decrease in vegetation biomass after the third cycle relative to the baseline case and did not have the positive benefit that was expected in a seasonal dry forest. This was due to the larger net loss of P that results from a mature/older secondary forest after biomass burning relative to a younger secondary forest. The higher net loss of P from the mature forest was due to the fact that more P is stored in the mature forest than the successional forest (Figure 4.14). Moreover, land use practices that increase the fraction of vegetation consumed by fire (thereby increasing nutrient losses) further reduce ecosystem resilience. Although a long fallow period benefits soil nutrient status via canopy enhanced P deposition and reduced leaching losses, unless losses can be reduced during biomass burning, the benefits of the long fallow period (up to twenty-five years) are not realized. This would likely not be the case for much longer fallow periods.

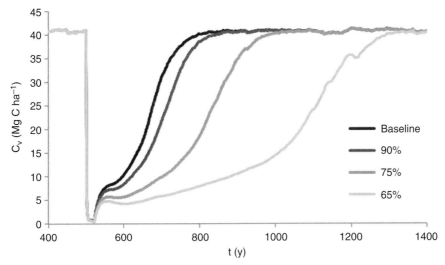

Figure 4.15. A change in the recovery rate of vegetation following deforestation at year t = 500 for decreasing values of postdisturbance recalcitrant P in the soil (humus) pool expressed as a percentage of a baseline case. Decreasing the recalcitrant P stocks to less than 65% of their baseline values leads to a state change where the system does not recover.
Source: Runyan and D'Odorico (2013).

The analysis of microbial-vegetation-P interactions suggests (Runyan and D'Odorico, 2013) that systems most susceptible to such state changes are those that have been previously deforested or those where the amount of P stored in the recalcitrant organic pool is low or becomes available very slowly (Figure 4.15). This occurred because once the recalcitrant organic P pool was depleted, there was no longer a source of P that could be accessed by the microbes and in turn be made available for plant uptake.

As discussed for the canopy trapping feedback, when there is a loss of the limiting nutrient and the land is being used for agricultural purposes, fertilizers will have to be used in order to maintain crop yield. For this feedback, deforestation results in a considerable loss of nutrients that have been built up and stored in the system over a long period. If fertilizers were cost prohibitive, this could also lead to a situation in which deforestation rates increased locally as a result of the increase in demand for land that could support agricultural production.

4.3 Feedbacks That Modify the Disturbance Regime

4.3.1 Landslides

Operating Principle. The deep roots of woody species aid in stabilizing the soil profile and binding the upper soil layer to saprolite substrates and bedrock (e.g.,

O'Loughlin and Pearce, 1976; Wu et al., 1979; Wu and Swanston, 1980). Vegetation can stabilize the soil as a result of (i) the presence of small and elastic roots that bind soil aggregates, thereby enhancing effective soil cohesion (Angers and Caron, 1998); (ii) the ability of thick and stiff roots to anchor the regolith to the bedrock, thereby stabilizing the soil mantle against mass wasting (i.e., mud flows, landslides, and debris flows); and (iii) the reduced total load (i.e., from water, soil, and vegetation) on the slope and reduced soil shear strength due in part to lower soil moisture levels as a result of transpiration (Sidle et al., 1985, 2006). Thus, the root strength of forest vegetation increases the shear strength – that is, the force required to cause hillslope failure and mass wasting – of the overall soil profile and contributes to preventing larger erosion events (Terwilliger and Waldron, 1991).

Extensive conversion of hillslope areas from native forest to pasture can reduce the shear strength of the regolith and render slopes more susceptible to landslides (e.g., Glade, 2003). After such events, it can be difficult for forest vegetation to reestablish because of the lack of an adequate soil substrate (Pimentel and Kounang, 1998). In turn, a lower return period for landslides on denuded hillslopes does not give seedlings an adequate time to grow and increase the stability of the soil (Dames and Moore, 1980; Brardinoni et al., 2002; Alcantara-Ayala et al., 2006; Stokes et al., 2008; Kuriakose et al., 2009). These dynamics could lead to the presence of the following two stable states: i) a state characterized by a soil-mantled slope stabilized by forest vegetation, and ii) a state where there is insufficient soil cover to allow for plant regrowth and to protect the soil from frequent erosion and landsliding.

Geographic Extent. An increase in landslides following deforestation has been observed in many mountain ranges in the world including the Himalayas (Haigh et al., 1995), Japan (Fujiwara, 1970), southern Italy (Wasowski, 1998), New Zealand (Glade, 2003; Crozier, 2005), Southeast Asia (Sidle et al., 2006), the Alps (Meusburger and Alewell, 2008), the Andes (Vanacker et al., 2003), the Pacific Northwest region of the United States (Swanson and Dyrness, 1975; Torres et al., 1998; Guthrie, 2002), and the Appalachian Mountains in the eastern United States (Wieczorek et al., 2000). Brabb and Harrod (1989) analyzed landslide occurrence and found that landslides have occurred in more than 100 countries. Areas that are at the highest risk for landslides are typically mountainous with steep slopes, have coarse soil, and/or lack vegetation to anchor the soil in place (Brabb and Harrod, 1989; Hong et al., 2006; Figure 4.16).

Observational Evidence. The occurrence of landslides following deforestation has been widely documented (DeGraff, 1979; Swanston, 1988; Prandini et al., 1977; DeGraff et al., 1989; Glade, 2003), with some of these landslides scouring the hillslope to bedrock (O'Loughlin and Pearce, 1976; Sidle and Swanston, 1982; Trustrum et al., 1983; Trustrum and De Rose, 1988). For instance, Swanson and Dyrness (1975) found that the quantity of material transported in landslides was 2.8 times greater in clear-cut areas than in forested areas of the western Cascades in Oregon. Similarly, prior to the

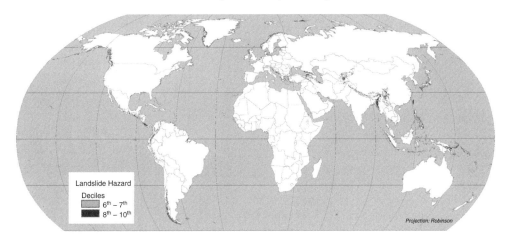

Figure 4.16. Landslide risk map from regional maps featuring some combination of coarse soil, land cover that was inadequate to stabilize the surface, and/or steep mountains. Areas where landslides occurred between 2003 and 2006 are marked with black dots on this map.
Source: Image obtained from http://earthobservatory.nasa.gov/IOTD/view.php?id=7783.

arrival of European settlers in New Zealand when hilly regions were only marginally influenced by human activity, erosion from these areas was relatively low (Glade, 2003). Bishop and Stevens (1964) observed a 4- to 5-fold increase in the number and area of landslides within 10 years after clear-cutting. An inventory of landslides in two forested areas in Japan revealed that the frequency and area affected by landslides increased 3- to 12-fold and 2.8- to 24-fold for two different forests after forest harvesting (Fujiwara, 1970). Similarly, Jakob (2000) found the frequency of landslides in logged terrain to be nine times higher than in natural forest, while Bruschi et al. (2013) found this frequency to increase up to twelve times in logged forests in northern Spain. Similarly, Beguería (2006) found that land use changes in the Pyrenees had a significant effect on landslides, even decades after human activities had ceased.

Modeling Techniques. Process based models to investigate under what conditions deforestation can lead to a state change (Runyan and D'Odorico, 2014) use a slope stability analysis coupled with a soil mass balance and a hydrologic model (e.g., Montgomery and Dietrich, 1994; Dietrich et al., 1995; D'Odorico and Fagherazzi, 2003) and a dynamic vegetation component. Results from using these coupled soil-vegetation dynamics can show the emergence of bistability; interestingly, the presence of this behavior can be strongly modified by rooting characteristics as well as by the effect of a stochastic (intermittent) forcing (Runyan and D'Odorico, 2014) (Figure 4.17).

Management Practices. Management practices that reduce resilience include clear-cutting of forest on unstable steep slopes and road development (e.g., Bruschi et al., 2013; Guthrie, 2002; Swanson and Dyrness, 1975). Roads overload and undercut slopes and intercept surface and subsurface drainage (Guthrie, 2002). Swanson

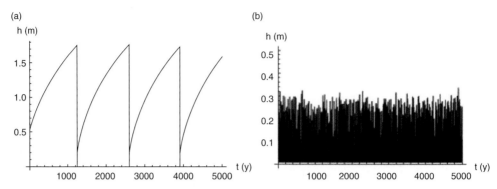

Figure 4.17. The effect of a change in rooting strength characteristics on the emergence of landslides (where h is soil depth) for supply limited steep hollows (similar results were shown for event limited steep hollows) with (a) stronger and deeper roots and (b) weaker and shallower roots.
Source: Runyan and D'Odorico, 2014.

and Dyrness (1975) analyzed the impact of clear-cutting and road construction on slide activity and found that at comparable levels of development (i.e., 8% roads, 92% clear-cut), road right-of-way and clear-cut areas contribute equally to the impact of management activity on erosion by landslides. They also found that the combined impacts of road construction and clear-cutting increased slide activity by about five times on road and clear-cut sites relative to forested areas over a 20-year period. Notably, Rice (1977) analyzed the effect of roads on landslides in a forested area of Oregon and found that the frequency of landslides was 315 times greater in areas affected by roads than in those without roads. Leaving a larger fraction of vegetation and using a longer logging period generally lead to enhanced resilience and a quicker and greater recovery of forest after each deforestation event (Sidle et al., 2006; Runyan and D'Odorico, 2014).

There are also natural factors that increase the susceptibility of an area to landsliding that could be taken into account by land managers/timber harvesters. These factors include slope, with steep, concave slopes being more susceptible to slides than shallow slopes (e.g., Montgomery and Dietrich, 1994) and geological settings with unstable bedrock such as limestone contributing to enhanced slope instability (e.g., Jakob, 2000; Guthrie, 2002). Rooting characteristics of vegetation (e.g., hardwood vs. conifers) such as tensile strength and the decline in root reinforcement as a function of depth also affect slope susceptibility to sliding (e.g., Sidle et al., 1985; Bischetti et al., 2005, 2009; Roering et al., 2003; Runyan and D'Odorico, 2014).

4.3.2 Fire

Operating Principle. Anthropogenic activities can significantly increase the susceptibility of landscapes to fire in areas that were historically unaffected by fire (Hoffman et al., 2003; Cochrane, 2003; Cochrane et al., 1999; Uhl and Kauffman, 1990). Forests that have been previously burned or selectively logged are more susceptible to fire because there are a more open canopy with higher understory temperatures, lower relative humidity, and more rapid fuel drying, which significantly increase the flammability of the forest (Hoffman et al., 2003; Uhl and Kauffman, 1990). Previously burned forests also have a more open canopy (50% to 70% cover) that allows greater solar heating and air movement to dry out the forest fuels (Nepstad et al., 2008; Hoffman et al., 2003). After the initial burn, thicker barked trees survive despite the low fire tolerance of species growing in areas that were historically unaffected by fire. The increase in the amount of fuel at the forest floor also leads to faster speeds and greater intensities of subsequent fires (Cochrane et al., 1999). Plant species in wet forests tend to be more vulnerable to fire than those in drier forests because of their lack of evolutionary adaptive traits to fire, which causes them to suffer higher mortality rates from fires (Cochrane, 2003; Cochrane and Barber, 2009). After deforestation, a transition to a grass state can occur because the increase in fire frequency (i) changes the availability of nutrients, water, and light to shift the competitive advantage in favor of grasses; (ii) preferentially selects for grasses because they tend to be dormant during the dry season when fires occur; and (iii) maintains selection for grasses by increasing future fire probability (Balch et al., 2009).

Geographic Extent. Human disturbances in tropical forests of the Amazon and Indonesian Borneo have significantly reduced the return interval of fires (Cochrane, 2003; Uhl and Kauffman, 1990; Malingreau et al., 1985; Leighton and Wirawan, 1986; Field et al., 2009). The role of fire in determining the emergence of woodland and grasslands as alternative stable states has been investigated in the case of the Serengeti National Park in Tanzania and other African ecosystems (Dublin et al., 1990). Modeling results also confirm the effect of deforestation on fire risk in tropical forests located in the Amazon, Congo, and Indonesia (Hoffman et al., 2003). Apart from tropical forests, this feedback has also been documented in semiarid to arid ecosystems (Anderies et al., 2002; D'Odorico et al., 2006).

Observational Evidence. Under undisturbed conditions, fires in the wet tropics are relatively rare, with return intervals of hundreds if not thousands of years (Cochrane, 2003; Hammond and ter Steege, 1998). However, human disturbance in tropical forests of the Amazon and Indonesian Borneo has significantly reduced this return interval (Cochrane, 2003; Uhl and Kauffman, 1990; Malingreau et al., 1985; Leighton and Wirawan, 1986; Field et al., 2009). For example, a single grass species now dominates approximately 300,000 km^2 in Southeast Asia that was once covered by moist tropical forest (MacDonald, 2004; Nepstad et al., 2008). Staver et al. (2011) and Hirota et al. (2011) recently analyzed remotely sensed tree cover data from tropical and subtropical zones in Africa, Australia, and South America and related these patterns to precipitation. Their results indicated that rather than seeing a gradual increase in tree cover with precipitation from savanna to forest, there was a greater

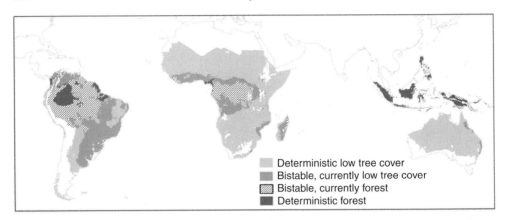

Figure 4.18. Biome distribution across South America, Africa, Southeast Asia, and Australia, based on whether the climate supports low tree cover/savanna (low precipitation, high variability), forest (high precipitation), or bistable vegetation dynamics presently either in the savanna or forest states.
Source: Modified from Staver et al., 2011.

likelihood of being in one of two (or even three) states, which they attributed to this feedback (Figure 4.18). Through a space-for-time substitution across entire continents, these authors highlighted that the savanna and forest states could exist under the same rainfall regime by differences in soil properties or other confounding factors. In semiarid to arid ecosystems, fires can also determine the relative abundance of woody vegetation versus grasses. A number of controlled burn experiments in semi-arid ecosystems have confirmed that in the absence of fire, shrubs and woody vegetation flourish, while grass growth is suppressed (e.g., du Toit et al., 2003; Briggs and Knapp, 2001); at the same time fire frequency and intensity decrease with decreasing grass cover (van Wilgen et al., 2003).

Modeling Techniques. Results from modeling studies tend to suggest that deforestation in the wet tropics leads to an increase in fire frequency, thereby inhibiting the regrowth of forest vegetation. Hoffman et al. (2003) used a GCM coupled with a land surface model to examine the effect of deforestation on fire risk in the Amazon, Congo, and Indonesia. They found that in all three regions, the largest relative increases in fire risk occurred in more humid areas with the lowest original fire risk. They estimated widespread increases in fire frequency of 44%, 80%, and 123%, in the Amazon, Congo, and Indonesia, respectively, after deforestation. Minimalist modeling frameworks have also shown that multiple stable states can exist in systems affected by this feedback (Anderies et al., 2002; van Langevelde et al., 2003; D'Odorico et al., 2006; Staver et al., 2011).

Management Practices. Fire has often been used as a tool for forest management. For instance, fire suppression favors a more rapid postfire recovery of forest vegetation. The timing of forest removal during a given season may affect the forest's susceptibility to fire. For instance, previously burned forests and selectively logged forests are more susceptible to fire during common dry season weather conditions (Meggers, 1994; Bowman et al., 2011; Balch et al., 2009). An increase in the frequency of logging or forest removal via slash and burn may also increase the susceptibility of a forest to fire.

4.3.3 Exposure to Freezing Events (Climate-Air-Temperature)

Operating Principle. Higher minimum temperatures have been documented to occur beneath forest canopies with respect to surrounding grasslands and tundra (Section 2.9). This phenomenon is due both to the lower albedo (particularly in boreal and alpine regions) of forest vegetation standing above the snow mantle as well as to the ability of the forest canopy to trap and back-radiate upwelling nocturnal long wave radiation from the ground (Bonan, 2008b; D'Odorico et al., 2013). Forested areas are typically less exposed to low temperatures than adjacent grasslands and tundra because of the effect of forest vegetation on both albedo and radiative cooling (Chen et al., 1993; Eugster et al., 2000; Bonan, 2008a; Grimmond et al., 2000; Renaud and Rebetez, 2009). Because woody plants have a relatively low cold tolerance – due to their susceptibility to frost damage, frost dissiccation, and reduction in photosynthesis and reproduction rates (Tranquillini, 1979) – a positive feedback exists between forest vegetation and the surrounding air temperature (D'Odorico et al., 2013). The presence of the forest canopy mitigates low temperatures, thereby providing a favorable habitat that further sustains the establishment and growth of forest vegetation. Forest vegetation also provides diurnal conditions favorable for seedling establishment because of the low tolerance of seedlings to climatic extremes (i.e., drought, high temperatures, low temperatures, and excessive sunlight) (e.g., Scholes and Archer, 1997; Germino and Smith, 1999, 2000; D'Odorico et al., 2013).

Geographic Extent. This feedback is present in climates where temperatures decrease below a species's cold tolerance and the presence of woody vegetation can influence the microclimate in a way that is favorable to the plants' growth (Wilson and Agnew, 1992). Such areas include (e.g., D'Odorico et al., 2013) the ecotone between boreal forest/woodland and tundra in the subarctic, which is found around the 10–12°C maximum temperature isotherm for July (e.g., Epstein et al., 2004; Figure 4.19); the margin between arid and semiarid shrubland and grassland (He et al., 2011); the tree line between alpine forests and high-elevation meadows in mountainous landscapes, which occurs around the 10°C summer isotherm (e.g., Grace, 1989; Korner, 1998); and, possibly, the transition between mangrove swamps and salt marshes (Figure 1.2) in subtropical intertidal environments that occurs around the 20°C winter seawater isotherm (McKee and Rooth, 2008; Feller et al., 2010; D'Odorico et al., 2013).

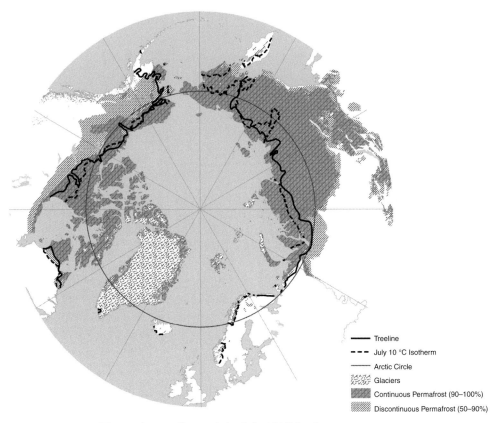

Figure 4.19. The arctic tree line and the July 10°C isotherm.
Source: D'Odorico et al., 2013.

Observational Evidence. Albedo estimates based on surface energy flux measurements and satellite observations indicate the existence of lower albedo in boreal forests and shrublands compared to higher albedo in tundra. In fact, the presence of boreal forests can increase local annual average temperatures by ~5°C (Betts, 2000). This confirms that the presence of these forests warms the climate in comparison to conditions where there are no trees (Chapin et al., 2000a). Evidence of these changes has also been documented from paleobotanical data, which indicate that during the middle Holocene, there was a northward expansion of boreal forests (Bryson et al., 1965; Webb et al., 1987; Foley et al., 1994).

Experimental evidence suggests that conversion from graminoid (i.e., grass or grasslike plants) tundra to shrub tundra increases radiation absorption and atmospheric heating as a result of the lower albedo of shrubs in comparison to graminoids (Chapin et al., 2000b; Eviner and Chapin, 2003). Chapin et al. (1979) showed that tussocks of *Eriphorum vaginatum* in the Alaskan Arctic tundra have a higher temperature during late spring and summer than does intertussock vegetation, which leads to a longer growing season (Wilson and Agnew, 1992). Because tussock growth is limited mainly by temperature and light, the plants' environmental modification can lead to increased plant growth. A recent expansion of shrub cover has been documented in the Alaskan Arctic (Sturm et al., 2001; Chapin et al., 2005). It has been suggested that the increase in shrub cover is mainly the result of the growth and expansion of dark colored alder, which has a lower albedo and, thus, a greater potential to increase the temperature of its surrounding microenvironment (Sturm et al., 2001). This warming effect can also be induced by the weaker nocturnal radiative cooling under forest canopies and has been documented in a wide variety of forest types. For instance, a temperature difference of up to 6°C on calm, clear-sky nights has been reported under relatively dense aspen stands with respect to nearby forest gaps (Groot and Carlson, 1996). Similarly, Scots pine canopies facilitate the growth of Norway spruce during clear-sky, calm nights (Langvall et al., 2002). Under these conditions, the minimum air temperature beneath Scots pine canopies can be up to 3°C higher than in clear-cut areas (Langvall et al., 2002). Collectively, these results confirm the existence of a positive feedback between forest vegetation and the surrounding air temperature.

Modeling Techniques. Both complex multidimensional models and minimalist process based models have been used to investigate this feedback. Coupled climate–land surface models have shown that this feedback has important impacts on climate at local, regional, and global scales (Bonan et al., 1992; Sturm et al., 2005). For instance, Foley et al. (1994) used a fully coupled global climate–vegetation model to estimate the potential for vegetation feedbacks on Arctic climate and concluded that the northward shift of trees and shrubs would raise spring temperatures by 1.1°C to 1.6°C. Coupled vegetation–atmospheric models can also be developed at the mesoscale to identify areas where vegetation and climate dynamics can be bistable because of microclimate feedbacks (He et al., 2015). Minimalist models have also been used to provide a process based understanding of the relation between microclimate feedbacks and the emergence of alternative stable states (D'Odorico et al., 2013) within a range of microclimatic conditions (Figure 4.20). Thus, gradual changes in temperature as brought about by regional or global warming may lead to abrupt changes in the state of the system.

Management Practices. In forests that are cut for commercial logging, some silvicutural techniques have historically been used to reduce the exposure to frost induced mortality in tree seedlings and therefore favor forest regeneration (Barg and Edmonds, 1999; Agestam et al., 2003; Pommerening and Murphy, 2004). For instance, *shelterwood forest regeneration* is a common tree harvesting method that leaves a residual tree canopy (Man and Lieffers, 1999) in the logged area. Forest managers rely on these residual trees to establish a nurse plant effect that favors forest regeneration by mitigating the extreme nocturnal temperature in their surroundings (e.g., Pommerening and Murphy, 2004; Paquette et al., 2006).

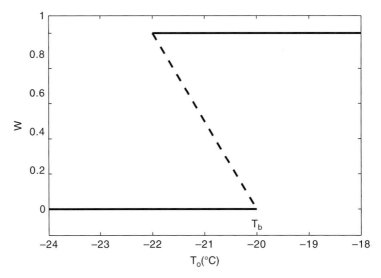

Figure 4.20. Stable (solid) and unstable (dashed) states of the northern Chihuahuan desert. In this figure, T_0 is the temperature that would exist with no woody plants and W is a relative value for woody vegetation.
Source: D'Odorico et al., 2013.

4.3.4 Salinity

Operating Principle. In semiarid areas, salts can accumulate in soils, particularly if ET, which leaves dissolved solutes behind, is high (Naumburg et al., 2005). Rainwater may partially leach these salts to deeper soil layers beneath the root zone or to groundwater (Naumburg et al., 2005). When the quantity of rainfall is not sufficient to leach salts accumulated beneath the root zone of native vegetation, this buildup of salt can accumulate over thousands of years (Rengasamy, 2006; Figure 4.21). The presence of salts in the unsaturated zone and groundwater will not constrain the productivity of vegetation when the groundwater table and capillary fringe remain beneath the rooting zone. However, after deforestation, a rise in the water table toward the surface can mobilize salts stored in the unsaturated zone and lead to an increase in groundwater salinity (Sorenson et al., 1991; Rengasamy et al., 2003). Once this saline groundwater rises into the shallow soil, exfiltration moves groundwater and its dissolved salts into the shallow rooting zone and up to the ground surface. When water evaporates from the soil surface, salts remain at the surface, where they can inhibit seedling survival and the reestablishment of vegetation because of the toxicity of salts to seedlings (Naumburg et al., 2005).

Geographic Extent. About 831 million hectares worldwide have salt-affected soils (Rengasamy, 2006; Figure 4.22). In many areas, the extent of soil salinization is increasing. For instance, 20% of irrigated land, or 45 million hectares, is affected by conditions of increasing salt content (Rengasamy, 2006). This trend has

Irreversibility and Ecosystem Impacts 139

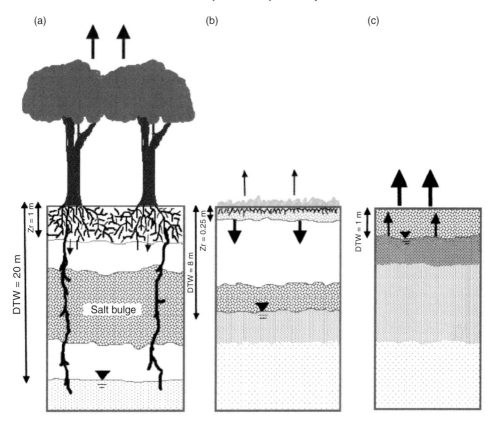

Figure 4.21. Conceptual representation of groundwater-vegetation-salinity dynamics as brought about by deforestation: (a) Forest vegetation with a dimorphic rooting system characterized by a tap root that reaches the saturated zone and a dense network of roots within the shallow rooting zone; (b) following deforestation, salts concentrated within the salt bulge are mobilized by the rising water table. Deforestation also increases percolation; (c) the resultant rise in the water table leads to the highest salinity water being found at the surface where exfiltration continues to concentrate salts from the groundwater at the surface via capillary rise. Downward pointing arrows represent flow from the rooting-zone to the groundwater table, the upward arrows into the atmosphere represent ET and arrows pointing from the groundwater table to the shallow rooting-zone represent exfiltration. The relative magnitude of these fluxes with respect to the total water budget is represented by the size of the arrow.
Source: Runyan and D'Odorico, 2010.

been documented in a number of major agricultural basins worldwide, including the Indo-Gangetic Basin in India (Gupta and Abrol, 2000), the Indus Basin in Pakistan (Aslam and Prathapar, 2006), the Yellow River Basin in China (Chengrui and Dregne, 2001), the Aral Sea Basin of Central Asia (Cai et al., 2003), the Euphrates Basin in Syria and Iraq (Sarraf, 2004), the Murray-Darling Basin in Australia (Rengasamy, 2006), and the San Joaquin Valley in the United States (Schoups et al., 2005; Qadir et al., 2007). Soil salinization also occurs under undisturbed conditions as a result of

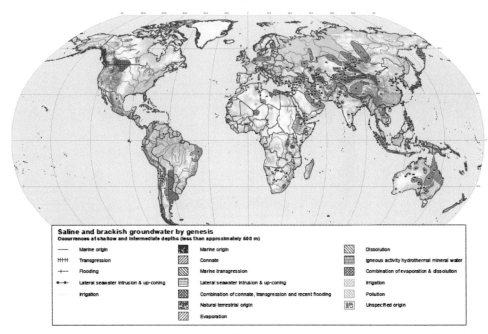

Figure 4.22. Global occurrence of saline groundwater at depths less than 500 m. Groundwater in these identified regions has a threshold limit of 1,000 mg/L TDS. Of the total land area on Earth 16% is affected by high salinity groundwater. The basins of West and Central Asia account for the largest area with high groundwater salinity contributing 14% to the total groundwater salinity area. The lowlands of South America and Europe, mountain belt of Central and Eastern Asia, and Eastern Australia all contribute individually at about 6%–7% to the total groundwater salinity area. See also color figures at the end of the book.
Source: Published by the International Groundwater Resources Assessment Centre (IGRAC).

geologic, hydrogeological, and hydromorphic characteristics of the watershed. For instance, some semiarid and arid areas such as Western Australia have naturally saline soils (Clarke et al., 2002).

Observational Evidence. One notable example where this feedback has been documented is the Murray-Darling Basin, where approximately 64% of woodlands and 63% of forests have been converted to agricultural use, with the majority of this conversion occurring over the last 50 years (Walker et al., 1993; Eberbach, 2003). The change in land use altered the water balance by reducing the amount of ET (in general, forests have an ET rate that is 1.6 times higher than that of grasslands (Zhang et al., 2001); decreasing interception (Cramer and Hobbs, 2002); and increasing groundwater recharge by 5% to 10% of the annual rainfall (Peck and Hurle, 1973; Allison et al., 1990; Eberbach, 2003). Thus, the change in vegetation cover caused the water table to rise, mobilizing salts stored in the vadose zone and leading to their accumulation in the rooting zone.

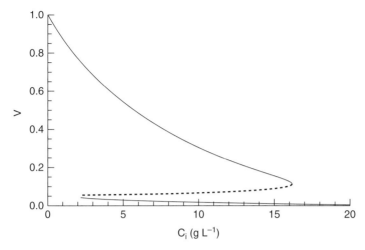

Figure 4.23. The location of the stable (solid lines) and unstable (dashed lines) states as a function of the salinity of irrigated water for the Murray-Darling Basin, Australia. *Source:* Runyan and D'Odorico, 2010.

Modeling Techniques. Mechanistic modeling frameworks have shown the emergence of bistability for systems affected by this feedback. Anderies (2005) developed a model specifically for the Goulburn Broken catchment in Australia that showed how feedbacks between vegetation and hydrologic dynamics lead to the emergence of two (alternative) stable states in salt-affected ecosystems. Bistable dynamics were demonstrated for these systems using a minimalist modeling framework in Runyan and D'Odorico (2010). In the resultant dynamics, two stable states potentially exist: (i) a state with vegetation cover, deep groundwater table, and low salinity; and (ii) a state with sparse or no vegetation, shallow water table, and high salinity. The presence of this feedback resulted in bistable dynamics for precipitation regimes ranging from 8 to 90 cm per year, an irrigated agriculture regime with salinity ranging from 2 to 17 g L^{-1}, and a range of groundwater salinities (Runyan and D'Odorico, 2010; Figure 4.23). Bistability reduces the resilience of these forest ecosystems and makes them prone to abrupt, irreversible, and largely unpredictable shifts to stable bare soil, saline conditions.

Management Practices. Management practices that reduce the amount of salt input to the soil via irrigation may include either switching between or combining saline irrigation water with fresh water, which is especially important in semiarid and arid areas, where water is a limiting resource (Hillel, 2000). Drainage systems can be used to reduce the amount of water that percolates beneath the rooting zone. Leaching of salts accumulated within the rooting zone with low salinity water is one common practice that is used to combat soil salinity (Qadir et al., 2000). However, in arid regions, where water is not available to leach these accumulated salts, soil salinization can be irreversible (Rozema and Flowers, 2008; Runyan and D'Odorico, 2010). Additionally, a lack of substantial drainage and the accumulation of salts may result in areas such as Western Australia with shallow, saline groundwater tables that have shallow topographic relief and long groundwater residence times and are located at long distances from potential discharge points (Clarke et al., 2002).

For agricultural systems, soil salinity not only reduces crop growth and yield but can also leave the soil in a permanently degraded state. Salt accumulation decreases ecosystem and crop productivity because elevated salt concentrations can inhibit plant establishment and growth (e.g., Munns and Tester, 2008). Once these salts have been transported into the rooting zone (via groundwater associated salinity), measures to mitigate soil salinity not only are costly, but often require fresh water, which may be scarce in arid and semiarid areas.

4.4 Noise Induced Transitions

Random drivers interacting with nonlinear dynamics may lead to the emergence of new states and bifurcations (or "noise induced transitions") that do not exist in the deterministic counterpart of the dynamics. While random forcings are commonly considered a source of disorganized random fluctuations in dynamical systems, the formation of ordered states and bifurcations is a rather counterintuitive effect of noise (Horsthemke and Lefever, 1984). The emergence of noise induced transitions has seldom been investigated in the context of environmental dynamics (Ridolfi et al., 2011). Models have shown that several of the dynamics discussed in this chapter (i.e., permafrost, erosion, and fire) may exhibit noise induced transitions. For instance, bistability was no longer observed when using a stochastic model of landslide occurrence with the same parameter set as in the (bistable) deterministic model (Runyan and D'Odorico, 2014). Similarly, for the permafrost feedback, the stable states of the deterministic system differed from those of the stochastic system as a result of the effect of noise on the states of the system (Runyan and D'Odorico, 2013). The opposite can also occur: Bistable dynamics can be induced by random environmental fluctuations (Horsthemke and Lefever, 1984). The significance of these noise induced phenomena to the dynamics of forest ecosystems, however, still needs to be investigated. While in the last few decades most of the research on the impact of climate change on ecosystem dynamics has concentrated on shifts in the mean and seasonal distributions of rainfall and temperatures, the effect of changes in the variance (e.g., interannual variability) of climate drivers has for the most part remained unexplored. Thus, it is unclear whether noise induced transitions might emerge as an effect of changes in climate variability or other environmental fluctuations (Ridolfi et al., 2011).

4.5 Leading Indicators of State Shifts

A major challenge in the management of bistable ecosystems is their susceptibility to state shifts. Because such transitions can be undesirable and hard to reverse, it is important to identify leading indicators of state shifts that could serve as early warning signs for ecosystem managers. Nonlinear dynamics theories indicate that as a system approaches its critical bifurcation point, the dynamics exhibit a *"critical slowing down"* (Strogatz, 1994); in other words, as the system approaches the bifurcation point its return to a stable equilibrium after a "small" perturbation becomes slower (van Nes and Scheffer, 2007). It has been shown that in systems forced by (additive) random drivers, this slowing effect causes an increase in the variance and autocovariance of the state variable as the system approaches the point of critical transition (Brock and Carpenter, 2006). Thus, in the proximity to this point, the system's recovery in response to a small perturbation is slowed (van Nes and Scheffer, 2007) and this slowing leads to an increase in the variance and autocorrelation of the system state (Scheffer et al., 2009). The fact that the variability of system behavior changes in advance of a regime shift can be used as a leading indicator of a state shift (Kleinen et al., 2003; van Nes and Scheffer, 2007; Carpenter and Brock, 2006; Dakos et al., 2008). Results from model simulations and experimental evidence suggest that regime shifts can be detected in advance using rising variance and autocorrelation as early warning signs (e.g., Dakos et al., 2008). For instance, using a model of lake eutrophication Carpenter and Brock (2006) found that the regime shift to eutrophication could be detected years in advance by studying the standard deviation of P concentration in the water. Experimental evidence based on manipulative experiments on lakes confirmed these findings (Carpenter et al., 2011): Early warning indicators such as increase in variance, changes in skewness, and elevated spectral ratio predicted the occurrence of the state shift more than a year before the transition. Rising variance has been found to be a leading indicator of state shift also in complex multivariate systems with several state variables, including social-ecological networks (Suweis and D'Odorico, 2014). Other indicators may precede a nonlinear transition, including extreme changes in skewness and shifts in variance spectra toward longer wavelengths (i.e., lower frequencies) prior to a regime shift (Kleinen et al., 2003; Carpenter and Brock, 2006; Scheffer et al., 2009; Carpenter et al., 2011). While these methods are being explored in the context of ecosystem ecology, particularly in laboratory experiments and lakes, applications to forest management are still missing. The main challenge in the case of deforestation is that this state shift typically does not occur as an effect of slow variations in environmental conditions that allow the system to remain close to equilibrium while approaching the bifurcation point (point V_1 in Figure 4.1). Rather, deforestation typically results from relatively abrupt vegetation removal that shifts the system away from equilibrium. Because most warning sign theories are based on the critical slowing phenomenon, which applies to systems close to equilibrium, we presently lack a framework to inform forest managers that the landscape is on the verge of a nearly irreversible transition.

4.6 Concluding Comments

Feedbacks discussed in this chapter involve a wide range of processes typical of ecosystems that span all continents containing forests. The previous sections have shown how such feedbacks may induce bistable ecosystem dynamics, with forest ecosystems occurring as metastable states of a system that also have an alternative stable configuration with no forest vegetation. The existence of such dynamics suggests that these forests are prone to abrupt and highly irreversible shifts to a stable and often "degraded" state with no trees. In most cases, some observational evidence in support of these feedbacks exists; however, the claim that the underlying dynamics are bistable could be enhanced via long-term datasets that are used to investigate suitable leading indicators of a state shift. Failure to account for bistable dynamics in systems affected by these feedbacks could have important economic and environmental implications as after the shift, the land no longer has the same capacity to support the original economic and ecosystem function. Moreover, once a transition to the bare state occurs, the restoration options will depend not only on the value of the ecosystem services that can be recovered once the system is restored, but also on the physical feasibility of restoration efforts and on the availability of adequate financial resources to spend on restoration. Importantly, in areas where restoration is either not feasible or more costly than the economic benefit provided by the restored forest, restoration will likely not be pursued and the land will remain in its alternative (or "degraded") state with no forest vegetation.

5
Economic Impacts and Drivers of Deforestation

5.1 Background

The classic economic issue surrounding deforestation is whether and how forest resources are allocated optimally in order to maximize net benefits to people (Sills and Pattanayak, 2006). Net benefits are the value of goods and services (people's willingness to pay in terms of money or some other valuable resource) minus their opportunity costs (what people have to pay or what resources they have available to invest to obtain the goods and services) (Sills and Pattanayak, 2006). The opportunity costs of forested land can vary considerably (Chomitz et al., 2006; Table 5.1). For instance, in Brazil's cerrado region, converting native woodlands to soy results in land worth more than $3,000 per hectare, whereas land in the Atlantic forest of Bahia, Brazil, is worth just $400 per hectare. As discussed by Sills and Pattanayak (2006), net benefits can be calculated from two perspectives. Private benefits and costs directly affect the families or companies making decisions about how resources are allocated. The private net benefits of deforestation are equal to the value of returns from agriculture minus the value of forgone future forest production. Future forest production includes timber that could be sold, as well as goods consumed directly from forest products (i.e., nontimber forest products). However, these private benefits and costs do not include adverse environmental effects, termed "environmental externalities," that result from deforestation (e.g., see Table 5.2). The other perspective on net benefits takes into account the cost of externalities to calculate social or public benefits and costs. Many socially valuable goods and services do not have prices because they are public goods and are not traded in markets. Public goods and services refer to those whose benefits cannot be denied to anyone and cannot be divided up and sold. For example, biodiversity and carbon sequestration are public goods provided by forests.

When forest is initially cleared, there are both immediate and future costs and benefits of the harvested timber. The future stream of net benefits from land uses such as agriculture and ranching is equal to the value of outputs minus the cost of

Table 5.1. *Variation in land values of forested area*

Location	Year	Land use or type	Price per hectare	Study
Bahia, Brazil	2000	Median land value	$400	Chomitz et al. (2005)
Goias, Brazil	2004	Savanna	$140–$1,290	FNP Consultoria & Agroinformativos
		High productivity agricultural land	$1,950–$3,150	
Para, Brazil	2004	Pasture	$200	Fundacao
Rondonia, Brazil			$318	Getulio Vargas
Cameroon	2001	Frontier land	$86	Kazianga and Masters (2006)
Bolivia	N/A	Pasture	$24–$500	Merry et al. (2002)
North Ecuador	2001	Grazing land	$150–$500	Olschewski and Benitez (2005)
Central Ecuador			$400–$1,000	
Quito, Ecuador			$800–$2,000	
Veracruz, Mexico	1998	Pasture	$210–$1,052	Ricker et al. (1999)

Source: Studies reviewed in Chomitz et al., 2006.

inputs. To compare costs and benefits at their equivalent present value, future values are adjusted by a discount rate. The discount rate can vary considerably in time and with location (e.g., Weitzman, 1994). For instance, interest rates in Madagascar have varied between 5% and 22.8% (Kremen et al., 2000). The sum of discounted benefits minus costs on a per acre basis is the "marginal net benefits of agriculture" (i.e., MNB_A; or the dollar value of one additional hectare of cleared land) (Figure 5.1). MNB_A declines as deforestation increases because costs to produce crops rise (for example, in response to greater labor inputs), and, thus, the net profit per crop sold decreases (Sills and Pattanayak, 2006). In contrast, MNB_F, or the "marginal net benefits of forest," rises as forest becomes more scarce (Sills and Pattanayak, 2006). When nearly all of the land is cleared, one remaining hectare of forest provides substantial benefits in terms of future production of fuelwood, vines, fruits, game, and other products obtained from the forest. The optimal level of deforestation (i.e., D_P^*) occurs when MNB_F equals MNB_A because it maximizes the total net benefits (Sills and Pattanayak, 2006).

MNB_A and MNB_F represent private net benefits. If the social benefits of forests are higher and/or the social benefits of agriculture are lower, then the optimal level of deforestation from a social perspective will be lower than D_P^* (i.e., see D_S^* in Figure 5.1). Social net benefits from forest ($SMNB_F$) are likely to be higher than the private net benefits because many outputs of forests are public goods (Sills and Pattanayak, 2006). Private landholders cannot capture these values and, thus, do not take them into account in decisions about deforestation. From a social perspective, forest outputs decades to

Table 5.2. *Ecosystem services, their function and examples of these services. Forests may contribute to the provision of all of these ecosystem services*

Ecosystem services	Ecosystem function	Examples
Atmospheric gas regulation	Regulation of atmospheric chemical composition	CO_2/O_2 balance, O_3 for UV_B protection
Climate regulation	Regulation of global temperature, precipitation and other biologically mediated climatic processes	Greenhouse gas regulation
Disturbance regulation	Capacitance, damping and integrity of ecosystem response to environmental fluctuations	Storm protection, flood control, and drought recovery
Water supply	Storage and retention of water	Provisioning of water by watersheds and aquifers
Erosion control and sediment retention	Retention of soil within an ecosystem	Prevention of soil loss by wind, runoff and mass wasting (e.g., landsliding)
Soil formation	Soil formation processes	Weathering of rock and the accumulation of organic material
Nutrient cycling	Storage, internal cycling, processing and acquisition of nutrients	Nitrogen fixation, retention of N, P and other nutrients against runoff and drainage losses
Waste treatment	Recovery of mobile nutrients (when in excess) and removal or breakdown of excess or xenic nutrients and compounds	Waste treatment, pollution control, detoxification and phytoremediation
Pollination	Movement of floral gametes	Provisioning of pollinators for the reproduction of plant populations
Biological control	Trophic-dynamics regulations of populations	Keystone predator control of prey species, reduction of herbivory by top predators
Refugia	Habitat for resident and transient populations	Nurseries, habitat for migratory species, regional habitats for locally harvested species or wintering grounds
Food production	Portion of gross primary production extractable as food	Production of fish, game, crops, nuts, fruits by hunting, gathering, subsistence farming or fishing
Raw materials	Portion of gross primary production extractable as raw materials	Production of lumber, fuel or fodder

(*continued*)

Table 5.2. (*cont.*)

Ecosystem services	Ecosystem function	Examples
Genetic resources	Sources of unique biological materials and products	Medicine, products for materials science, genes for resistance to plant pathogens and crop pests
Recreation	Providing opportunities for recreational activities	Eco-tourism, sport fishing and other outdoor recreational activities
Cultural	Providing opportunities for non-commercial uses	Aesthetic, artistic, educational, spiritual and/or scientific values of ecosystems

Source: Costanza and Folke, 1997.

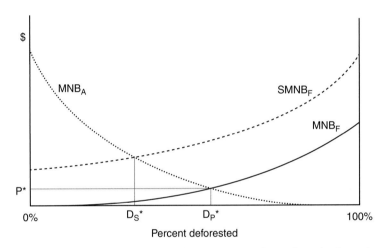

Figure 5.1. Marginal net benefits of agriculture versus forest from both a private and a social perspective. Total net benefits are equal to the area under MNB_A from the left to D_P^* plus the area under MNB_A from D_P^* to the right. The total value of agriculture is the area below the MNB_A curve up to the chosen level of deforestation, while the total value of forest is the area below the MNB_F curve up to chosen level of deforestation.
Source: Redrawn from Sills and Patanayak, 2006.

centuries from now may be important, but for individual landholders, these distant benefits are often outweighed by current concerns (Sills and Pattanayak, 2006).

5.2 Economic Uses of Forested Land

There are many economic uses of forests, of which one of the most common is harvesting timber. Forested land can be used for agricultural purposes, in conjunction with its use for timber or exclusively. Nontimber forest products such as rattan, resins, honey, fruit, and nuts are additional sources of economic uses derived from the forest. There are also displacement uses (i.e., its use for other purposes) of forested land, such as the construction of hydro dams and the building of roads (Sills and Pattanayak, 2006). Last, there are the uses of forests in a conserved state, which include i) maintaining ecological functions such as watershed protection, nutrient cycling, and microclimatic regulation; ii) indirect values, such as recreation and tourism (Table 5.2); iii) the option value of the future use of forest resources, particularly biodiversity; and iv) the existence value derived from the willingness of people to pay for the existence of these resources, irrespective of whether they use them or not (Sills and Pattanayak, 2006).

5.3 Factors Driving Deforestation

Causes of deforestation can be distinguished at several different levels (Angelsen and Kaimowitz, 1999). Deforestation is driven by both proximate causes and underlying driving factors (Figure 5.2). The choices made by deforestation agents (i.e., individuals, households, or companies) are influenced by proximate causes such as prices, market outlets, technologies, and agroecological conditions (Angelsen and Kaimowitz, 1999). These actions are in turn affected by broader national and international macrolevel and policy instruments, known as the underlying causes. Notably, the distal factors that shape the proximate ones can be difficult to connect empirically to land outcomes, typically owing to the number and complexity of the linkages involved (Turner et al., 2007b; Turner and Robbins, 2008). Often, deforestation can be explained by multiple proximate and causal factors (Geist and Lambin, 2002). Additionally, numerous factors tend to operate in chain-linked ways and their specific configuration and interaction may lead to dissimilar outcomes (Turner et al., 2007b).

5.3.1 Proximate Causes

Proximate causes are human activities or immediate actions at the local level that originate from intended land use and directly impact forest cover (Geist and Lambin, 2002). Proximate causes of deforestation include infrastructure expansion, agricultural expansion, and wood extraction. The relative importance of these different causes as a driver of deforestation can vary both in time and with geographic location.

5.3.1.1 Infrastructure Extension

One of the primary proximate causes cited as driving deforestation in rural areas is infrastructure extension. Establishing new or improving existing roads opens up new

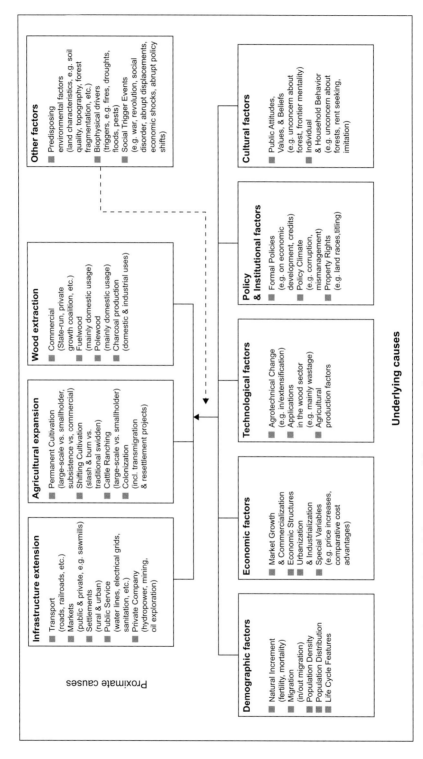

Figure 5.2. Proximate and underlying causes of forest removal.
Source: Geist and Lambin, 2002.

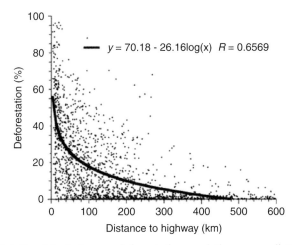

Figure 5.3. Relationship between deforestation and the mean distance to paved highways (Def and Dist in the equation in this figure, respectively) in the Brazilian Amazon.
Source: Laurance et al., 2002.

areas, reduces transport costs, provides market access, and thereby makes deforesting activities more profitable (Pfaff, 1996; Angelsen, 2010). Roads are among the most powerful factors contributing to deforestation across the tropics (Geist and Lambin, 2002); highways have considerably larger-scale impacts than roads because they provide efficient, year-round access to forests and reduce transportation costs (Laurance et al., 2002; Figure 5.3). In the Brazilian Amazon, 95% of all deforestation occurs within 50 km of highways or roads (Laurance, 2001). Similar results have been found in remote temperate forests in developed countries. For instance, ~30% of the boreal forest in Canada is within 1 km of a road or access route (Global Forest Watch, 2000), and 63% of Canada's original frontier forest has been lost to or is at risk of deforestation (Burton et al., 2003). Notably, road construction may not lead to an increase in deforestation rates when deforestation occurs in areas that already have an extensive transportation infrastructure (e.g., as in the case of the semiarid Chaco in Argentina; Grau et al., 2005).

5.3.1.2 Agricultural Expansion

Agricultural expansion includes forest conversion for permanent cropping, cattle ranching, shifting cultivation, and colonization agriculture (Naylor, 2011). An analysis of 152 cases of deforestation by Geist and Lambin (2002) spanning tropical forests in Asia, Africa, and the Americas showed that agricultural expansion drives 96% of all deforestation cases. More recently, Hosonuma et al. (2012) found that agricultural expansion drives ~80% of deforestation across 46 developing countries worldwide. The geographic importance of commercial versus subsistence agriculture differs worldwide. For instance, Geist and Lambin (2002) found that commercial agriculture was the most important driver in Latin America – associated with 68% of deforestation cases – whereas in Africa and Asia it was associated with 35% of deforestation cases. Local and subsistence agriculture drives deforestation quite equally among the continents (i.e., 27%–40%; Hosonuma et al., 2012). Traditional modes of small-scale clearing in frontier landscapes to support subsistence needs or local markets are no longer the dominant driver of deforestation in many places (Chomitz et al., 2006; Nepstad et al., 2006; Rudel, 2005). In fact, DeFries et al. (2010) found that higher rates of forest loss in 41 countries across the humid tropics in the period 2000–2005 were strongly associated with demands for agricultural products in distant urban and international locations. Agricultural expansion is also a cause of forest decline in temperate regions. For instance, agricultural development in western Canada is largely unregulated, unencumbered with provincial operating rules, and more likely to result in the permanent conversion of forest habitat (Hobson et al., 2002).

5.3.1.3 Wood Extraction

Wood extraction involves both commercial timber harvesting as well as the harvesting of fuelwood for domestic purposes. Although commercial timber harvesting is generally not cited as the predominant cause of tropical deforestation (i.e., Burgess, 1993; Geist and Lambin, 2002), commercial wood extraction is frequent in both mainland and insular Asia (Hosonuma et al., 2012). Tropical timber plays a relatively minor role in the global timber market, accounting for approximately 15% of the total volume of global timber production (Burgess, 1993). In 2013, China, Russia, Germany, the United States, and Canada accounted for 45% of the global exports of wood and wood pulp (UN comtrade data, 2014). Only 17% of the total tropical timber production is used for industrial purposes with the remainder being consumed for fuelwood and other nonindustrial uses (Burgess, 1993). The harvesting of fuelwood for domestic uses dominates cases of deforestation associated with wood extraction in Africa (Geist and Lambin, 2002). While less is known about the causes of deforestation in temperate regions than tropical regions (Zipperer, 1993; Kress et al., 1996; Zheng et al., 1997), commercial timber harvesting plays a greater role in boreal forest extraction. For instance, approximately 50% of Canada's boreal forest is under tenure to a few large-scale forest companies (Burton et al., 2003).

5.3.2 Underlying Causes

Underlying causes of deforestation are fundamental social processes, such as human population dynamics or agricultural policies, that drive the proximate causes of deforestation and either occur at the local level or have an indirect impact at the national or global level (Geist and Lambin, 2002; Figure 5.2). Underlying causes of deforestation include demographic factors, technological factors, economic factors, policy and institutional factors, and environmental factors.

5.3.2.1 Demographic Factors

Population growth has frequently been cited as an underlying cause influencing deforestation (e.g., Myers, 1980). Deforestation may increase after population growth because more land is needed for food, fibers, fuelwood, timber, or other forest products. Increases in growing populations of shifting cultivators and small-holder colonists along pioneer fronts have driven deforestation across a wide range of settings (Carr, 2009). Even in areas where the dominant land use conversion is increasingly due to capital-intensive crops, as has been documented recently with soybean expansion in the Amazon (Hecht, 2005; Fearnside, 2007), much of the cropped land was originally cleared by family-scale colonist farmers (Laurance et al., 2002). In many rural areas in recent decades, deforestation has accelerated despite a deceleration in rural population increase, suggesting that agriculturalists in areas of lower population density may be increasingly responsible for greater forest clearing (Carr, 2009). DeFries et al. (2010) found that urban population growth was positively associated with forest loss in tropical forests located within Africa, Asia, and Latin America. Urbanization raises consumption levels and increases demand for agricultural products (DeFries et al., 2010; Godfray et al., 2010). Urban consumers generally eat more processed foods and animal products than rural consumers, thereby inducing commercial production of crops and livestock (Kennedy et al., 2004; Mendez and Popkin, 2004; Kastner et al., 2012). Thus, although the initial forest clearing may be conducted by a relatively small number of rural agriculturalists at the forest frontier, it may be driven by demand for timber and agricultural products in urban areas. The conversion to large-scale commercial farming can lead to a disconnection between urban populations and the land they rely on for food products (Chapin et al., 2010). In fact, consumers that are not directly affected by the environmental impacts of their choices are less likely to value a responsible and sustainable management of forest ecosystems located far from where they live (Chapin et al., 2010).

5.3.2.2 Technological Factors

A common policy prescription for reducing deforestation is the development and dissemination of technologies that allow farmers to increase production on land that is already deforested. Agricultural intensification has been identified as a viable alternative to agricultural expansion in order to meet increased food demand from a growing and increasingly affluent human population (e.g., Foley et al., 2011; Godfray, 2011). Technological progress lowers the farmers' average costs and therefore shifts the supply curve outward, leading to increased output and lower prices. This argument is referred to as the *Borlaug hypothesis*, named after Norman Borlaug (2007), who claimed that the intensification of cereal production between 1950 and

2000, partly as a result of Green Revolution technologies, saved more than 1 billion hectares of land from being put into agricultural production. Rudel et al. (2009a) examined this hypothesis by calculating the percentage change in the ratio of yields to cultivated area for a range of crops between 1990 and 2005 using national-level data from FAOSTAT. These authors concluded that for the great majority of crops, yield growth did not lead to land sparing because increases in productivity from new technologies increase the profitability of agriculture in comparison with alternative land uses (such as forests), thereby encouraging expansion of agricultural land. Vosti et al. (2002) addressed this same question and found by using a simulation model of farms in the western Brazilian Amazon that improved ranching and farming technology would increase the rate of deforestation. Similarly, Morton et al. (2006) found that agricultural intensification in the Brazilian Amazon did not lead to a decline in deforestation rates. A review of modeling studies examining these two different views found that they are not mutually exclusive, but rather coexist (Villoria, 2014). Technological progress in one region can influence production decisions elsewhere, as productivity-induced price changes are transmitted through product markets (Villoria, 2014). Such readjustments in global production could relieve or encourage pressures in forests elsewhere.

Deforestation may be reduced by some forms of improved technology including more profitable techniques for sustainable forest management; infrastructure such as irrigation equipment that can lead to a boom in the labor market resulting from higher cropping intensities and alleviating land pressure on the forest frontier, thereby reducing cropland expansion (Shively and Pagiola, 2004); and labor-intensive agricultural techniques. In general, technologies that use more scarce resources such as labor will constrain deforestation in the short run, although in the long run, they may attract more of the scarce resource to the region (e.g., through migration). For instance, Maertens et al. (2006) studied the impact of rice production in Indonesia on deforestation rates and found that the use of tractors (labor-saving) encourages land expansion. Grau et al. (2005) found that the introduction of transgenic soybean cultivars, which increase yields, contributed to increased deforestation rates because they reduce production costs (Perez et al., 2002; Perez and Gonzáles-Lelong, 2003). Thus, yield-increasing and labor-saving technology may increase deforestation, especially if the crop is planted extensively (for example, maize, wheat, and soybean) rather than intensively (for example, coffee and fruits) (Angelsen and Kaimowitz, 2001).

5.3.2.3 Economic Factors

Economic factors are prominent underlying forces of tropical deforestation (Geist and Lambin, 2002). Higher national income and economic growth may reduce the pressure on domestic forests by improving off-farm employment opportunities but increase it (also through international trade) by stimulating demand for agricultural

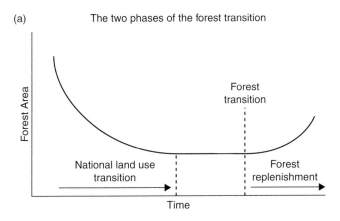
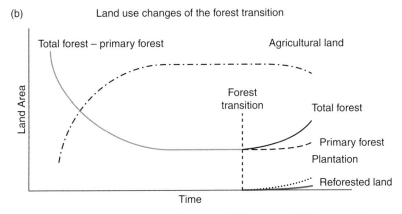

Figure 5.4. Land use change and the forest transition curve: (a) the two phases of the forest transition and (b) land use changes of the forest transition.
Source: Barbier et al., 2010.

and forest products and improving access to previously unharvested forests and markets (Angelsen and Kaimowitz, 1999). This trend of an initial increase in income accelerating the rate of deforestation, followed by a reduction in the rate of deforestation beyond a certain level of income is referred to as the environmental Kuznet's curve (EKC; e.g., Lopez, 1994). Some empirical evidence may support this relationship (e.g., Cropper and Griffiths, 1994; Koop and Tole, 1999; Bhattarai and Hammig, 2001; and Culas, 2007). For instance, Culas (2007) found significant evidence of an EKC relationship for deforestation in Latin American countries. The transition from prevalent deforestation to reforestation as countries become economically more developed is also known as *forest transition* (Mather, 1992), and this trajectory is termed the *forest transition curve* (Figure 5.4). This phenomenon has been explained as an effect of industrialization, agricultural intensification, and international trade. The latter allows for an international displacement of land use whereby reforestation in more affluent countries occurs at the expense of forests in developing countries

(Meyfroidt et al., 2010). If induced by conservation policies, this "outsourcing of deforestation" is known as *leakage* and is often an unintended consequence of reforestation (Meyfroidt and Lambin, 2011). Part of the "regrowth" contributing to the forest transition curve may be due to commercial plantations (Rudel, 2012) rather than "proper" reforestation (i.e., regrowth of a natural forest in previously forested areas). The nature of the regrowth (i.e., second growth forest vs. commercial plantation) depends on the region: In Latin America the increase in forest cover is mostly the result of a spontaneous regeneration of secondary forests or scrubs; in Africa it is mainly due to agroforestry by small-holders planting trees in upland settlements; while in Asia it is contributed to by both spontaneous regrowth and forest plantations. The drivers of the regrowth also vary from region to region and may include new conservation policies, land tenure reforms, changes in global patterns of agricultural production, and migrations (Rudel, 2012).

Economic variables such as low domestic costs (for land, labor, fuel, or timber) and an increase in the price of agricultural crops have been shown to stimulate deforestation rates (Geist and Lambin, 2002). For instance, high soybean prices stimulated an increase in deforestation rates in northwestern Argentina (Grau et al., 2005) and Brazil (Hargrave and Kis-Katos, 2013). Morton et al. (2006) found that in the southern area of the Brazilian Amazon, the area deforested for cropland and mean annual soybean price in the year of forest clearing were directly related, suggesting that deforestation rates could increase with a rebound of crop prices in international markets (Figure 5.5). Availability of subsidized rural credit may increase deforestation rates by resulting in a rise in expected profits from reduced costs of clearing (Hargrave and Kis-Katos, 2013). In contrast, an increase in the price of rural wages (Angelsen and Kaimowitz, 1999), a shortage of off-farm employment opportunities, and an increase in the price of agricultural inputs such as seeds, pesticides, and hand tools reduce forest clearing rates (Ruben et al., 1994; de Almeida and Campari, 1995; Monela, 1995). Another important economic factor affecting deforestation rates is access to international markets, which often results in more stable demand and prices. This factor was cited by Grau et al. (2005) as contributing to increased deforestation rates in northwestern Argentina. Commercialization and the growth of timber markets (as driven by national and international demands) are frequently reported to drive deforestation because they increase the profitability of logging. Rising profits can also result from currency devaluation, as evidenced by Brazil in 1999, thereby making exports more attractive (Chomitz et al., 2006).

5.3.2.4 Policy and Institutional Factors

In this section, we examine the potential impact of the property rights regime, policies, and political instability on deforestation rates. These factors may play a substantial role in driving deforestation rates. For instance, in their analysis of 152 cases of tropical deforestation, Geist and Lambin (2002) found that institutional factors

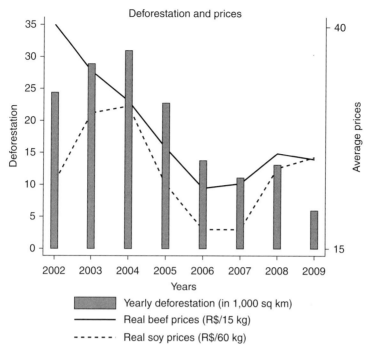

Figure 5.5. Deforestation rates in the Brazilian Amazon and average price movements (in Brazilian reals) in 2002–2009.
Source: Hargrave and Kis-Katos, 2013).

drive 78% of deforestation cases. These factors included measures such as policies on land use and economic development as related to colonization, transportation, or subsidies for land-based activities. Land tenure arrangements and policy failures such as corruption or mismanagement in the forestry sector are other institutional factors affecting deforestation rates (Geist and Lambin, 2002).

5.3.2.4.1 Property Rights Regime
The impact of property rights on deforestation rates is unclear because property rights can affect forest cover in two different ways with opposite implications (Liscow, 2013; Figure 5.6). Notably, the world's most carbon-rich forests are often found in regions where ownership is poorly defined, contested, or insecure; thus, understanding the effect of property rights on deforestation rights is of critical importance (Robinson et al., 2013; Figure 5.7). In one way, property rights could lead landholders to discount the future less (i.e., place greater emphasis on the future stream of benefits) and reap the long-term benefits of forestry instead of the short-term benefits of agriculture (Barbier and Burgess, 2001; Bohn and Deacon, 2000; Mendelsohn, 1994). Recent metaanalyses seem to suggest that empirical studies may not support this view, although exceptions exist. For instance, Robinson et al. (2013) reviewed 36

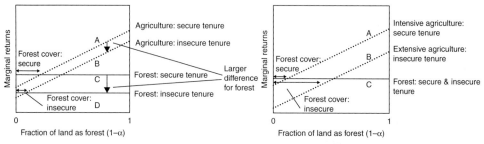

Figure 5.6. The two different impacts of property rights on deforestation rates. (left) Because forestry's returns are weighted toward the future, higher effective discount rates disproportionately reduce a forest's value, as evidenced by agriculture's value decreasing relatively less (i.e., from line A to line B) with greater insecurity than the forest cover's value (i.e., from line C to line D). Thus, a larger amount of forest with higher marginal returns than agriculture results with secure than with insecure tenancy. This theory predicts that landholders with insecure tenure will have less forest. (right) Because of the greater likelihood of capturing future benefits from investment in a secure-tenancy regime, landholders will invest more, increasing agricultural productivity by changing from more extensive to more intensive agriculture increasing agricultural returns from line B to line A. Returns to forest, on the other hand, are not directly affected by investment and are the same in both high-risk and low-risk regimes (i.e., line C). Thus, increases in agricultural productivity increase the returns to agriculture and, therefore, deforestation. This theory predicts that landholders with insecure tenure, or a high risk of expropriation, will have more forest.
Source: Liscow, 2013.

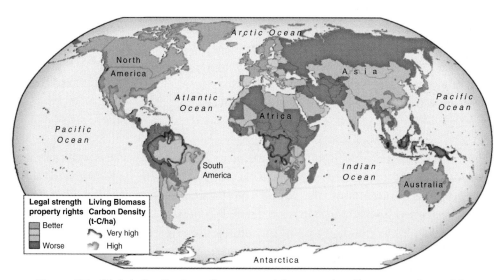

Figure 5.7. Global distribution of property rights security. Forests are located in areas with high live plant biomass and tend to be found in regions with poorly defined property rights. See also color figures at the end of the book.
Source: Bruce et al., 2010.

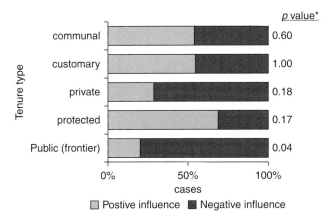

Figure 5.8. The effect of land tenure on positive (light gray) versus negative (dark gray) deforestation outcomes. Land tenure is categorized into one of five categories: public (unmanaged public frontier or open access), protected (managed public land with restrictions on forest clearing), private, communal, and customary/traditional. For each deforestation case in the review of Robinson et al. (2013), a positive or negative forest outcome was assigned with a 1 or 0, respectively. Positive outcomes included slowed deforestation rates (relative to other local sites), maintained forest cover, or regenerated forest cover. Negative outcomes included increased deforestation rates or overall forest loss.
Source: Robinson et al., 2013.

studies across South America, Central America, and Africa linking land cover change to tenure conditions. They found that protected land is associated with low deforestation rates in all regions, and public/frontier land as well as private land seem to be particularly vulnerable to negative forest outcomes in South America (Figure 5.8). Community-managed forests are also associated with lower and less variable deforestation rates relative to protected forest (Figure 5.8). A recent literature review of community-managed forests by Porter-Bolland et al. (2012) supports this trend. There are several reasons why communities might be more effective land managers. They have better knowledge of the local forest and its users and uses compared with the state, thereby making policing of deforestation easier. Local management is often crucial to the ethic of land stewardship, which favors a more sustainable use of natural resources, including forests (Chapin et al., 2010). Communities may also apply a different set of sanctions, as resource management is embedded in larger social systems (Ostrom, 1990; Ensminger, 1996; Aoki, 2001). However, achieving collective outcomes is difficult, particularly when the user group is large, heterogeneous, and poor; the forest is an open "access" resource; no ostracism mechanism can be adopted (Tavoni et al., 2012); or the forest benefit flow and economic environment are unstable (Agrawal and Angelsen, 2009). Moreover, when a common pool forest resource exhibits a positive feedback (see Chapter 4), where a transition to a bare state can follow deforestation (or continued deforestation), the inability to switch harvesting strategies in response to changing resource levels can make the system more susceptible to a state change and, thus, reduced economic benefits (Lade et al., 2013). Rules (Hayes, 2006), enforcement of rules (Gibson et al., 2005), flexibility (Lade et al., 2013), monitoring and maintenance (Van Laerhoven, 2010), and metrics of governance (Hyde et al., 2003; Persha et al., 2011) have all emerged as important factors for positive forest outcomes in community-managed forests.

On the other hand, more secure property rights increase investment in land, increasing the value of agriculture relative to that of forest and, in turn, leading to a decrease in forest cover (Liscow, 2013; Farzin, 1984; Figure 5.6). Investment increases for two reasons: (1) Landholders can use their land as collateral for credit, allowing investment, and (2) landholders have a greater ability to recoup future returns, encouraging investment (Liscow, 2013). Because maximizing profits – not minimizing costs for a fixed output – is the landholders' goal, investments that increase yield could raise production area (Liscow, 2013). This has been observed in temperate regions such as the boreal forest of western Canada, where privately owned lands are more likely to have been cleared than provincial government or native lands (Hobson et al., 2002). These trends are partially due to the implementation of policies that encourage the conversion of forest to agricultural lands and sustain low costs for logging (Hobson et al., 2002). Similar findings have been detected in tropical forests as well. Liscow (2013) examined the effect of property rights on forest cover in Nicaragua. He found that insecure land tenure substantially increases forest cover and that titling decreases the fraction of a landholding as forest by 14 percentage points. This occurred because the owners of titled land receive more credit, use more fertilizer, and practice more resource-intensive forms of agriculture, suggesting that property rights increase the returns to deforestation, thereby encouraging landholders to reduce the forest cover on their land.

5.3.2.4.2 Policies
Policies aimed at agricultural development such as credits, low taxation, incentives for cash cropping, and legal land titling can increase deforestation rates and lead to the expansion of commercial crops and pastures (Geist and Lambin, 2002). For instance, tax incentives and subsidized credit for large cattle ranches were important drivers of deforestation in the Brazilian Amazon during the 1970s (Mahar, 1979). In Saskatchewan, Canada, programs such as the Grazing Lease Improvement Program from the 1970s provided direct payment for felling trees and clearing or burning bush (Hobson et al., 2002). Commercial forestry policies in Saskatchewan have also encouraged logging via high provincial subsidies that allow forestry companies in the region to transport wood over large distances at relatively little cost (Hobson et al., 2002). In contrast, policies that police illegal logging activities via fines tend to reduce

deforestation rates (e.g., Hargrave and Kis-Katos, 2013). For instance, the Brazilian Forestry Code prescribes that properties in the Amazon retain 80% of their area in original forests. All instances of deforestation must be authorized by the environmental agency with any violation potentially resulting in fines and the prohibition of agricultural production in the deforested area. Hargrave and Kis-Katos (2013) analyzed yearly data from 457 municipalities in the Brazilian Amazon in 2002–2009 and found that increases in fining intensity are associated with decreases in deforestation rates.

Deforestation rates are also driven by national development policies (e.g., Klepeis, 2003). For instance, Klepeis (2003) performed a historical analysis of the impact of centralized (i.e., national) versus decentralized (i.e., local) approaches to development in the southern Yucatan Peninsula, Mexico. He found that while both approaches to development emphasize natural resource exploitation, the rates of deforestation tend to be faster, the patterns of forest clearing more pronounced, and land use decision making less democratic under systems of centralized control. Similarly, Pfaff (1996) analyzed satellite data on deforestation for the Brazilian Amazon and found that government-subsidized development projects result in greater deforestation.

Deforestation is also favored by the recent escalation in large-scale land acquisitions by domestic or foreign investors (Anseeuw et al., 2012). In a number of cases agribusiness corporations buy forested areas and convert them into farmland (e.g., Naylor, 2011), as reported in the cases of Papua New Guinea (Nelson et al., 2014), Myanmar (Webb et al., 2014), and Brazil (Oliveira, 2013; Hermele, 2014). The assessment of deforestation induced by large-scale land acquisitions is a difficult task because official records of the acquired land and its geographic location are seldom available. A recent study (Davis et al., 2015) concentrated on Economical Land Concessions (ELCs) in Cambodia, which account for 36% of Cambodia's agricultural land and 12% of its forests. ELCs exhibited deforestation rates (4.3%–5.2% yr^1) about 29%–105% greater than those of comparable land outside ELC areas, indicating that large-scale land acquisitions can be significant drivers of deforestation (Davis et al., 2015). To date there are no international policies to regulate large-scale investments in agriculture and their environmental impacts. There are some guidelines for responsible land investments (FAO, 2011c); this is a code of conduct, which, however, is voluntary and cannot be enforced.

In recent years, deforestation has also been induced by the demand for biofuels. By mandating a certain degree of reliance on renewable energy, policies adopted by the United States and the European Union have increased the demand for biofuel crops, thereby favoring agricultural expansion and deforestation, often abroad (e.g., Meyfroidt et al., 2013). Thus, policies aimed at reducing atmospheric CO_2 concentrations may inadvertently increase CO_2 emissions (e.g., Fargione et al., 2008; Carlson et al., 2012). To prevent a potentially undesirable "leakage effect," European Union policies specify that biofuel produced at the expense of ecosystems of "high biodiversity value" (including woodlands and primary forests) does not count toward the renewable energy requirement of that country (E.U., 2009). The direct and indirect effects of biofuel production on these ecosystems, however, remain difficult to assess (Hermele, 2014).

5.3.2.4.3 Political Stability

Institutional instability leads to high discount rates, which favor the allocation of resources to the present at the expense of the future. This can result because dictators are concerned with the immediate need of securing their tenure in office, and, thus, resources at their disposal such as forests can be exploited for this purpose without any regard to future needs or efficiency considerations (Didia, 1997). Since dictators are less certain of their tenure in office than elected officials in democracies, the pressure from this uncertainty leads to actions and policies that are detrimental to tropical forest conservation (Didia, 1997). For instance, during 1976–1980, the tropical countries with the highest rates of deforestation were Ghana, Ivory Coast, and Nigeria, all of which experienced dictatorships for a considerable portion of this period (Didia, 1997).

5.3.2.5 Environmental Factors

Multiple environmental factors may influence an area's susceptibility to deforestation. In the tropics, drier, more deciduous forests are most vulnerable to deforestation because they are easier to burn, a characteristic that reduces the effort needed to clear forests and maintain pastures and croplands (Laurance et al., 2002). Similarly, in the Amazon, deforestation is most concentrated in drier, more seasonal forests (Laurance et al., 2002; Steininger et al., 2001). However, in semiarid areas such as the Chaco of Argentina, relatively higher deforestation rates have been documented in areas receiving more than 600 mm of precipitation per year (Grau et al., 2005). Grau et al. (2005) found that 40% of forests have been removed in areas with more than 600 mm of rainfall annually, while sectors with less than 600 mm annually have lost less than 20% of their forest cover. Soybean expansion into forested areas in northwestern Argentina has been partially attributed to the increase in rainfall that occurred in this area during the late 20th century (Minetti and Lamelas, 1997; Minetti and Vargas, 1997). Other environmental factors driving deforestation rates may include soil fertility (e.g., Pfaff, 1996) because more fertile and higher quality soils could lead to increased productivity of cleared land. Thus, farmers and ranchers on low fertility soils could be forced to clear larger areas of forest than those on better soils in order to remain viable (Laurance et al., 2002). Historical analyses of environmental factors influencing deforestation (and continued forest removal) suggest that relatively low rainfall, temperature, and soil nutrient content all lead to a permanent reduction in forest cover (Rolett and Diamond, 2004). Steep land slope is another environmental factor that has been cited as leading to reduced deforestation rates (e.g., Rolett and Diambond, 2004; Mas et al., 1996). Rolett and Diamond (2004) suggest that agriculture (hence land clearance) decreases with elevation because of low temperatures that may be unfavorable for crops, steep slopes, and difficult access. Apart from environmental characteristics of the forest, ecological attributes such as the presence of a positive feedback can affect the value of the standing forest and, thus, potentially make forest conservation more economically attractive. For instance, Runyan et al. (2015) coupled a biophysical model accounting for a positive feedback with an economic model and found that the optimal quantity of vegetation was considerably greater when the forest exhibited a positive feedback. While environmental variables alone cannot be used as predictors of deforestation rates, they may influence an area's susceptibility to deforestation.

5.4 Modeling Frameworks to Examine Deforestation

There are three primary spatial scales of economic modeling frameworks that are used to examine deforestation: microeconomic, regional, and macroeconomic. The decisions made by deforestation agents about how much and when to deforest are influenced by immediate and underlying causes (Angelsen and Kaimowitz, 1999). Immediate causes driving deforestation are characteristics such as background, preferences, and resources and decision parameters based on microeconomic factors such as prices, technology, institutions, new information, and access to services and infrastructure. Underlying macrocauses of deforestation influencing the agents' characteristics and decision parameters are the market, the dissemination of new technologies and information, the development of infrastructure, and institutional factors such as the property regime (Angelsen and Kaimowitz, 1999). Thus, microeconomic models focus on immediate causes, whereas macroeconomic models focus on underlying causes.

Three different methodologies are generally used with these modeling frameworks: analytical, simulation, and empirical. Analytical models are theoretical constructs that use no empirical data but seek to clarify the implications regarding different assumptions about how agents behave and how the economy operates (e.g., Runyan et al., 2015). Simulation models at the level of economic agents are becoming increasingly popular where environmental change is simulated as an emergent property of interactions between agents (Turner et al., 2007b). Empirical models quantify the relationship between variables on the basis of empirical data primarily using regression analysis (Angelsen and Kaimowitz, 1999). Analytical models produce results that are more general in nature and can shed light on the mechanisms underlying observed trends, whereas results from simulation and empirical models may be limited to the area in which they are applied. Notably, challenges confront these modelling frameworks because of the need for validation data on decision-making processes in the case of multiagent simulations and interactions between actors as well as how the scale of analysis affects modeling results (Turner et al., 2007b).

5.4.1 Microeconomic models

Microeconomic models seek to explain how individuals allocate their resources, using standard economic variables such as background and preferences, prices, institutions, access to infrastructure and services, and technological alternatives (Angelsen and Kaimowitz, 1999). A major distinction is that between models that assume all prices are market determined where decisions are guided by market prices and can be studied as a profit-maximizing problem (Southgate, 1990; Mendelsohn, 1994; Bluffstone, 1995; Angelsen, 1999), and those that do not (Dvorak, 1992; Holden, 1993; Angelsen, 1999) where decisions are based, in part, on farmers' subjective shadow prices (i.e., the price for an increment of additional agricultural production). Analytical types of models are useful in highlighting the role played by the underlying market and behavioral assumptions. Farm-scale simulation and regression models are informative because they use data regarding the magnitude of deforestation and farmers' behavior (Angelsen and Kaimowitz, 1999). However, their conclusions may have limited validity outside the region for which the model is constructed, and correlations obtained from statistical analyses do not establish causal relationships among variables (e.g., Lambin et al., 2000).

5.4.2 Regional models

Regional models are specifically applied to regions or areas with a distinct set of ecological characteristics, agrarian structure, institutional and political history, set of trade networks, and pattern of settlement and land use (Lambin, 1994). Most regional models are regression models, which may be spatial or nonspatial. Spatial models assess the impact of land use variables such as how far the forest is from markets and roads, topography, soil quality, precipitation, population density, and zoning categories (Angelsen and Kaimowitz, 1999). For instance, Megevand (2013) developed a spatially explicit regional economic model to analyze the potential drivers of deforestation in the Congo Basin over the next several decades. Their results showed that future deforestation in the Congo Basin depends critically on national strategies to be implemented in the next years and on local investments. The scenarios with the highest negative impact on forest cover resulted from improvement in transportation infrastructure and technological change. Nonspatial models use data obtained at the provincial or regional level in a manner similar to that used by multicountry regression models (Angelsen and Kaimowitz, 1999).

5.4.3 Macroeconomic

National and multicountry models emphasize the relationship among underlying variables, decision parameters, and deforestation. Analytical, simulation, and regression models are all used at this scale. Both analytical and simulation models at the national level add two important components that are often absent from smaller-scale models (Angelsen and Kaimowitz, 1999). First, they make some prices endogenous. Thus, they move beyond examining how decision parameters influence agents and look at how the underlying variables determine certain prices. Second, most models include the interactions among different sectors, such as agriculture, forestry, and manufacturing; as a result, they are useful for analyzing the underlying causes of deforestation (Angelsen and Kaimowitz, 1999). Multicountry regression models compose the largest category of deforestation models. They rely on national data to make global generalizations regarding the major processes affecting tropical deforestation. However, problems with methodology and data can make their usefulness and validity questionable. In order to produce meaningful cross-country results, the variables included must affect deforestation in roughly the same manner across countries (Angelsen and Kaimowitz, 1999).

5.5 Economic Effects of Deforestation

5.5.1 Benefits

Deforestation produces benefits to society because it provides essential resources (e.g., food, timber, and other raw materials), as well as much-needed jobs and income to many developed and developing countries worldwide (e.g., Foley et al., 2007). The FAO estimates that forest industries contribute more than U.S. $450 billion to national incomes, contributing nearly 1% of the global GDP in 2008 and providing formal employment to 0.4% of the global labor force (Agrawal et al., 2013). It is estimated that 735 million rural people live in or near tropical forests and savannas, relying on them (or the crops and pasture from removing them) for much of their fuel, food, and income (Chomitz et al., 2006). For instance, the oil palm sector in Indonesia employs 1.7–2 million people (Wakker, 2006; Zen et al., 2006). These benefits are also received by people dependent on forest products in temperate regions as well. In Canada, it is estimated that forest products account for $26.6 billion in exports, $1.38 billion in government revenues from the sale of timber, and 211,200 direct jobs nationwide in 1999 (CCFM, 2002). Moreover, more than half a million Canadians are employed directly or indirectly in the processing of wood fibers from Canada's boreal forests (Burton et al., 2003). Both wealthy and poor populations benefit from products derived from deforestation. For instance, rural populations may rely on low-productivity agriculture for subsistence while urban populations can have high demands for commodities such as beef, palm oil, coffee, soybeans, and chocolate (in addition to wood and pulp).

The people and communities benefiting from deforestation can also differ across spatial scales. Kremen et al. (2000) estimated conservation costs of forests from local, national, and global perspectives using a case study from Madagascar. They found that conservation generated significant benefits over logging and agriculture locally and globally. Nationally, however, financial benefits from industrial logging were larger than conservation benefits (by providing benefits between $6 million to $265 million to the state). Thus, when Madagascar conserves forests, it pays most of the costs of the global benefit of preventing transfers of carbon from the biosphere to the atmosphere. Similarly, Kumari (1994) quantified different human uses of tropical forest areas in Sleangor, Malaysia, and found that high-intensity logging was associated with greater private benefits through timber harvesting, but reduced social and global benefits (through loss of nontimber forest products, flood protection, carbon stocks, and endangered species). Although many of the benefits of deforestation can be private, such as income from agriculture and timber sales from areas being cleared, deforestation can offer local benefits for securing food, income, fresh water, and/or raw materials (DeFries et al., 2004). For instance, Rodrigues et al. (2009) analyzed levels of income at different stages of deforestation across the Brazilian Amazon. They found that the relative levels of income improve very quickly during the early stages of deforestation. Levels of literacy and life expectancy also increase, possibly because improved income and government investment result in better living conditions while improved transportation infrastructure provides better access to education and medical care. However, if natural resources that sustained the initial boom are exhausted, a "bust" can result, leading to a decline in the standard of living, literacy, and life expectancy to levels similar to those in the prefrontier municipalities (Rodrigues et al., 2009).

5.5.2 Costs

5.5.2.1 Loss of Nontimber Forest Products

In case studies conducted worldwide, economists have quantified the value of nontimber forest products (NTFPs) used as food (e.g., nuts, honey, bush meat, and fruits), medicine, fuel, construction materials (e.g., rubber), bioprospecting (i.e., value for new pharmaceutical products), and agricultural products such as fodder for livestock. In many cases, the value of these products, estimated at market prices or in terms of willingness to trade with market products, is quite low per hectare of forest, compared to the value of timber and agricultural use of the land. However, studies that have quantified the value of tropical forests as potential storehouses of undiscovered pharmaceutical products can indicate very high potential values (Simpson et al., 1996; Goeschl and Swanson, 2002; Craft and Simpson, 2001; Rausser and Small, 2000). For instance, Rausser and Small (2000) estimated the bioprospecting value of 20 biodiversity "hotspots" and found incremental values as high as $9,177/ha (in western Ecuador).

The value of NTFPs as a percentage of household income can be significant, especially for the poorest households (Box 5.1). The demand for NTFPs is also widespread.

> **Box 5.1 Empowering Communities through the Valuation of Communal Forests**
>
> In many regions of the developing world, rural communities are losing access to their forests because of land concessions granted to commercial agribusinesses potentially without adequate informed consent by local villagers. In many cases communities are not aware of the value of the land and have little sense of how their livelihoods depend on these forests and the services they provide. To empower the local communities, groups of "community paralegals" are trying to level off the information asymmetries existing between investors and villagers (Knight, 2015). They teach local communities how to evaluate the replacement costs of communal forests. The notion of replacement cost is used to express the impact of forest concessions on rural livelihoods without claiming that their natural capital can/should be sold or replaced by man-made capital. Using a grassroots approach, teams of legal advocates meet with each community to establish the forest's replacement cost. To that end, the community is asked to list the different ways its members use the communal forest, identify the resources and other services they extract from it, and indicate their monetary value (e.g., how much would it cost to buy firewood at the local market). These estimates are then used to determine how much the typical family would have to pay in order to buy the resources it takes from the community forest. An analysis of 14 communities across Mozambique, Liberia, and Uganda has shown (Knight, 2015) that the "typical family" uses common forest resources with a replacement cost in the range ≈$300–$3,000 per household per year (on average $1,500 yr^{-1}), which is a nontrivial amount for communities often living below the poverty line of $1.25 cap^{-1} d^{-1}. These values are comparable to those determined by other authors for Tanzania and South Africa. On average a community uses the equivalent of $1.4 million yr^{-1} in forest resources and services. In Uganda the size of concession areas ranges between 500 and 1,000 hectares; thus a typical rent of the order of $2–5 ha^{-1} yr^{-1} (e.g., Cotula, 2011) is several orders of magnitude smaller than the replacement cost of the goods and services provided by the forest (Knight, 2015).

For instance, it has been estimated that in Mexico, around 25 million people prepare their meals with fuelwood (Masera et al., 2005). An estimated 1.5 million people in the Brazilian Amazon utilize and depend on a wide array of products from the forest (Chibnik, 1994; Clay and Clement, 1993). NTFPs may also help rural households smooth income fluctuations (Pattanayak and Sills, 2001). Chopra (1993) estimated the value of NTFPs for tropical deciduous forests in India and found that forest products (e.g., dyes, bidi leaves, sal leaves, tassan cocoons, and lacquer) provided a value of U.S. $601.20 per hectare while fuelwood and fodder provided average values of U.S. $123.41 and U.S. $263.17 per hectare, respectively.

The potential economic loss of NTFPs may be substantial when the future commercialization of these products is considered (Box 5.1). NTFP commercialization is expected to increase income and employment opportunities, especially for poor and otherwise disadvantaged people because of the emergence of new markets for natural products, and the development of new marketing mechanisms (e.g., green marketing, fair trade) (Belcher and Schreckenberg, 2007). When conducted sustainably, NTFP commercialization can provide opportunities for benign forest utilization (Myers, 1988) and even create incentives for forest conservation (Belcher and Schreckenberg, 2007). Despite this, several factors can limit the commercialization of NTFPs, and once they are achieved, mixed results can ensue. Lack of proximity to markets is considered one of the most limiting factors for NTFP commercialization (Neumann and Hirsch, 2000). However, global growth has increased market access (and will continue to do so) in remote locations, thereby increasing potential opportunities for NTFP commercialization. Mixed results can follow NTFP commercialization as a result of both supply and demand related problems. Many of these products are considered "luxury" items, meaning that change in demand is particularly difficult to predict and variable between years. These products are often traded and produced in relatively small volumes that are dispersed over wide areas, leading to market inefficiencies. Market inefficiencies result under these circumstances because they do not function as markets for major agricultural commodities where producers can purchase needed inputs and sell their produce, often with many options for sales outlets (Belcher and Schreckenberg, 2007). Barriers to market entry may also be difficult in that very few low-income countries have the high degree of infrastructural and institutional development, strict quality control, and sophisticated supply-chain management practices necessary to enter the international market with a new product (Belcher and Schreckenberg, 2007). Sustaining supply can also be a difficult problem. Once producers can access markets for the product, one of the most important challenges is to increase the quality and quantity of production at competitive prices. The Center for International Forestry Research analyzed 61 case studies of NTFP commercialization in Asia, Africa, and Latin America and found that in many cases, the initial response to increased demand was more intensive harvesting, leading to overexploitation of the species (Marshall et al., 2006). Overexploitation can result from open access conditions where increased value leads to uncontrolled competition for resources and inefficient and damaging harvesting practices, which in turn lead to highly variable supply of the product.

While different constraints exist on NTFP commercialization, many factors may contribute to the successful commercialization of an NTFP. These factors include i) biological attributes of the plant species such that fast-growing, high-yielding species are less likely to be overexploited than slower-growing, low-yielding species and the returns on investment are quicker; ii) national level policies that encourage investment in processing and trade to support raw material producers; and iii) improvement in the quality and quantity of their product by investing in postharvest storage and processing to extend the economic life of the harvest, reduce some of the urgency for selling, and allow for the collection of larger volumes at one time and place (Belcher and

Schreckenberg, 2007). Thus, successful NTFP development may be achieved in the form of better markets, improved infrastructure, and higher product demand and/or prices, which provide a strong incentive for increased production.

5.5.2.2 Loss of Ecosystem Services after Deforestation

Ecosystem services consist of flows of materials, energy, and information from natural capital stocks (such as trees and forests) that are combined with manufactured and human capital services to produce human welfare (Costanza and Folke, 1997). These services are expansive (see Table 5.2 for a list of these services and their function) and have local to regional (e.g., flood control) to global (e.g., carbon sequestration) benefits (Foley et al., 2007). For instance, the Amazon provides an important ecosystem service to the planet by storing organic carbon in biomass and soil, thereby preventing the release of greenhouse gases into the atmosphere (Foley et al., 2007). Ecosystems also regulate many hydrological functions (Brauman et al., 2007). For instance, deforestation in the Amazon would substantially impact the regulation of river discharge. Costa and Foley (1997) estimated that widespread deforestation in the Amazon would increase runoff and river discharge by about 20%. One other notable ecosystem service provided by tropical forests is moderating the risk of infectious disease by regulating the populations of disease organisms (i.e., viruses, bacteria, and other parasites), their animal hosts, or the intermediary disease vectors (most often insects or rodents) (Foley et al., 2007). For instance, Vittor et al. (2006) found that heavily deforested areas in the Peruvian Amazon can see up to a 300-fold increase in the risk of malaria infection, compared to areas of intact forest. Deforestation can also degrade basic ecosystem services especially those tied to the long-term functioning of the ecosystem (see Chapter 4; Foley et al., 2007) and subsequent land use to sustain farming and cattle grazing (Müller et al., 2004). For instance, climate model simulations of large-scale deforestation in the Amazon Basin generally show a considerable reduction in evapotranspiration as tropical forest vegetation is replaced with grasses and shrubs. This reduction in evapotranspiration has the effect of substantially warming the surface and inhibiting convection, regional precipitation, and cloud cover, thereby suppressing the regrowth of forest vegetation (e.g., Nobre et al., 1991; Costa and Foley, 2000; Chapter 4). The loss of these ecosystem services that follows deforestation shows that while many of the costs of deforestation are shared by all of the world's citizens, the benefits may accrue only to a relatively small group of agents.

The loss of ecosystem services is hard to account for in the economic analysis of different forest management decisions (Kinzig et al., 2011). The valuation and discounting of ecosystem services remain major tasks for environmental economists. In recent years the notion of valuating these services (i.e., assigning a "dollar value" to them) has also been challenged on the grounds that some of them cannot be replaced and "bought" with the commodification of natural capital being detrimental to long-term economic development (e.g., Fairhead et al., 2012; Dasgupta,

2013; D'Odorico and Rulli, 2014). Using markets to induce people to take account of the environmental costs of their behavior is one mechanism to allow governments and nongovernmental organizations to pay for environmental public goods, such as habitat provision, watershed protection, or carbon sequestration (Kinzig et al., 2011). Markets have not developed for many of these services partly because of the public good nature of those services, the lack of well-defined property rights and the various costs of forming markets (Kinzig et al., 2011).

Although ecosystem services are often not quantified in terms of their economic value, that value can be appreciable (Box 5.1). For instance, Xiao et al. (2000) quantified the indirect economic value of tropical forests in Hainan Island, China, which cover 44,667 ha. Indirect economic value of these services added up to 664 million yuan (Chinese RMB; 2000 mean exchange rate of 8.28 yuan/USD), which is equal to $1,795 USD/ha. Approximately 60% of this value was due to the forests' ability to hold water, while 27% was from air purification. Forests can also aid in purifying water reaching drinking water sources. One study in Chile calculated that native forest is worth more than U.S. $200 per hectare per year because of its effects on drinking water supply (Guo et al., 2000; Nunez et al., 2006). Pattanayak and Kramer (2001) quantified the value of forests to downstream farmers in Indonesia by comparing agricultural profits across watersheds with different levels of water flow and amounts of deforestation. They found that farmers were willing to pay ~10% of annual agricultural costs, 75% of annual irrigation fees, and 3% of annual food expenditures for drought mitigation. The presence of forests can also aid in enhancing groundwater recharge. For instance, the net present value of one forested watershed for enhancing groundwater recharge in Hawaii was calculated to range between $1.5 and $2.5 billion (Kaiser and Roumasset, 2002).

5.6 Alternative Policies Aimed at Deriving Value from Forested Lands (i.e., REDD)

About one-fifth of global CO_2 emissions arise from tropical deforestation – and the costs of abating some of these emissions are low. In Latin America, dense tropical forest is often cleared to create pastures worth a few hundred dollars a hectare, while releasing 500 tons of CO_2 per hectare. Though volatile in price, when other land uses are considered, the price for CO_2 abatement generally ranges between $1 and $3 per ton. Thus, deforesters are destroying a carbon storage asset potentially worth $1,500–$10,000 to create a pasture worth $200–$500 per hectare (Chomitz et al., 2006;). Paying the opportunity cost for forest conservation is an economical mechanism for securing reduced atmospheric greenhouse gas concentration (Pearce and Brown, 1994), while promoting other benefits including the protection of a threatened biota (Mittermeier et al., 1998), maintenance of ecosystem services, and enhancing human welfare (Kremen et al., 2000). As noted previously, there is often a discrepancy between immediate private benefits gained from forest conversion and external benefits from standing forest that tend to be global in nature (Balmford et al., 2002). Hence, conserving relatively intact forests in developing countries may require compensatory mechanisms that are financed at the global level to mitigate the impact of forgone private and local benefits.

Programs such as REDD (i.e., Reduced Emissions from Deforestation and Degradation) aim to reduce deforestation through incentives to maximize agricultural production on already-cleared lands while minimizing new forest clearing (Matson and Vitousek, 2006). International programs that meet the dual goals of maintaining forest carbon and increasing agricultural production are needed as countries face pressures to clear more forests. However, several issues surround the potential effectiveness of REDD and more recent programs such as REDD+ that also consider the role of conservation, sustainable management of forests, and enhancement of forest carbon stocks in developing countries. These issues include *additionality*, which is the risk that payments for reduced emissions might be provided for reductions that would have occurred even without payments; *permanence*, which is the risk that current emissions may be offset via higher levels of deforestation in the future; and *leakage*, which is the risk that emissions reduced in one location may simply be shifted to higher levels of deforestation at another location (Agrawal et al., 2011). Additionally, future conservation costs may increase as a result of higher agricultural land rents resulting from productivity increases (Phelps et al., 2013). Despite difficulties with implementing REDD, such a program could generate significant funding to reduce deforestation. Dutschke and Wolf (2007) estimated the market potential of REDD at U.S. $10 billion, which is substantial when considering that forestry exports from the developing world were worth ~U.S. $39 billion in 2006 (FAO, 2008).

5.7 Conclusion

Deforestation produces economic benefits to society because it provides essential resources such as food and timber as well as jobs and income to societies worldwide. However, maximizing the economic benefits from deforestation can be challenging because many externalities resulting from deforestation are public, while benefits and direct costs incurred by deforestation are private. Deforestation results from choices made by small-scale farmers at the forest frontier in the tropics as well as commercial loggers in the boreal. These choices regarding when and how much to deforest are influenced by multiple proximate and underlying drivers. Proximate causes of deforestation are human activities or immediate actions at the local level that originate from intended land use with agricultural expansion frequently being cited as a proximate underlying driver of deforestation globally. Underlying causes of deforestation are fundamental social processes that underpin the proximate causes. While some of these causes have a predictable effect on deforestation rates such as the increase in the price of agricultural commodities leading to an increase in deforestation rates, other causes such as the effect of property rights regime have a less predictable effect. There is a range of costs, which are sometimes unquantified, associated with the loss of forests, many of which are global in nature such as carbon storage and greenhouse gas regulation. Because of the potential discrepancy between private benefits from deforestation and global benefits from standing forest, programs such as REDD+ have been pursued as an economical mechanism to minimize forest conversion while creating a financial market to compensate developing countries for their standing forests.

6

Synthesis and Future Impacts of Deforestation

6.1 Benefits of Preserving Forests

Forests provide an expansive range of environmental benefits (Table 5.2) that have local to regional (e.g., flood control) and global (e.g., carbon sequestration) relevance (Foley et al., 2007). Hydrological benefits of forests include regulating water supply and river discharge (i.e., moderating high and low flows) by increased transpiration, water storage beneath the forest, and increased travel time for water to reach streams/rivers. Climate benefits of forests include maintaining available precipitation via precipitation recycling (in areas where this feedback exists) and regulating local and global temperature both directly – by reducing diurnal sensible heat fluxes and nocturnal radiative cooling – and indirectly – by taking up atmospheric CO_2 during photosynthesis. For instance, a review by Lawrence and Vandecar (2015) highlighted that complete deforestation of the tropics would lead to a 0.1–1.3°C increase in temperature across the tropics and drying of approximately –270 mm yr^{-1} (or up to 10%–15% decrease of annual rainfall). The presence of forests can also affect edaphic processes by reducing soil erosion and enhancing soil formation. Biogeochemical benefits of forests include enhancing nutrient availability and reducing nutrient losses, thereby increasing the amount of nutrients available for plant uptake and aiding in sustaining forest growth. Forests also provide many ecological services such as maintaining biodiversity and regulating a range of dynamical trophic relationships. As discussed in Chapter 4, forests can be important for maintaining their own habitat, possibly reducing the occurrence of disturbances such as fire or landsliding, and, in turn, maintaining the wide span of benefits described previously. Forests provide many economic benefits to societies from nontimber forest products (NTFPs) that are harvested from them. These NTFPs provide food (e.g., nuts, honey, bush meat, and fruits), medicine, construction materials (e.g., rubber), bioprospecting (i.e., value for new pharmaceutical products), and agricultural products such as fodder for livestock. Forests also have recreation, cultural, intellectual, aesthetic, and spiritual values that are important to society. Thus, deforestation strongly affects the environment and society. In the following sections we review and summarize some of its major impacts.

6.2 Ecohydrological and Climate Impacts of Deforestation

Most of the past research on the hydrological impacts of deforestation has focused on changes in water yields and flow regulation. The removal of forest vegetation causes a reduction in soil infiltration and evapotranspiration and, consequently, an increase in infiltration-excess runoff and soil erosion. Overall, deforestation leads to an increase both in water yields and in peak flows. Large-scale deforestation of continental regions is expected to reduce precipitation and alter the timing of the rainy season in the same region, while smaller-scale deforestation may lead to an enhancement of precipitation by mesoscale circulations (*canopy breezes*). The possible existence of teleconnections between land cover change in one region and the precipitation and temperature regime in another remains poorly understood (see Chapter 2). Likewise the hydrological impacts of large-scale reforestation and forest restoration still need to be adequately investigated. In fact, some authors argue that most studies have concentrated on small-scale watersheds, and their results cannot be extended to larger regions because they do not account for important feedbacks on precipitation (Ellison et al., 2012). As explained in Chapter 2, forests may enhance water availability either by enhancing fog or cloud water deposition or by increasing regional precipitation. Both effects need to be accounted for while evaluating the hydrological benefits of large-scale reforestation efforts (Palmer et al., 2015).

6.3 Effect of Forest Loss on Biogeochemical Processes

Carbon (C). Carbon emissions from deforestation account for an estimated 12% of global anthropogenic CO_2 emissions (Van der Werf et al., 2009). Deforestation is largely driven by agricultural expansion. Hosonuma et al. (2012) found that agricultural expansion drives ~80% of deforestation across 46 developing countries worldwide. Whether pasture and agricultural soils are a net sink or source of C after deforestation depends substantially on their management. Carbon emissions can be reduced after deforestation by adopting management practices that increase the soil C content. In the tropics, increasing C inputs to the soil are obtained by improving the fertility and productivity of cropland and pastures using practices such as no-till and, for systems practicing shifting cultivation, using planted fallows and cover crops (Paustian et al., 1997). In temperate regions, key management strategies involve increasing cropping frequency and reducing bare fallow, increasing the use of perennial forages (including N fixing species) in crop rotations, retaining crop residues, and reducing or eliminating tillage (i.e., no-till; Paustian et al., 1997).

Nitrogen (N). After deforestation, there is an increase in leaching and runoff losses of nitrate to groundwater and stream water that increase when the harvesting method is intensified (Sollins and McCorison, 1981). Nitrate export to stream water increases after deforestation because of reduced plant uptake, increased rates of N mineralization and nitrification, and increased leaching through the soil profile. Deforestation can increase soil temperature and moisture, thereby enhancing conditions for mineralization. Subsequently, nitrification may occur so rapidly that uptake by vegetation and immobilization by microbes are insufficient to prevent large losses of NO_3^- to stream water and groundwater (Schlesinger, 1997).

N is also lost during deforestation as a result of the removal of organic N contained in logged wood. In contrast to the effects on N cycling in forests that are logged, fire differently alters patterns of N cycling. During burning there is a loss of both N stored in vegetation (via smoke and airborne ash) and soil N lost to wind and water erosion (Kauffman et al., 1993).

Atmospheric losses of N result both during and after deforestation. The quantity of N lost is dependent on the mechanism of forest removal. Fire volatilizes N from vegetation and litter, reducing N stored in the burned ecosystem but often increasing mineralization of the remaining organic matter (e.g., Turner et al., 2007a). Increased fluxes of N to the atmosphere also result over relatively short periods from denitrification (i.e., the reduction of nitrogen oxides to dinitrogen gas by microorganisms).

Phosphorus (P). In P limited forests, the availability of P for plant uptake is dependent on mechanisms and symbioses with trees that aid in increasing P availability. For instance, plant roots may produce extracellular phosphatases, microorganisms and fungi mineralize unavailable forms of P, cluster roots maximize P acquisition, while symbiotic structures such as mycorrhizae and root nodules increase scavenging volumes. In P limited forests, P is cycled very conservatively with minimal losses from the system. However, after deforestation, P losses are high, often leaving the system in an unproductive and highly nutrient limited state. If vegetation growth is constrained by nutrient availability, less carbon will be fixed to structural biomass, and larger amounts of CO_2 will be respired and released to the atmosphere (Körner, 2006; Canadell et al., 2007).

It has been suggested that 80% of deforestation will occur in tropical forests, which are often P limited (Gibbs et al., 2010). Thus, if the cleared land is used for agriculture, it will most likely be able to sustain only a few crop cycles. It has been estimated that 30%–40% of the world's arable land is primarily P limited (Runge-Metzger, 1995; von Uexkull and Mutert, 1995). Short-term P limitations can be overcome in agricultural systems by adding fertilizer to stimulate crop yields; however, the supply of rock phosphate reserves used to create P fertilizer could be depleted in as little as 50–100 years (Vance et al., 2003; Cordell et al., 2009). Moreover, excess fertilizer application of N and P can reach surface water and groundwater bodies, causing problems such as eutrophication, and once dissolved oxygen falls below a critical level, creating dead zones. Globally, dead zones affect a total area greater than 245,000 square kilometers and have been reported for more than 400 marine systems, a number of which are major fishery areas (Diaz and Rosenberg, 2008).

6.4 Economic Impacts of Deforestation

The choices regarding when and how much to deforest are influenced by multiple proximate and underlying drivers. Proximate causes of deforestation are human activities or immediate actions at the local level that originate from intended land use while the underlying factors are fundamental social processes that drive the proximate causes (Geist and Lambin, 2002). As noted previously, agricultural expansion is frequently cited as the primary proximate driver of deforestation globally, accounting for 80% of deforestation across a number of developing countries (Hosonuma et al., 2012). A common policy for reducing deforestation is to develop and disseminate technologies that allow farmers to increase production on land that is already deforested. While it is hypothesized that intensification would lead to a decrease in deforestation, this trend is often not observed (Rudel et al., 2009b; Vosti et al., 2002; Morton et al., 2006). A recent review by Villoria (2014) shed light on this paradox. They found that deforestation may be reduced by some forms of improved technology including more profitable techniques for sustainable forest management and infrastructure such as irrigation equipment that can lead to a boom in the labor market (i.e., employment), while sustaining higher cropping intensities, alleviating land pressure on the forest frontier, and reducing cropland expansion (Shively and Pagiola, 2004). In general, technologies that use more labor will constrain deforestation in the short term (for rural areas that are more remotely located and have smaller populations), although over the long term, they may attract more labor to the region (e.g., through migration). In contrast, yield increasing and labor saving technology may increase deforestation. Agroeconomic variables such as low domestic costs (for land, labor, fuel, or timber) and an increase in the price of agricultural crops can stimulate deforestation rates (Geist and Lambin, 2002). In contrast, an increase in the price of rural wages (Angelsen and Kaimowitz, 1999), a shortage of off-farm employment opportunities, and an increase in the price of agricultural inputs such as seeds and pesticides can reduce deforestation rates (Ruben et al., 1994; de Almeida and Campari, 1995).

While some of these causes have a more predictable effect on deforestation rates, such as the increase in the price of agricultural commodities leading to an increase in deforestation rates, other causes such as the effect of the property rights regime have a less predictable effect. For instance, the impact of property rights on deforestation could affect forest cover in two different ways with opposite implications (Liscow, 2013). Property rights could lead landholders to discount the future less (i.e., place greater emphasis on the future stream of benefits) and reap the long-term benefits of forestry instead of the short-term benefits of agriculture (see Chapter 5). On the other hand, more secure property rights increase investment in land because farmers are more motivated to "develop" the land and are able to use it as a collateral for borrowing money for such investments. As a result, the value of agriculture increases relative to that of forest, thereby leading to a decrease in forest cover.

While there are multiple underlying and proximate factors driving the decision to deforest, both the quantity and timing of deforestation activities will be determined by maximizing the economic benefits resulting from deforestation. However, maximizing the economic benefits from deforestation can be challenging because many externalities (e.g., the environmental cost of the loss of services provided by the forest) from deforestation are public, while benefits and direct costs incurred by deforestation are private. Thus, immediate benefits gained from forest conversion tend to be private while external benefits from standing forest tend to be public and global in nature (Balmford et al., 2002). In other words, the typical economic impacts of deforestation include both the provision of private benefits to the individuals or corporations that clear the land (e.g., for logging, mining, or agriculture) and the emergence of public costs associated with the loss of important services provided by the forest. There are also unquantifiable and irreplaceable losses for the local and global community associated with the environmental, cultural, and ethical value of forests. Thus, there is a separation between those who are affected by the costs and the benefits of land conversion. Regardless of whether the benefits are evenly distributed within the community affected by forest loss, deforestation may lead to the irreversible loss of natural resources, and, as such, it may raise issues of intergenerational equity and weak sustainability: that is, treating natural capital (the forest) and man-made capital as substitutable and interchangeable goods (Solow, 1974; Hartwick, 1978). If land clearing causes environmental degradation, the decision to deforest for short-term profits reduces the amount of natural capital available to future generations.

6.5 Irreversible Changes Induced by Deforestation

Positive feedbacks may induce bistable ecosystem dynamics, with forest ecosystems occurring as metastable states of a system that has an alternative stable configuration with no forest vegetation. The existence of such dynamics suggests that some forests are prone to abrupt and highly irreversible shifts to a stable and often "degraded" state with no trees. These feedbacks span many different environmental processes and are found across forests globally.

In general, deforestation that leads to complete forest removal (i.e., clear-cutting) results in the greatest reduction in system resilience (i.e., the ability of the forest to recover after disturbance). In contrast, selective logging tends to have the smallest effect on system resilience. Once a shift to the bare state occurs, it could endure if the underlying dynamics are bistable. Therefore, it is important for land managers to be cognizant of when a system may be close to its tipping point. Moreover, once a transition to the bare state occurs, the restoration options will depend not only on the value of the ecosystem services that could be recovered once the system is restored, but also on the physical feasibility of restoration efforts and on the availability of adequate financial resources to spend on restoration. Importantly, in areas where restoration is either not feasible or more costly than the economic benefit provided by the restored forest, restoration will likely not be pursued and the land will remain in its degraded state with no forest vegetation. Early warning signs of state change could provide useful tools for land managers to prevent the shift to a degraded state. To date, however, the existing theories refer to the case of systems that remain near equilibrium under the effect of a "slowly" changing environmental parameter (e.g., Scheffer et al., 2009). In the case of deforestation, the system undergoes an abrupt shock, and there are currently no tools to predict whether or not the forest will be able to recover.

6.6 Biodiversity Loss

Biodiversity is defined as the sum of all of the plants, animals, fungi, and microorganisms on Earth; their genetic and phenotypic variation; and the communities and ecosystems of which they are a part (Dirzo and Raven, 2003). Fossil records of terrestrial multicellular eukaryotes (i.e., nonbacterial organisms) indicate maximum biodiversity at the present time with a nearly exponential increase that has occurred since organisms first emerged on land about 440 million years ago (Dirzo and Raven, 2003; Figure 6.1). The number of organisms has continually increased over the past 570 million years despite five major extinction events (i.e., the Ordovician, Devonian, Permian, Triassic, and Cretaceous events). Estimates of the total number of species on Earth range between 2 million and 100 million species (Costello et al., 2012), though a new review of estimates leans toward more conservative estimates of ~5 million ± 3 million species (Costello et al., 2013). Mora et al. (2011) estimated that there are 8.7 million eukaryotic species globally (including 2.2 million marine species, of which ~20% of this total have been described; Table 6.1). These authors derived this estimate from their discovery that the higher taxonomic classification of species (i.e., the assignment of species to phylum, class, order, family, and genus) follows a consistent and predictable pattern from which the total number of species in a taxonomic group can be estimated and validated against well-known taxa for all domains of life. Their approach predicted ~7.77 million species of animals, ~298,000 species of plants, ~611,000 species of fungi, ~36,400 species of protozoa, and ~27,500 species of chromists (Table 6.1).

The current distribution of biodiversity is largely concentrated in the tropics. In broad geographic terms, species densities range from more than 5,000 species/10,000 km^2 in tropical regions to fewer than 100 species/10,000 km^2 in the highest latitudes (e.g., Dirzo and Raven, 2003). Tropical forests represent 70% of the world's floristic and faunistic species (Wassenaar et al., 2007). However, these species are not

Table 6.1. *Currently catalogued and predicted total number of species on Earth*

Species	Catalogued	Predicted	±SE
Eukaryotes			
Animalia	953,434	7,770,000	958,000
Chromista	13,033	27,500	30,500
Fungi	43,271	611,000	297,000
Plantae	215,644	298,000	8,200
Protozoa	8,118	36,400	6,690
Total	1,233,500	8,740,000	1,300,000
Prokaryotes			
Archaea	502	455	160
Bacteria	10,358	9,680	3,470
Total	10,860	10,100	3,630
Grand Total	1,244,360	8,750,000	1,300,000

Source: Mora et al., 2011.

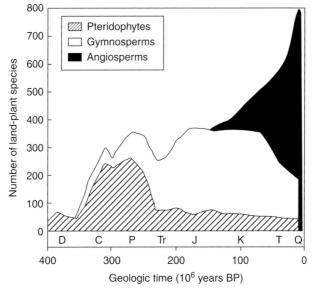

Figure 6.1. The number of land-plant fossil species, including three major groups, Angiosperms (i.e., flowering plants), Gymnosperms (i.e., plants that reproduce by means of an exposed seed or ovule), and Pteridophytes (i.e., vascular plants that reproduce and disperse via spores), through the last 400 million years. D, Devonian; C, Carboniferous; P, Permian; Tr, Triassic; J, Jurassic; K, Cretaceous; T, Tertiary; Q, Quaternary.
Source: Drawing based on plot by Dirzo and Raven, 2003.

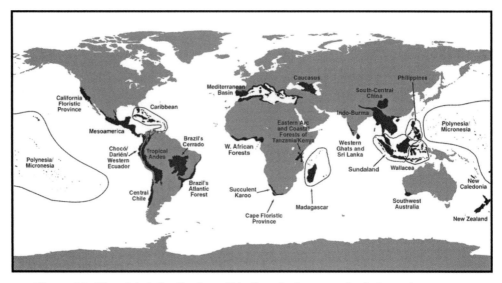

Figure 6.2. The global distribution of biodiversity hotspots (in dark gray).
Source: Myers et al., 2000.

distributed evenly among the tropics. As many as 44% of all species of vascular plants and 35% of all species in four vertebrate groups are confined to 25 hotspots comprising only 1.4% of the Earth's land surface (Myers et al., 2000; Figure 6.2). Species number analyses of plants suggest that about 90,000 species occur in the Neotropics, which are approximately twice as many as in Africa south of the Sahara and 50% greater than in Asia (Dirzo and Raven, 2003). Similar trends have been observed in other groups, including butterflies (Neotropics > Southeast Asia > Africa), frogs (Neotropics > Africa/Asia > Papua/Australia), and birds (Neotropics > Africa > Asia/Pacific > Australopapuan) (Dirzo and Raven, 2003). However, biodiversity may currently be threatened by a major extinction event (which is defined when the Earth loses more than three-quarters of its species in a geologically short interval) as indicated by higher than would be expected extinction rates (Barnosky et al., 2011).

Land use change is one of the most severe drivers of changes in biodiversity, with estimates of global losses of intact ecosystems ranging between 0.5% and 1.5% annually (Sala et al., 2000; Jenkins, 2003). One extreme consequence of habitat loss is extinction. Habitat loss can cause extinctions directly by removing all individuals over a short period but can also be indirectly responsible for lagged extinctions by facilitating invasions, improving hunter access, eliminating prey, altering biophysical conditions, and increasing inbreeding depression (Brook et al., 2008). Although extinction is a natural process, it may be occurring at a higher rate relative to background rates as a consequence of human activities. Already we may have caused the extinction of 5%–20% of the species in many groups of organisms, and current rates of extinction are estimated to be 100–1,000 times greater than average prehuman rates (Chapin et al., 2000c). For instance, estimates of mammalian extinction rates over the past four hundred years are between 36 and 78 times background levels (Regan et al., 2001). While extinction rates vary for different groups of organisms and different fauna and flora, estimates of extinction rates vary widely (i.e., 0.04%–30%), yet are more likely to be in the range of <1% per decade (Stork, 2010). Species most at risk of extinction are those with small ranges located in areas experiencing high levels of habitat loss (Pimm and Raven, 2000). Additional traits such as ecological specialization, low population density, and low reproduction rates elevate extinction risk, because rarity imparts higher risk and specialization reduces the capacity of a species to adapt to habitat loss by shifting range or changing diet (Davies et al., 2004). A geographical analysis of the 811 recorded extinctions that occurred between 1500 and 2000 AD show that they have primarily taken place on islands (Dirzo and Raven, 2003). For instance, since human arrival on the Hawaiian Islands 1,500 to 2,000 years ago, 60 endemic species of land birds have become extinct while another 20 to 25 bird species were lost in the past two centuries (Steadman, 1995). The most comprehensive list of globally threatened species (i.e., species facing a high risk of extinction in at least the near future resulting both directly and indirectly from human activities) is the IUCN 2014 edition, which lists 22,413 threatened species (IUCN, 2014).

In addition to considering species extinctions, assessing biodiversity change includes considering other metrics such as species abundance and community structure, habitat loss and degradation, and shifts in the distribution of species and biomes (Balmford et al., 2003; Pereira et al., 2010). Examining the risk of species extinction addresses the irreversible component of biodiversity change, yet species extinctions have weaker links to ecosystem function (e.g., the last few individuals of a threatened species are unlikely to make a major, measurable contribution to ecosystem functioning) and respond less rapidly to global change than do other metrics (Balmford et al., 2003). Extinction often lags behind partial habitat clearance because it can take a century or more for the smaller community to adjust to its reduced habitat area, during which time the persistence of very small populations can give the misleading impression that species losses are not as severe as predicted (Pimm and Brooks, 2000; Turner et al., 1994; Brooks et al., 1999). To understand and mitigate better the impact of humans on different ecosystems, the long-term perspective provided by examining extinction rates needs to be supplemented with measures of the ecosystem state that are sensitive to short-term changes (Balmford et al., 2003).

More-responsive metrics of biodiversity change include examining changes in species abundances and community structure, shifts in the distribution of species, and, at a higher organizational level, habitat loss or biome changes (Pereira et al., 2010). It is important to understand these changes as globally, the average annual loss of habitats and the populations they support leads to a loss of goods and services worth approximately U.S. $250 billion annually (Balmford et al., 2002). The GLOBIO model is one tool that has been used to estimate the effect of land use change on mean species abundance for terrestrial systems (Alkemade et al., 2009). Scenarios examined using GLOBIO project a decline of 9% to 17% in mean species abundance by 2050 relative to 2000 (UNEP, 2007; Brink et al., 2007). Hooper et al. (2012) conducted a metaanalysis of published data from 192 studies examining species loss and ecosystem productivity and found that intermediate levels of species loss (i.e., 21%–40%) reduced plant production by 5%–10% while a 50% species loss reduced plant production by approximately 13%. The impact of habitat loss and degradation on biodiversity decline over the next 50 years could be substantial as forest loss is expected to be highest in the tropics, where biodiversity is greatest (Pereira et al., 2010). Gibson et al. (2011) used 2,200 pairwise comparisons of biodiversity in tropical forests from a metaanalysis of 138 studies to assess the impact of disturbance and land conversion on biodiversity. They found that birds were the most sensitive taxonomic group to primary forest removal while mammals were the least sensitive to disturbance. However, the results varied, depending on the type of disturbance (for instance, birds were the most sensitive group to forest conversion into agriculture, whereas plants were the most sensitive group to forest burning and conversion to shaded plantations). Sala et al. (2000) estimated that by the year 2100, land use change will have the greatest impact on biodiversity (in comparison to rising CO_2 concentrations, N deposition, rise in atmospheric temperatures, and biotic exchanges, i.e., deliberate or accidental introduction of plants and animals to an ecosystem) as a result of effects on habitat availability and consequent species extinctions. They also found that biomes such as tropical and southern temperate forests were largely influenced by changes in land use and were less sensitive to the effect of other drivers. In contrast, Arctic ecosystems were largely influenced by climate change while other biomes such as northern temperate forests, deserts, Mediterranean ecosystems, savannas, and grasslands were influenced by all drivers. Habitat loss can also drive species range shifts. For instance, Jetz et al. (2007) evaluated the impact on all 8,750 land bird species of projected changes in climate and land use. They found that the mean expected range contraction across a variety of scenarios by the year 2050 is 21%–26% with roughly 170–260 species projected to experience substantial (i.e., a greater than 50%) range declines and the total number of threatened species to increase 19%–30% by 2050. Species most at risk are predominantly narrow-ranged and endemic to the tropics, where projected range contractions are driven by anthropogenic land conversions. One other metric to assess biodiversity change is to examine changes in population size because many populations of vertebrates, plants, and invertebrates have been monitored over long periods (Balmford et al., 2003). For example, the World Wildlife Fund initiated the Living Planet Index (LPI), which is a measure of the state of the world's biological diversity based on population trends of 10,380 populations of more than 3,038 vertebrate species (e.g., fishes, amphibians, reptiles, birds, and mammals). The index shows that terrestrial populations declined by 39% in 1970–2010, coinciding with a decline in protected areas for these populations of 18% (McRae et al., 2014).

Table 6.2. *Biodiversity and the cross-sectoral ecosystem services*

Beneficiary	Ecosystem service	Important taxa
Climate change	Carbon storage	Trees, decomposers
Agriculture	Pollination	Insects (e.g., bees and flies), bats, birds
Agriculture	Biological control	Parasitic microorganisms, invertebrates (e.g., insects), birds, small mammals
Forestry	Seed dispersal	Birds, rodents
Forestry	Soil formation and nutrients	Invertebrates (e.g., arthropods), microorganisms
Forestry	Productivity	Mostly trees, but also fully functioning ecosystems
Forestry	Reduction or prevention of invasive species	All species in the ecosystem
Life quality	Water retention	Plants, soil microorganisms

Source: Modified from Thompson et al., 2011.

The results also indicate that vertebrates are declining in both temperate and tropical regions, but that the average decline is greater in the tropics (i.e., 56% versus 36%). Thus, examining multiple metrics of biodiversity loss including measuring extinction rates in addition to monitoring populations and habitats provides a complete picture of the impacts humans have had and are likely to have on biodiversity (Balmford et al., 2003).

6.6.1 Role of Biodiversity in Ecosystem Processes

Certain species can strongly affect ecosystem processes by directly mediating energy and material fluxes or by altering abiotic conditions that regulate these processes (Table 6.2; Chapin et al., 2000c). For instance, Balvanera et al. (2005) found that 13% of forest tree species can contribute to 90% of the carbon storage. Species's alteration of the availability of limiting resources, the disturbance regime, and the climate can strongly affect ecosystem processes. For instance, the introduction of deep-rooted salt cedar to the Mojave and Sonoran Deserts of North America increased the amount of soil water and solutes accessed by vegetation, enhancing productivity of the salt cedar and increasing surface litter and salts at the surface through the presence of salt exudates on the surfaces of leaves. These changes inhibited the regeneration of many native species and led to an overall reduction in biodiversity (Berry, 1970). Species can also influence the disturbance regime. For example, several species of palatable but flammable grasses were introduced to Hawaii to support cattle grazing. Some of these grasses spread into protected woodlands, where they caused a 300-fold increase in the extent of fire and a substantial decrease in woody plant extent because fire eliminates woody plants, whereas grasses rebound quickly (D'Antonio and Vitousek, 1992).

Trophic interactions can have large effects on ecosystem processes either by directly modifying fluxes of energy and materials or by influencing the abundances of species that control those fluxes. When top predators are removed, prey populations can increase rapidly and deplete their food resources, leading to a trophic cascade (i.e., a perturbation at one trophic level that propagates through lower levels with alternating positive and negative effects) of ecological effects (Chapin et al., 2000c). While there is a debate between whether terrestrial ecosystems are regulated in a top-down manner (i.e., predators limit herbivores and thereby prevent them from overexploiting vegetation) or bottom-up manner (i.e., plant chemical defenses limit plant depredation by herbivores), removal of predators has been shown to alter ecosystem dynamics substantially (Terborgh et al., 2001). For instance, Terborgh et al. (2001) examined a set of islands in Venezuela where predators of vertebrates are absent. They found densities of rodents, howler monkeys, iguanas, and leaf-cutter ants to be 10 to 100 times greater than on the nearby mainland, where predators of vertebrates are present. Similarly, Dunham (2008) conducted an experiment in a tropical rain forest, where she excluded top predators of the forest floor and then measured changes in nutrient levels and herbivory rates in above- and belowground communities. Exclusion of top predators of the forest floor increased seedling herbivory rates and macroinvertebrate (>5 mm) densities with the strongest effects on herbivorous taxa, spiders, and earthworms suggesting that mammalian and avian terrestrial insectivores directly control densities of understory invertebrate herbivores through predation. In contrast, densities of microbivores showed the opposite trend, which may have led to the substantial decrease in inorganic soil phosphorus (i.e., inorganic soil P was 1.2 times higher in the control versus experimental plots). As an extreme example of the effect of predator loss on plant productivity, wolves have been absent from the Scottish island of Rùm for 250 to 500 years. Over this same period, Rùm has transitioned from being a forested environment to a treeless island, providing evidence of a trophic cascade in the absence of top-down regulation and illustrating that the ultimate effect of predator loss and elevated herbivory in a temperate forest may be the transition to a deforested state (Estes et al., 2011).

6.6.2 Societal Impacts of Biodiversity Loss

Changes in biodiversity resulting from deforestation impact the provision of ecosystem goods and services utilized by society. Changes in biodiversity can directly reduce the availability of food, fuel, structural materials, medicine, or genetic resources (Chapin et al., 2000c). Changes in community composition can affect ecosystem goods or function either by directly reducing abundances of beneficial species (by predation or competition) or by altering controls on ecosystem processes (Chapin et al., 2000c). These impacts can be wide-ranging and costly. For example, increased evapotranspiration due to the invasion of *Tamarix* in the United States costs an estimated $65–$180 million per year from reduced municipal and agricultural water supplies (Zavaleta, 2000). In addition to raising water costs, *Tamarix* stands trap sediment, an effect that has led to narrowed river channels and obstructed over-bank flows, increasing flood damage by as much as $50 million annually (Zavaleta, 2000).

6.6.3 Strategies for Protecting Biodiversity

Only 9.8% of the entire tropical forest biome lies within strictly protected areas (Schmitt et al., 2008), and the long-term viability of existing reserves is strongly affected by patterns of human activity in adjacent areas (Wittemyer et al., 2008). Thus, the future of tropical forest biodiversity depends on effectively mitigating adverse anthropogenic impacts on landscapes (Harvey et al., 2008; Perfecto and Vandermeer, 2008) as well as preserving areas with high concentrations of biodiversity. One approach to conservation is to identify "hotspots," or areas featuring exceptional concentrations of endemic species and experiencing exceptional loss of habitat (Figure 6.2; Myers et al., 2000). As noted above, as many as 44% of all species of vascular plants and 35% of all species in four vertebrate groups are confined to 25 hotspots comprising only 1.4% of the Earth's land surface (Myers et al., 2000). Because the number of species threatened with extinction far surpasses available conservation resources, this places a premium on identifying priority areas to protect the most species per dollar invested (Myers et al., 2000). Hence, the greater the number of endemic species in a region (i.e., the proportion of taxa in a specified geographical area that are found nowhere else), the more biodiversity is lost if that region is disturbed. Using a hotspot-type approach to conservation allows for better use of conservation funds as the area of these hotspots covers a little more than 1 million square kilometers (Pimm and Raven, 2000; Myers et al., 2000).

While preserving areas with high biodiversity (i.e., hotspots) is a critical strategy, the health of the ecological preserve also needs to be maintained in order to serve the goal of protecting biodiversity. Laurance et al. (2012) assessed the health of 60 pantropical protected areas and found that approximately half are experiencing a relatively large loss of biodiversity that is often widespread taxonomically and functionally. Maintaining *functional diversity* (i.e., the spectrum of ecosystem functions fulfilled by different species) as well as *response* (or *disturbance*) *diversity* (i.e., providing multiple species that fulfill similar functions but have different responses to human landscape modification) make the system more resilient in the face of poor management (Box 4.2). Thus, functional diversity and response diversity are two important ecosystem properties to consider in conservation management (Elmqvist et al., 2003). Ecosystem functioning can be altered after land use change by the emergence of invasive or aggressive species that become overly abundant after deforestation and negatively affect other species through aggressive behavior, competition, or predation (Fischer et al., 2006). For example, in southeastern Australia, widespread land clearing for agriculture has led to expanded populations of the noisy miner bird, which is highly aggressive and has led to a decline in insectivorous birds. This decline has been linked to insect outbreaks and reduced tree health in many agricultural landscapes (Grey et al., 1998).

A number of other general management principles have been proposed to guide biodiversity conservation. For instance, Lindenmayer et al. (2006) proposed a series of guiding principles at several spatial scales. At the regional scale, management should establish large, structurally complex ecological reserves (Fischer et al., 2006). Maintaining primary forest is important as the biodiversity of degraded or secondary forest substantially differs from that of primary forest. For instance, Gibson et al. (2011) analyzed 2,200 pairwise comparisons of biodiversity values in primary forests and disturbed forests and found that disturbed forests had significantly lower biodiversity. Conserving keystone species whose presence or abundance has a disproportionate effect on ecosystem processes is another important ecosystem property to maintain (Fischer et al., 2006). While reserves and protected areas are critical for conserving forest biodiversity, reserves alone are insufficient to conserve forest biodiversity adequately (Daily et al., 2001; Lindenmayer and Franklin, 2002) because 92% of the world's forests are located outside formally protected areas (Commonwealth of Australia, 1999; Lindenmayer et al., 2006). At the landscape scale, off-reserve conservation measures should include having (1) protected areas within production forests, (2) buffers for aquatic ecosystems, (3) appropriately designed and located road networks, (4) careful spatial and temporal arrangement of harvest units, and, (5) appropriate fire management practices. Fischer et al. (2006) also suggested maintaining corridors to link patches of native vegetation and creating buffers around sensitive areas to help lessen negative edge effects from strongly altered conditions. At the stand level, off-reserve conservation measures should include (1) the retention of key elements of stand structural complexity (e.g., large living and dead trees with hollows, understory thickets, and fallen logs), (2) long rotation times (coupled with structural retention at harvest), (3) silvicultural systems alternative to traditional high impact ones (e.g., clear-cutting in some forest types), and (4) appropriate practices to manage fire and other kinds of disturbances. It is also important to establish protected areas and forest reserves while protecting rural livelihoods from dispossession and the economic and societal consequences of their losing access to the forest (Fairhead et al., 2012).

6.7 Impact of Deforestation on Human Health

Most human diseases originated from animals and have been affected in various ways by animals' interactions with the environment. Deforestation – and environmental change in general – can explain the emergence of infectious diseases in wildlife (e.g.,

Nipah virus, West Nile virus disease, HIV) and the consequent losses of biodiversity and human (*zoonotic*) infections. Some of the emerging infectious diseases are a major threat to global health because of their high fatality rates or lack of prevention and treatment therapies (Daszak et al., 2001). Wildlife populations often play a crucial role in the emergence of infectious diseases because they provide a reservoir for pathogens that are then transmissible to humans (zoonotic infection). While pathogens seldom cause disease in their reservoir hosts, they can reach humans who enter into close contact with them (Daszak et al., 2001). In turn, land use change and the associated wildlife habitat destruction can favor the flow of pathogens from reservoir species to humans (Wilby et al., 2009). This phenomenon can also be enhanced by the loss of biodiversity resulting from human encroachment into forested areas (Sala et al., 2009).

Deforestation affects human health by increasing exposure to zoonotic infections. Forest destruction enhances interactions with wildlife, which result from human encroachment into previously forested areas, bush meat consumption, and loss of the buffering effect of biodiversity. Deforestation can improve the habitat for reservoir species and increase their populations. Overall the chances of contact with wildlife and the spillover of the zoonotic pathogens from wildlife reservoirs to humans greatly increase as an effect of deforestation (Myers et al., 2013). Likewise, deforestation has been found to increase exposure to vector-borne diseases such as dengue (Froment, 2009), lishmaniasis (Coimbra, 1991), and malaria, both in Africa (Coluzzi, 1994; Cohuet et al., 2004; Guerra et al., 2006; Yasuoka and Levins, 2007) and in South America (de Castro et al., 2006; Singer and de Castro, 2006).

6.8 Food Security as a Major Future Driver of Deforestation

The growth of human population to 9 billion people around 2050 and the increase in per capita food consumption will likely result in the growth of global food production by 60% (Alexandratos and Bruinsma, 2012). Future projections suggest that by 2050, the world's average daily calorie demand could rise to 3,070 kcal per person, an 11% increase over its level in 2005/2007 (i.e., 2,772 kcal per person) (Alexandratos and Bruinsma, 2012). The continuation of current yield trends until 2050 will not suffice to meet the rising global food demand without an accelerated increase in yields or further growth of cropland areas, likely at the expense of forests (Ray et al., 2013). To achieve this growth (given an annual projected increase in crop yields of 1.1%), arable land would need to expand by 70 Mha with an expansion of about 120 Mha in developing countries being offset by a decline of 50 Mha in developed countries (Alexandratos and Briunsma, 2012). The environmental impacts of increasing global crop production by 60% will depend on how increased production is achieved (Tilman et al., 2002; Foley et al., 2011). Societies could reduce the need to increase food production by decreasing their consumption rates through changes in diets or waste reduction (Kummu et al., 2012; Davis et al., 2014).

6.8.1 Reduce Food Losses

Reducing the substantial losses that occur in the pre- and postharvest production stream represents an important strategy to meet the increasing demand for agricultural products without having to intensify agriculture or expand croplands. Roughly one-third of the edible parts of food produced for human consumption is lost or wasted globally as a result of preharvest losses, transport, preprocessing, storage, processing, packaging, marketing, and plate waste losses (Kummu et al., 2012; Popp et al., 2014). Food losses in industrialized countries are as high as in developing countries, but more than 40% of the food losses in developing countries occurs at postharvest and processing levels. Focusing agricultural expansion on close proximity to markets where there is demand for these products could reduce these food losses. However, much of the suitable land not yet in use for agricultural production is concentrated in a few countries in Latin America and sub-Saharan Africa, which are often removed from areas where food is most needed leading to food being increasingly accessed through international trade (D'Odorico et al., 2013). For instance, some 60% of the world's population is in Asia, where only limited additional arable land is available (FAO, 2011b). In contrast, in industrialized countries, more than 40% of the food losses occur at retail and consumer levels (FAOSTAT, 2011) and food waste could be reduced by improving food marketing and distribution strategies, and consumer behavior.

Reducing preharvest losses caused by pests, pathogens, and weeds is one of the major challenges to increasing yields in agricultural production. Globally, an average of 35% of potential crop yield is lost to preharvest pests (Oerke, 2006). These losses occur even with the use of herbicides and insecticides because of the evolution of strains of weeds/insects resistant to these products. For instance, within about one or two decades of introducing each of the seven major herbicides, herbicide-resistant weeds were observed (IEA, 2011). Similarly, insects often evolve resistance to insecticides within a decade. As discussed previously, reducing these losses could in part be achieved by obtaining greater crop diversity and sequencing because dynamic cropping systems have been shown to result in reduced weed and disease infestations, as well as greater nutrient and precipitation use efficiency (Hanson et al., 2007). In contrast, monocultures are more prone to the development of weed, insect, and disease infestations over time and are less responsive to external stressors such as fluctuating climatic conditions (Krupinsky et al., 2002).

6.8.2 Intensification versus Extensification

As noted in Chapter 1, an increase in agricultural production could result from *agricultural extensification* (i.e., clearing additional land for crop production) or *intensification* (i.e., achieving higher yields by using fertilizers, herbicides, pesticides, improved agronomic practices, and crop varieties) or a combination of both. Although intensifying food production can cause multiple environmental problems, several studies suggest that these impacts are smaller (in terms of biodiversity loss, carbon emissions, permanent habitat degradation) than those of introducing new land into agricultural use (e.g., Godfray, 2011). Burney et al. (2010) computed the net effect of historical agricultural intensification on greenhouse gas (GHG) emissions between 1961 and 2005. They found that the net effect of higher yields has prevented emissions of up to 161 gigatons of carbon (Gt C) (590 $GtCO_2e$) since 1961, corresponding to 34% of the total 478 Gt C emitted by humans between 1850 and 2005. In addition, yield increases saved approximately 1,514 Mha of cropland (an area slightly smaller than the size of Russia). Thus, although there are environmental problems with intensification as currently practiced, their impacts are smaller than would have occurred if yields had remained at pre–Green Revolution levels and production had been raised by land clearance (Godfray and Garnett, 2014). *Sustainable intensification* has recently been proposed as one solution that can meet the increase in food demand by limiting the expansion of agricultural lands and forest loss and increasing yield, while reducing environmental problems associated with agricultural intensification. Sustainable intensification uses a combination of ideas from biotechnology, conventional farming, agroecology, and organic farming that best meet the goals of increasing yields while minimizing environmental damage (Godfray and Garnett, 2014).

Initially, intensified production provides farmers with higher yields per hectare and growth in their gross income, which may induce them to expand the area that they have under cultivation. However, if demand for crops is relatively inelastic, the increase in supply will result in a decline in crop prices. In turn, price declines will dissuade farmers from expanding the area they use to cultivate these crops (Wunder and Dermawan, 2007). Over time, cropland may decline as farmers abandon their less-productive fields. The area abandoned by farmers has the potential to become land where forests regenerate. Thus, policies to promote intensification may lead to forest regrowth. Such a trend was observed in eastern North America beginning in the 1890s (Pfaff and Walker, 2010). However, if the increase in supply does not result in a large price decline (i.e., the demand for the crop is relatively elastic), the overall incentive for higher production by using more land remains in place and agricultural intensification will not deter cropland expansion and forest loss (Rudel et al., 2009b). For instance, Angelsen (2010) found that increases in local yield often stimulate rather than reduce agricultural encroachment because increasing local profits promote expansion. Rudel et al. (2009b) analyzed country-level agricultural production data on a global scale from the FAO and concluded that cropland abandonment occurred only with food imports, political upheaval, or specific, targeted conservation set-aside programs.

6.8.3 Mitigating the Environmental Impacts of Intensification

Substantial gains in crop yield have occurred since the start of the Green Revolution as a result of large increases in the use of nitrogen (N) and phosphorus (P) fertilizers as well as increases in the area under irrigation globally. If past trends in N and P fertilization and irrigation use continue, Tilman et al. (2001) forecast that global N fertilization will be 2.7 times 2000 levels by 2050. By 2050, N fertilization would add $236*10^6$ MT of N yr^{-1} to terrestrial ecosystems (Tilman et al., 2001), compared with $140*10^6$ MT yr^{-1} of N from natural sources (Vitousek et al., 1997b). Moreover, it is estimated that only 30%–50% of applied N fertilizer (Smil, 1999) and ~45% of P fertilizer (Smil, 2000) is taken up by crops. At a regional level, Ju et al. (2009) examined the effects of N fertilization in intensive agricultural areas of China and found that more efficient use of N fertilizer would allow current N application rates to be reduced by 30% to 60% while still maintaining crop yields.

Much of the excess applied N and P travels to lakes, rivers, estuaries, and coastal waters via overland and subsurface flow. The excess nutrients result in overenrichment, eutrophication, and low-oxygen conditions in the water bodies they reach, adversely impacting aquatic biota and humans dependent on these water sources. Nitrogen inputs to agricultural systems contribute to emissions of the greenhouse gas nitrous oxide (4.4 Tg N_2O yr^{-1}, or 15.8% of total N_2O emissions) (Skiba et al., 2012) and other nitrogen oxides (NO_x).

These adverse environmental impacts could be reduced by applying fertilizer more efficiently. Nutrient use efficiency could be improved by better matching temporal and spatial nutrient supply with plant demand. Applying fertilizers during periods of greatest crop demand, at or near the plant roots, and in smaller and more frequent applications all have the potential to reduce losses while maintaining or improving yields and quality (Matson et al., 1996; Matson et al., 1998; Cassman et al., 1993; Peng et al., 1996; Tilman et al., 2002). Such "precision agriculture" has typically been used in large-scale intensive farming but is possible at any scale (Cassman, 1999).

Water is a critical resource and irrigation has been an important contributor to yield growth during the Green Revolution. Forty percent of crop production results from irrigated agriculture, which accounts for 16% of agricultural land and 70% of global water withdrawals (Tilman et al., 2002). The world's irrigated areas are estimated to be 300 Mha, more than twice the irrigated area in the early 1960s. However, the potential for further expansion of irrigation is limited (Falkenmark and Rockstrom, 2006) because irrigation is most needed in water scarce regions such as the Near East/North Africa and northern China. Developing countries have a relatively high potential for irrigation expansion: Their irrigated land (presently, about 68 Mha) could increase by another 180 Mha. However, it is likely that only a small portion (i.e., 20 Mha) will be equipped for irrigation by 2050 because of lack of investments in irrigation technology (Alexandratos and Bruinsma, 2012).

Despite the increase in yields resulting from irrigating crops, there can be substantial environmental impacts resulting from diverting water for such use. Irrigation can alter many hydrological processes including depleting surface water and groundwater and cause decline in the quality of water bodies, aquifers, and soils via salinization and transport of herbicides, pesticides, and fertilizers. Yield increases in water limited environments could also be obtained by cultivating crops with high water use efficiency and developing crops with greater drought tolerance (Charles, 2001; DeVries and Toenniessen, 2001).

6.8.4 Crop Selection and Sequencing

Intensive monoculture agriculture tends to rely heavily on synthetic fertilizers and pesticides that, in turn, increase soil physical degradation, reliance on agricultural chemicals, and nutrient leaching, while causing loss of biodiversity (Karlen et al., 1994; Giller et al., 1997; Tilman et al., 2002; Malezieux et al., 2009; Bullock, 1992). In contrast, using crop rotations (i.e., growing different crops on the same land in a successive pattern) can reduce soil erosion and enhance soil structure, permeability, microbial activity, water storage capacity, and organic matter content (Bullock, 1992; Gebremedhin and Schwab, 1998). Crop rotation can also reduce the use of external inputs through internal nutrient recycling, maintenance of the long-term productivity of the land, avoidance of pest accumulation associated with monoculture, consequently leading to an increase in crop yields (Bullock, 1992; Gebremedhin and Schwab, 1998; Krupinsky et al., 2002). These beneficial effects can further be improved by combining crop rotations with cover crops and reduced or no-tillage practices (Gebremedhin and Schwab, 1998; Zentner et al., 2002).

Similar to using crop rotations, using mixed cropping systems whereby two or more crops are planted simultaneously in the same field can reduce pests and diseases (Trenbath, 1993; Hauggaard-Nielsen et al., 2001), aid in suppressing weed growth (Liebman and Altieri, 1986; Bulson et al., 1997; Hauggaard-Nielsen and Jensen, 2005), and significantly reduce populations of harmful insects in comparison with monocultures of the same species (Nickel, 1973; Perrin, 1977; Vandermeer, 1989). Other environmental benefits documented with mixed cropping systems include i) nutrient recycling (species with different rooting depths have better nutrient access throughout the soil profile); ii) soil conservation and improved water quality (by enhancing surface cover and root density, mixed cropping increases infiltration and soil water retention, thereby decreasing erosion) (Swift et al., 2004); iii) enhanced carbon sequestration over that of monocultures (Malezieux et al., 2009); and iv) biodiversity conservation (greater diversity of crop plants may increase the diversity of plants, insects, birds, and mammals) (Brussaard et al., 2007; Perfecto et al., 2003). Species diversity has been shown to reduce outbreaks by pest species by diluting the availability of their hosts. Jactel and Brockerhoff (2007) conducted a metaanalysis comparing insect herbivory in mixed forests with that in monoculture forests and found that the more diverse stands were significantly less affected by herbivory (Thompson et al., 2011). Similarly, Koh (2008) showed that excluding birds from patches of oil palm resulted in a significant increase in insect damage to crop plants. Diversity also decreases host specific diseases (Burdon, 1993), plant feeding nematodes (Wasilewska, 1995), and consumption of preferred plant species (Bertness and Leonard, 1997).

Multispecies systems also offer a number of economic benefits. For instance, they allow for *increased profitability* when the growth rate of the plant is slow. In fact, when setting up perennial plantations such as oil palms or rubber trees, crop mixing is widely used to generate income in the first years of the plantation, when the palms or trees are still unproductive and occupy a confined field space without generating any economic returns. Mixed cropping also enhances *income stability* by providing several crops that can maintain a more stable income, particularly if price variations for those crops are not related to variations in the other crops' prices. Moreover, multispecies cropping provides ecological services either for the internal benefit of the fields in which they are cultivated (e.g., pollination or soil conservation) or for external benefits, such as water quality, biodiversity, or aesthetic value. For example, hay yield has been shown to be higher (up to 60%) in species rich (25–41 plant species) than species poor sowings (6–17 species) (Bullock et al., 2001). Finally, these cropping systems enhance labor productivity because the demand for labor is often distributed more evenly over the year (Tscharntke et al., 2005).

Despite these benefits, diversified systems are generally more difficult to manage than monocultures (Malézieux et al., 2009). Crop rotations may not be adopted because of the increased need for diversified farm activities and required information, as well as the requirement for more agricultural equipment and additional storage facilities. It also requires a more flexible approach to crop selection and the need for an expanded knowledge base regarding multiple crops. In addition, farmers may prefer to remain with their established systems because of the potential lower risk with planting their current crops and their unfamiliarity with the crop rotations' management requirements (Ikerd, 1991; Zegada-Lizarazu and Monti, 2011). Crop diversification by itself is of limited use without knowledge of how individual crops affect each other in a sequence. Nonetheless, achieving biodiversity in agroecosystems could contribute to constant biomass production; reduce the risk of crop failure in unpredictable environments; restore disturbed ecosystem services, such as water and nutrient cycling; and reduce the risks of invasions, pests, and diseases (Gurr et al., 2003).

6.8.5 Location of Agricultural Expansion

Incentives to locate agriculture on previously cleared (i.e., abandoned and degraded) land could be a resourceful use of land for agricultural production while curtailing greenhouse gas emissions and reducing further expansion into forested areas. Reliable data on the extent of abandoned and degraded lands are scarce, but estimates indicate there are ≈385–472 Mha of abandoned agricultural land globally (Campbell et al., 2008). For instance, Murdiyarso et al. (2010) examined oil palm production in peat swamp forests of Indonesia and concluded that intensifying existing production and locating new plantations in degraded secondary forests and grasslands could satisfy the demand for increased palm production while reducing greenhouse gas emissions. Such a strategy would also reduce deforestation of primary forests and lead to a more productive land use. The increasing prevalence of large-scale commercial agriculture also presents an opportunity to locate new production on land that has already been cleared or land with relatively low carbon stocks. Such changes could be feasible for newly establishing commercial enterprises, but may be more difficult to implement for small-scale agriculture in which farmers are not able to shift locations readily (DeFries and Rosenzweig, 2010).

There is also the potential to increase the use of marginal and degraded lands for bioenergy crops. There are three primary types of bioenergy crops: i) *grain seed and high-sugar crops* composed of annual and perennial crops. Annual crops such as maize and wheat currently make the largest contribution to biofuels, matched only by perennial sugarcane and oil crops; ii) *perennial rhizomatous grasses* that have little or no woody tissue and mostly consist of bunch-type grasses (e.g., elephantgrass, kleingrass, buffalograss, switchgrass, and miscanthus); and iii) *fast growing trees* that have a great range of adaptability and good disease resistance (e.g., poplar, willow, sweetgum, and cottonwood) (Lemus and Lal, 2005; Karp and Shield, 2008). While most biofuels are grown from grain seed and high-sugar crops, one alternative is to produce *second-generation* (or *advanced*) biofuels using perennial grasses or fast growing trees planted on marginal lands (Tilman et al., 2006). Marginal lands are those poorly suited to field crops because of low crop productivity due to inherent edaphic or climatic limitations or because of location in areas that are vulnerable to erosion or other environmental risks when cultivated. Often such lands are suitable for grasses, short-rotation tree crops, or other perennial vegetation with persistent roots that are better adapted to low-nutrient, erodible, or drought prone soils. As long as the conversion of these lands to biofuel production in the short term removes more CO_2 from the atmosphere than the preexisting vegetation would have stored (Searchinger et al., 2009; Gelfand et al., 2011) and therefore does not create large carbon debts (Fargione et al., 2008), planting biofuel crops on marginal lands might provide substantial GHG emission mitigation without the risk of indirect carbon costs due to displaced food and feed production (Searchinger et al., 2009). However, even though the use of the so-called marginal lands for the production of advanced biofuels neither directly nor indirectly causes deforestation, it can lead to irreversible land degradation, habitat loss, and impaired ecosystem functioning.

6.9 Concluding Comments

In this book, we have examined and discussed the variety of ways in which deforestation affects environmental processes as well as the societal implications of these changes. We have also reviewed historical and current drivers of deforestation and considered future drivers of deforestation to understand better the areas that may be susceptible to future loss of forest. The long-term loss of forest resources affects not only the functioning of ecosystems but also the societies, health, and livelihoods of those dependent upon them. Understanding how best to manage these forest resources in order to preserve their unique aspects and qualities for future generations will be a challenge that requires an integrated and concentrated effort from scientists and policy makers alike.

References

Aber, J. D., Botkin, D. B., and Melillo, J. M. (1978). Predicting the effects of different harvesting regimes on forest floor dynamics in northern hardwoods. *Canadian Journal of Forest Research*, 8(3), 306–315.

Aber J. D., and Melillo, J. M. (1991). *Terrestrial Ecosystems*. Saunders College Publishing, Orlando, FL.

Aber, J. D., Nadelhoffer, K. J., Steudler, P., and Melillo, J. M. (1989). Nitrogen saturation in northern forest ecosystems. *BioScience*, 39(6), 378–286.

Aber, J. D., Ollinger, S. V., Driscoll, C. T., Likens, G. E., Holmes, R. T., Freuder, R. J., and Goodale, C. L. (2002). Inorganic nitrogen losses from a forested ecosystem in response to physical, chemical, biotic, and climatic perturbations. *Ecosystems*, 5(7), 0648–0658.

Achard, F., Eva, H. D., Mayaux, P., Stibig, H. J., and Belward, A. (2004). Improved estimates of net carbon emissions from land cover change in the tropics for the 1990s. *Global Biogeochemical Cycles*, 18(2).

Aerts, R. (1997). Climate, leaf litter chemistry and leaf litter decomposition in terrestrial ecosystems: a triangular relationship, *Oikos*, 79, 439–449.

Aerts, R., Wallen, B., and Malmer, N. (1992). Growth-limiting nutrients in Sphagnum-dominated bogs subject to low and high atmospheric nitrogen supply. *Journal of Ecology*, 80, 131–140.

Agrawal, A., and Angelsen, A. (2009). Using community forest management to achieve REDD+ goals. In A. Angelsen (ed.) *Realising REDD+: National Strategy and Policy Options*, 201–212.

Agrawal, A., Cashore, B., Hardin, R., Shepherd, G., Benson, C., and Miller, D. (2013). Economic contributions of forests. In Background paper prepared for the United Nations Forum on Forests, Available at http://www. un.org/esa/forests/pdf/session_documents/unff10/EcoContrForests.pdf (accessed August 15, 2013).

Agrawal, A., Nepstad, D., and Chhatre, A. (2011). Reducing emissions from deforestation and forest degradation. *Annual Review of Environment and Resources*, 36, 373–396.

Agarwal, S. K. (2008). *Fundamentals of Ecology*. APH Publishing Corporation, New Delhi.

Agestam, E., Ekö, P. M., Nilsson, U., and Welander, N. T. (2003). The effects of shelterwood density and site preparation on natural regeneration of Fagus sylvatica in southern Sweden. *Forest Ecology and Management*, 176, 61–73.

Alcantara-Ayala, I., Esteban-Chavez, O., and Parrot, J. F. (2006). *Landsliding related to land-cover change: A diachronic analysis of hillslope instability distribution in the Sierra Norte*, Puebla, Mexico, *Catena*, 65(2), 152–165.

Aldrich, M., Billington, C., Edwards, M., and Laidlaw, R. (1997). Tropical montane cloud forests: An urgent priority for conservation (No. 2). Cambridge, UK, World Conservation Monitoring Centre.

Alexandratos, N., and Bruinsma, J. (2012). World agriculture towards 2030/2050: The 2012 revision. ESA Working paper No. 12-03. Rome, FAO.

Alkemade, R., van Oorschot, M., Miles, L., Nellemann, C., Bakkenes, M., and Ten Brink, B. (2009). GLOBIO3: A framework to investigate options for reducing global terrestrial biodiversity loss. *Ecosystems*, 12(3), 374–390.

Allison, G. B., Cook, P. G., Barnett, S. R., Walker, G. R., Jolly, I., and Hughes, M. (1990). Land clearance and river salinization in the Western Murray Basin, Australia. *J. Hydrol.*, 119, 1–20.

Amiro, B. D. et al. (2006). Carbon, energy and water fluxes at mature and disturbed forest sites, Saskatchewan, Canada. *Agricultural and Forest Meteorology*, 136(3), 237–251.

Anderies, J. M. (2005). Minimal models and agroecological policy at the regional scale: An application to salinity problems in southeastern Australia, *Reg. Environ. Change*, 5, 1–17.

Anderies, J. M., Janssen, M. A., and Walker, B. H. (2002). Grazing management, resilience, and the dynamics of a fire-driven rangeland system. *Ecosystems*, 5, 23–44.

Andréassian, V. (2004). Waters and forests: From historical controversy to scientific debate. *Journal of hydrology*, 291(1), 1–27.

Angelsen, A. (1999). Agricultural expansion and deforestation: Modelling the impact of population, market forces and property rights. *Journal of Development Economics*, 58(1), 185–218.

Angelsen, A. (2010). Policies for reduced deforestation and their impact on agricultural production. *Proceedings of the National Academy of Sciences*, 107(46), 19639–19644.

Angelsen, A., and Kaimowitz, D. (1999). Rethinking the causes of deforestation: Lessons from economic models. *The World Bank Research Observer*, 14(1), 73–98.

Angelsen, A., and Kaimowitz D. (2001). Introduction: The role of agricultural technologies in tropical deforestation. In A. Angelsen and D. Kaimowitz (eds.), *Agricultural Technologies and Tropical Deforestation*. CABI. Wallingford.

Angers, D. A., and J. Caron. (1998). Plant-induced changes in soil structure: Processes and feedbacks. *Biogeochemistry*, 42(1/2), 55–72.

Anseeuw W., Boche M., Breu T., Giger M., Lay J., Messerli P., and Nolte. (2012). Transnational land deals for agriculture in the global south K., available at http://www.landcoalition.org/fr/publications/transnationalland-deals-agriculture-global-south.

Aoki, M. (2001). *Toward a Comparative Institutional Analysis*. MIT Press, Cambridge, MA.

Armstrong, R. L. et al. (2001). State of the cryosphere: Response of the cryosphere to global warming. Glaciological Data, 18.

Ashraf, A., Maah, M. J., and Yusoff, I. (2011). Introduction to remote sensing of biomass, biomass and remote sensing of biomass, Dr. Islam Atazadeh (ed.), InTech, doi:10.5772/16462 available at: http://www.intechopen.com/books/biomass-and-remote-sensing-of-biomass/introduction-to-remote-sensing-of-biomass.

Aslam, M., and Prathapar, S. A. (2006). Strategies to mitigate secondary salinization in the Indus Basin of Pakistan: A selective review. Research report 97. International Water Management Institute (IWMI), Colombo, Sri Lanka.

Asner, G. P. (2001). Cloud cover in Landsat observations of the Brazilian Amazon. *International Journal of Remote Sensing*, 22(18), 3855–3862.

Attiwill, P. M., and Adams, M. A. (1993). Nutrient cycling in forests. *New Phytologist*, 124(4), 561–582.

Australian Bureau of Rural Sciences. (2010). *Australia's Forests at a Glance 2010*. Canberra, Australia, Australian Government Department of Agriculture, Fisheries and Forestry.

Avissar, R., and Liu, Y. (1996). Three-dimensional numerical study of shallow convective clouds and precipitation induced by land surface forcing. *Journal of Geophysical Research*, 101(D3), 7499–518.

Avissar, R., Silva Dias, P. L., Silva Dias, M. A. F., and Nobre, C. (2002). The large-scale biosphere-atmosphere experiment in Amazonia (LBA): Insights and future research needs. *Journal of Geophysical Research*, 107(D20), 8086, doi:10.1029/2002JD002704.

Baath, E., and Soderstrom, B. (1979). Fungal biomass and fungal immobilization of plant nutrients in Swedish coniferous forest soils. *Revue d'Ecologie et de Biologie du Sol* 16, 477–489.

Bader, M., Rietkerk, M., and Bregt, A. (2007). Vegetation structure and temperature regimes of tropical alpine treelines. *Arctic, Antarctic, and Alpine Research*, 39, 353–364.

Baidya Roy, S. (2009). Mesoscale vegetation-atmosphere feedbacks in Amazonia. *Journal of Geophysical Research*, 114, D20111, doi:10.1029/2009JD012001.

Baidya Roy, S., and Avissar, R. (2002). Impact of land use/land cover change on regional hydrometeorology in Amazonia. *J. Geophys. Res.*, 107, 8037, doi:10.1029/2000JD000266.

Balat, M. (2007). Global bio-fuel processing and production trends. *Energy Explor Exploit*, 25, 195–218.

Balat, M., and Balat, H. (2009). Recent trends in global production and utilization of bio-ethanol fuel. *Applied Energy*, 86(11), 2273–2282.

Balch, J. K., Nepstad, D. C., and Curran, L. M. (2009). Pattern and process: Fire-initiated grass invasion at Amazon transitional forest edges. In *Fire Ecology of Tropical Ecosystems*, (M. Cochrane, ed.), 481-502.

Balesdent, J., Manotti, A., and Guillet, B. (1987). Natural I3C abundance as a tracer for studies of soil organic matter dynamics. *Soil Biol. Biochem.* 19, 25–30.

Ball, M. C., Hodges, V. S., and Laughlin, G. P. (1991). Cold-induced photoinhibition limits regeneration of snow gum at tree-Line. *Functional Ecology*, 5, 663–668.

Balmford, A. et al. (2002). Economic reasons for conserving wild nature. *Science*, 297(5583), 950–953.

Balmford, A., Green, R. E., and Jenkins, M. (2003). Measuring the changing state of nature. *Trends in Ecology and Evolution*, 18(7), 326–330.

Balvanera, P., Kremen, C., and Martínez-Ramos, M. (2005). Applying community structure analysis to ecosystem function: Examples from pollination and carbon storage. *Ecological Applications*, 15(1), 360–375.

Barbier, E. B. (2004). Agricultural expansion, resource booms and growth in Latin America: Implications for long-run economic development. *World Dev.*, 32, 137–57.

Barbier, E. B., and Burgess, J. C. (2001). Tropical deforestation, tenure insecurity, and unsustainability. *Forest Science*, 47(4), 497–509.

Barbier, E. B., Burgess, J. C., and Grainger, A. (2010). The forest transition: Towards a more comprehensive theoretical framework. *Land Use Policy*, 27(2), 98–107.

Barg, A. K., and Edmonds, R. L. (1999). Influence of partial cutting on site microclimate, soil nitrogen dynamics, and microbial biomass in Douglas-fir stands in western Washington. *Canadian Journal of Forest Research*, 29, 705–713.

Barnosky, A. D. et al. (2011). Has the Earth/'s sixth mass extinction already arrived? *Nature*, 471(7336), 51–57.

Barreto, P., Arima, E., and Brito, M. (2006). Cattle ranching and challenges for environmental conservation in the Amazon. State of the Amazon No. 5. Belem, Brazil: IMAZON.

Barron, A. R. (2007). Patterns and controls of nitrogen fixation in a lowland tropical forest, Panama, Ph.D. diss. Princeton, NJ, Princeton University.

Barroso, C. B., and Nahas. E. (2005). The status of soil phosphate fractions and the ability of fungi to dissolve hardly soluble phosphates. *Applied Soil Ecol.*, 29, 73–83.

Barson, M., Randall, L., and Bordas, V. (2000). *Land Cover Change in Australia. Results of the Collaborative Bureau of Rural Sciences – State Agencies' Project on Remote Sensing of Land Cover Change*. Kingston ACT, Australia: Bureau of Rural Sciences.

Barton, L., McLay, C. D. A., Schipper, L. A., and Smith, C. T. (1999). Annual denitrification rates in agricultural and forest soils: A review. *Soil Research*, 37(6), 1073–1094.

BassiriRad, H., Gutschick, V. P., and Lussenhop, J. (2001). Root system adjustments: Regulation of plant nutrient uptake and growth responses to elevated CO_2. *Oecologia*, 126(3), 305–320.

Bauen, A., Chudziak, C., Vad, K., and Watson, P. (2010). A Casual Descriptive Approach to Modelling the GHG Emissions Associated with the Indirect Land Use Impacts of Biofuels. Final Report. A Study for the UK Department for Transport, E4tech, London.

Bauhus, J., and Barthel, R. (1995). Mechanisms for carbon and nutrient release and retention in beech forest gaps. II. The role of soil microbial biomass. *Plant and Soil*, 168–169, 585–592.

Beare, M. H., Hendrix, P. F., and Coleman, D. C. (1994). Water-stable aggregates and organic matter fractions in conventional and no-tillage soils. *Soil Sci. Soc. Am. J.*, 58, 777–786.

Beguería, S. (2006). Changes in land cover and shallow landslide activity: A case study in the Spanish Pyrenees. *Geomorphology*, 74(1), 196–206.

Belcher, B., and Schreckenberg, K. (2007). Commercialisation of non-timber forest products: A reality check. *Development Policy Review*, 25(3), 355–377.

Bengtsson, J., and Wikström, F. (1993). Effects of whole-tree harvesting on the amount of soil carbon: model results. *New Zealand Journal of Forest Science* 23, 380–389.

Berg, A. S., and Joern B. C. (2006). Sorption dynamics of organic and inorganic phosphorus compounds in soil. *J. Environ. Qual.*, 35, 1855–1862.

Berglund, B. E. (2006). Agrarian landscape development in northwestern Europe since the neolithic: Cultural and climatic factors behind a regional/continental pattern. In Hornburg, A., and C. L. Crumley (eds.), *The World System and the Earth System: Global Socioenvironmental Change and Sustainability since the Neolithic*. Left Coast Press, Walnut Creek, CA.

Berhe, A. A., Harte, J., Harden, J. W., and Torn, M. S. (2007). The significance of the erosion-induced terrestrial carbon sink. *Bioscience*, 57, 337–346.

Beringer, J., Chapin, F. S., Thompson, C. C., and McGuire, A. D. (2005). Surface energy exchanges along a tundra-forest transition and feedbacks to climate. *Agricultural and Forest Meteorology*, 131, 143–161.

Berndes, G., Bird, N., and Cowie, A. (2010). *Bioenergy, Land Use Change and Climate Change Mitigation*, IEA Bioenergy, Aadorf, Switzerland

Bernhardt, E. S., Likens, G. E., Buso, D. C., and Driscoll, C. T. (2003). In-stream uptake dampens effects of major forest disturbance on watershed nitrogen export. *Proceedings of the National Academy of Sciences*, 100(18), 10304–10308.

Berry, W. L. (1970). Characteristics of salts secreted by Tamarix aphylla. *American Journal of Botany*, 57, 1226–1230.

Bertness, M. D., and Leonard, G. H. (1997). The role of positive interactions in communities: Lessons from intertidal habitats. *Ecology*, 78(7), 1976–1989.

Best, A., L. et al. (2007). A critical review of paired catchment studies with reference to seasonal flows and climatic variability, Murray Darling Basin Comission Publication 11/03, Murray-Darling Basin Commission, Canberra, ACT, Australia.

Betts, A. K., and Ball, J. H. (1997). Albedo over the boreal forest. *Journal of Geophysical Research*, 102, 28,901–28,909.

Betts, R. A. (2000). Offset of the potential carbon sink from boreal forestation by decreases in surface albedo. *Nature*, 408, 187–190.

Bever, J. D., Schultz, P. A., Pringle, A., and Morton, J. B. (2001). Arbuscular mycorrhizal fungi: More diverse than meets the eye and the ecological tale of why. *BioScience*, 51(11), 923–932.

Bhattarai, M., and Hammig, M. (2001). Institutions and the environmental Kuznets curve for deforestation: A crosscountry analysis for Latin America, Africa and Asia. *World Development*, 29(6), 995–1010.

Biggs, T. W., Dunne, T., and Martinelli, L. A. (2004). Natural controls and human impacts on stream nutrient concentrations in a deforested region of the Brazilian Amazon basin. *Biogeochemistry*, 68(2), 227–257.

Billington, C., Kapos, V., Edwards, M. S., Blyth, S., and Iremonger, S. (1996). *Estimated Original Forest Cover Map – a First Attempt*. WCMC, Cambridge, UK.

Bischetti, G. B., Chiaradia, E. A., Epis, T., and Morlotti, E. (2009). Root cohesion of forest species in the Italian Alps. *Plant and Soil*, 324(1–2), 71–89.

Bischetti, G. B., Chiaradia, E. A., Simonato, T., Speziali, B., Vitali, B., Vullo, P., and Zocco, A. (2005). Root strength and root area ratio of forest species in Lombardy (Northern Italy). *Plant and Soil*, 278, 11–22.

Bishop, D. M., and Stevens, M. E. (1964). Landslides on logged areas in southeast Alaska. USDA Forest Service, Northern Forest Experiment Station Research Paper NOR-1.

Blaike, P., and Brookfield, H. (1987). *Land Degradation and Society*. Methuen, London.

Bluffstone, R. A. (1995). The effect of labor market performance on deforestation in developing countries under open access: An example from rural Nepal. *Journal of Environmental Economics and Management*, 29(1), 42–63.

Boeken, B., and Orenstein, D. (2001). The effect of plant litter on ecosystem properties in a Mediterranean semi-arid shrubland. *J. Veg. Sci.*, 12(6), 825–832.

Bohn, H., and Deacon, R. T. (2000). Ownership risk, investment, and the use of natural resources. *American Economic Review*, 526–549.

Bonan G. (2008a). *Ecological Climatology*. Cambridge University Press, New York.

Bonan, G. B. (2008b). Forests and climate change: Forcings, feedbacks, and the climate benefits of forests. *Science*, 320, 1444–1449.

Bonan, G. B., Pollard, D., and Thompson, S. L. (1992). Effects of boreal forest vegetation on global climate. *Nature*, 359, 716–718.

Bonan, G. B., and Shugart, H. H. (1989). Environmental factors and ecological processes in boreal forests. *Annu. Rev. Ecol. Syst.*, 20, 1–28.

Borg, H., Stoneman, G. L., and Ward, C. G. (1988). The effect of logging and regeneration on groundwater, streamflow and stream salinity in the southern forest of Western Australia. *J. Hydrol.*, 99, 253–271.

Borlaug, N. (2007). Feeding a hungry world. *Science* 318(5849): 359–359.

Bormann, F. H., and Likens, G. E. (1979). *Pattern and Process in a Forested Ecosystem*. Springer-Verlag, New York.

Borneman, J., and Triplett, E. W. (1997). Molecular and microbial diversity in soils from Eastern Amazonia: Evidence for unusual microorganisms and microbial population shifts associated with deforestation. *Applied and Environmental Microbiology*, 63(7), 2647–2653.

Bosch, J. M., and Hewlett, J. A. (1982). A review of catchment experiments to determine the effect of vegetation changes on water yield and evapotranspiration. *J. Hydrol.*, 55, 3–23.

Bossio, D. A. et al. (2005). Soil microbial community response to land use change in an agricultural landscape of western Kenya. *Microbial Ecology*, 49, 50–62.

Bouwman, A. F., Lee, D. S., Asman, W. A. H., Dentener, F. J., Van Der Hoek, K. W., and Olivier, J. G. J. (1997). A global high-resolution emission inventory for ammonia. *Global biogeochemical cycles*, 11(4), 561–587.

Bowman, D. M. et al. (2011). The human dimensions of fire regimes on Earth. *J. Biogeogr.*, 1–14, doi:10.1111/j.1365-2699.2011.02595.x.

Brabb, E. E., and Harrod, B. L. (1989). Landslides: Extent and economic significance. In Proceedings of the International Geological Congress (No. 28).

Bradshaw, C. J. (2012). Little left to lose: Deforestation and forest degradation in Australia since European colonization. *Journal of Plant Ecology*, 5(1), 109–120.

Brady, N., and Weil, R. (2008). *The Nature and Properties of Soils*, 14th ed. Prentice Hall, Upper Saddle River, NJ.

Braithwaite, L. W. (1996). Conservation of arboreal herbivores: The Australian scene. *Aust. J. Ecol.*, 21, 21–30.

Brannstrom, C. 2002. Rethinking the 'Atlantic forest' of Brazil: new evidence for land cover and land value in western Sao Paulo, 1900–1930. *Journal of Historical Geography* 28, 420–439.

Brannstrom, C., and Oliveira, A. M. S. (2000). Human modification of stream valleys in the western plateau of Sao Paulo, Brazil: Implications for environmental narratives and management. *Land Degrad. Dev.* 11, 535–548.

Brardinoni, F., Hassan, M. A., and Slaymaker, H. O. (2002). Complex mass wasting response of drainage basins to management in coastal British Columbia. *Geomorphology*, 49, 109–124.

Brauman, K. A., Daily, G. C., Duarte, T. K. E., and Mooney, H. A. (2007). The nature and value of ecosystem services: An overview highlighting hydrologic services. *Annu. Rev. Environ. Resour.*, 32, 67–98.

Breman, H., and Kessler, J. J. (1995). *Woody Plants in Agro-Ecosystems of Semi-Arid Regions*. Springer, New York.

Briggs, J. M., and Knapp, A. K. (2001). Determinants of C3 forb growth and production in a C4 dominated grassland. *Plant Ecology*, 152, 93–100.

Brink, B. T. et al. (2007). Cross-roads of planet earth's life: Exploring means to meet the 2010 biodiversity target: Solution-oriented scenarios for Global Biodiversity Outlook 2. CBD technical series (31).

Brock, W. A., and Carpenter, S. R. (2006). Variance as a leading indicator of regime shift in ecosystem services. *Ecology and Society*, 11(2), 9.

Broich, M., Hansen, M. C., Potapov, P., Adusei, B., Lindquist, E., and Stehman, S. V. (2011). Time-series analysis of multi-resolution optical imagery for quantifying forest cover loss in Sumatra and Kalimantan, Indonesia. *International Journal of Applied Earth Observation and Geoinformation*, 13(2), 277–291.

Brolsma, R. J., and Bierkens, M. P. F. (2007). Groundwater – soil water – vegetation dynamics in a temperate forest ecosystem along a slope. *Water Resour. Res.*, 43, W01414, doi:10.1029/2005WR004696.

Brook, B. W., Sodhi, N. S., and Bradshaw, C. J. (2008). Synergies among extinction drivers under global change. *Trends in Ecology and Evolution*, 23(8), 453–460.

Brooks, T. M., Pimm, S. L., and Oyugi, J. O. (1999). Time lag between deforestation and bird extinction in tropical forest fragments. *Conservation Biology*, 13(5), 1140–1150.

Brovkin, V. et al. (1998). On the stability of the atmosphere – vegetation system in the Sahara/Sahel region. *Journal of Geophysical Research – Atmospheres*, 103, 31613–31624.

Brown, A. E., Zhang, L., McMahon, T. A., Western, A. W., and Vertessy, R. A. (2005). A review of paired catchment studies for determining changes in water yield resulting from alterations in vegetation. *Journal of Hydrology*, 310(1), 28–61.

Brown, J. C., Jepson, W., and Price, K. P. (2004). Expansion of mechanized agriculture and land-cover change in Southern Rondônia, Brazil. *J. Latin Am. Geogr.*, 3, 96–102.

Brown, S., and Lugo, A. E. (1990). Tropical secondary forests. *Journal of Tropical Ecology*, 6(01), 1–32.

Bruce, J. P., Frome, M., Haites, E., Janzen, H., Lal, R., and Paustian, K. (1999). Carbon sequestration in soils. *Journal of Soil and Water Conservation*, 54(1), 382–389.

Bruce, J., Wendland, K., Naughton-Treves, L., 2010. Whom to Pay? Key Concepts and Terms regarding Tenure and Property Rights in Payment-based Forest Ecosystem Conservation, Land Tenure Center Policy Brief. Land Tenure Center, Madison, WI.

Bruijnzeel, L. A. (2002). Hydrology and water management in the humid tropics, Proceedings of the Second International Colloquium, 22–26 March 1999, International Hydrology Programme, Technical Documents in Hydrology: No. 52 UNESCO, Paris.

Bruijnzeel, L. A., and Veneklaas, E. J. (1998). Climatic conditions and tropical montane forest productivity: The fog has not lifted yet. *Ecology* 79(1), 3–9.

Bruschi, V. M. et al. (2013). Land management versus natural factors in land instability: Some examples in northern Spain. *Environmental Management*, 52(2), 398–416.

Brussaard, L., De Ruiter, P. C., and Brown, G. G. (2007). Soil biodiversity for agricultural sustainability. *Agriculture, Ecosystems and Environment*, 121(3), 233–244.

Bryson, R. A., Irving, W. N., and Larsen, J. A. (1965). Radiocarbon and soil evidence of former forest in the southern Canadian tundra. *Science*, 147, 46–48.

Bubb, P., May, I., Miles, L., and Sayer, J. (2004). *Cloud Forest Agenda*. UNEP-WCMC, Cambridge, UK, available at: http://www.unep-wcmc.org/resources/publications/UNEP_WCMC_bio_series/20.htm.

Buchner, O., and Neuner, G. (2011). Winter frost resistance of Pinus cembra measured in situ at the alpine timberline as affected by temperature conditions. *Tree Physiology*, 31, 1217–1227.

Bullock, D. G. (1992). Crop rotation. *Critical Reviews in Plant Sciences*, 11(4), 309–332.

Bullock, J. M., Pywell, R. F., Burke, M. J., and Walker, K. J. (2001). Restoration of biodiversity enhances agricultural production. *Ecology Letters*, 4(3), 185–189.

Bulson, H. A. J., Snaydon, R. W., and Stopes, C. E. (1997). Effects of plant density on intercropped wheat and field beans in an organic farming system. *Journal of Agricultural Science*, 128(1), 59–71.

Burdon, J. J. (1993). The structure of pathogen populations in natural plant communities. *Annual Review of Phytopathology*, 31(1), 305–323.

Burgess, J. C. (1993). Timber production, timber trade and tropical deforestation. *Ambio*, 136–143.

Burgess, S. S., Adams, M. A., Turner, N. C., and Ong, C. K. (1998). The redistribution of soil water by tree root systems. *Oecologia*, 115(3), 306–311.

Burney, J. A., Davis, S. J., and Lobell, D. B. (2010). Greenhouse gas mitigation by agricultural intensification. *Proceedings of the national Academy of Sciences*, 107(26), 12052–12057.

Burton, P. J., Messier C., Weetman G. F., Prepas E. E., Adamowicz W. L., and R. Tittler (2003). The current state of boreal forestry and the drive for change. In *Towards Sustainable Management of the Boreal Forest*. In Burton P. J., C. Messier, D. W. Smith and W. L. Adamowicz. NRC Research Press, Ottawa, pp. 1–40.

Butt, N., de Oliveira, P. A., and Costa, M. H. (2011). Evidence that deforestation affects the onset of the rainy season in Rondonia, Brazil. *J. Geophys. Res*.116,D11120.

Byerlee, D., and Rueda, X. (2015). From public to private standards for tropical commodities: A century of global discourse on land governance on the forest frontier. *Forests*, 6(4), 1301–1324.

Cai, X., McKinney, D. C., and Rosegrant, M. W. (2003). Sustainability analysis for irrigation water management in the Aral Sea region. *Agric. Syst.*, 76, 1043–1066.

Caldwell, B. A., Griffiths, R. P., and Sollins, P. (1999). Soil enzyme response to vegetation disturbance in two lowland Costa Rican soils. *Soil Biology and Biogeochemistry*, 31, 1603–1608.

Camacho Villa, T. C., Maxted, N., Scholten, M. A., and Ford-Lloyd, B. V. (2005). Defining and identifying crop landraces. *Plant Genetic Resources: Characterization and Utilization*, 3(3), 373–384.

Camargo, A. M. M. P. (2008). Dynamics and tendency of sugarcane expansion over other agricultural activities. State of São Paulo 2001–2006. *Informações Econômicas*, 38(3), 47–66.

Campbell, G. S., and Norman, J. M. (1998). *An Introduction to Environmental Biophysics*, 2nd ed. Springer, New York, doi:10.1007/978-1-4612-1626-1.

Campbell, J. E., Lobell, D. B., Genova, R. C., and Field, C. B. (2008). The global potential of bioenergy on abandoned agriculture lands. *Environmental Science and Technology*, 42(15), 5791–5794.

Canadell, J. G. et al. (2007). Saturation of the terrestrial carbon sink. In *Terrestrial Ecosystems in a Changing World*. Springer, Berlin and Heidelberg, pp. 59–78.

Canadian Council of Forest Ministers (CCFM). (2002). Criteria and indicators of sustainable forest management in Canada: National status 2005. Ottawa, Natural Resources Canada.

Carle, J., and Holmgren, P. (2008). Wood from planted forests: A global outlook 2005–2030. *For. Prod. J.*, 58, 6–18.

Carlson, K. M. et al. (2012). Expanding oil palm plantations in West Kalimantan, Indonesia: Impacts on land cover change and carbon emissions. *Proceedings of the National Academy of Sciences*, doi/10.1073/pnas.1200452109.

Carlyle-Moses, D. E. (2004). Throughfall, stemflow, and canopy interception loss fluxes in a semi-arid Sierra Madre Oriental matorral community. *Journal of Arid Environments*, 58(2), 181–202.

Carnicer. J., Barbeta, A., Sperlich, D., Coll, M., and Peñuelas, J. (2013). Contrasting trait syndromes in angiosperms and conifers are associated with different responses of tree growth to temperature on a large scale. *Front. Plant Sci.* 4:409, doi:10.3389/fpls.2013.00409.

Carpenter, S., and Brock, W. A. (2006). Rising variance: A leading indicator of ecological transition. *Ecol. Lett.*, 9, 308–315.

Carpenter, S. R. et al. (2011). Early warnings of regime shifts: A whole ecosystem experiment. *Science*, 10.1126/science.1203672.

Carr, D. (2009). Population and deforestation: Why rural migration matters. *Progress in Human Geography*, 33(3), 355–378.

Carter, C., Finley, W., Fry, J., Jackson, D., and Willis, L. (2007). Palm oil markets and future supply. *European Journal of Lipid Science and Technology*, 109(4), 307–314.

Cassman, K. G. (1999). Ecological intensification of cereal production systems: Yield potential, soil quality, and precision agriculture. *Proceedings of the National of Sciences*, 96(11), 5952–5959.

Cassman, K. G., Kropff, M. J., Gaunt, J., and Peng, S. (1993). Nitrogen use efficiency of rice reconsidered: What are the key constraints? *Plant and Soil*, 155(1), 359–362.

Cavelier, J., and Goldstein, G. (1989). Mist and fog interception in elfin cloud forest in Colombia and Venezuela. *Journal of Tropical Ecology*, 5, 309–322.

Cavelier, J., Solis, D., and Jaramillo, M. A. (1996). Fog interception in montane forests across the Central Cordillera of Panama. *Jounral of Tropical Ecology*, 12, 357–369.

Cavender-Bares, J., Cortes, P., Rambal, S., Joffre, R., Miles, B., and Rocheteau, A. (2005). Summer and winter sensitivity of leaves and xylem to minimum freezing temperatures: A comparison of co-occurring Mediterranean oaks that differ in leaf lifespan. *New Phytologist*, 168, 597–612.

Cernusak, L. A. et al. (2013). Tropical forest responses to increasing atmospheric CO_2: Current knowledge and opportunities for future research. *Functional Plant Biology*, 40(6), 531–551.

Cerri, C. C., Volkoff, B., and Andreaux, F. (1991). Nature and behaviour of organic matter in soils under natural forest, and after deforestation, burning and cultivation, near Manaus. *Forest Ecology and Management*, 38(3), 247–257.

Cerri, C. C., Volkoff, B., and Eduardo, B. P. (1985). Efeito do desmatament sobre a biomassa microbiana em latossolo amarelo da Amazônia. *Revista Brasileira Ciên. Solo*, 9(1), 1–4.

Chaer, G., Fernandes, M., Myrold, D., and Bottomley, P. (2009). Comparative resistance and resilience of soil communities and enzyme activities in adjacent native forest and agricultural soils. *Microb. Ecol.*, 58, 414–424.

Chagnon, F. J. F., and Bras, R. L. (2005). Contemporary climate in the Amazon. *Geophys. Res. Lett.*, 32, L13703.

Chagnon, F. J. F., Bras, R. L., and Wang, J. (2004). Climatic shift in patterns of shallow cumulus clouds over the Amazon. *Geophys. Res. Lett.*, 31, L24212, doi:10.1029/2004GL021188.

Chamberlain, A. C., and Little, P. (1981). Transport and capture of particles by vegetation. In Grace, J., E. D. Ford, and P. G. Jarvis (eds.), *Plants and Their Atmospheric Environment*. Blackwell, Oxford, UK.

Chambers, J. C., and Linnerooth, A. R. (2001). Restoring riparian meadows currently dominated by Artemisia using alternative state concepts – the establishment component. *Appl. Veg. Sci.*, 4, 157–166.

Chang, M., (2002). *Forest Hydrology: An Introduction to Water and Forests*. CRC Press, Boca Raton, FL.

Chapin, F. S., Eugster, W., McFadden, J. P., Lynch, A. H., and Walker, D. A. (2000a). Summer differences among Arctic ecosystems in regional climate forcing. *Journal of Climate*, 13, 2002–2010.

Chapin, F. S., Matson, P. A., and Mooney, H. A.(2002). *Principles of Terrestrial Ecosystem Ecology*. Springer-Verlag, New York.

Chapin, F. S., and Shaver, G. R. (1985). Individualistic growth response of tundra plant species to environmental manipulations in the field. *Ecology*, 66(2), 564–576.

Chapin, F. S., van Cleve, K., and Chapin, M. C. (1979). Soil temperature and nutrient cycling in the tussock growth form of *Eriophorum vaginatum*. *Journal of Ecology*, 67(1), 169–189.

Chapin, F. S. et al. (2000b). Arctic and boreal ecosystems of western North America as components of the climate system. *Global Change Biology*, 6, 211–223.

Chapin, F. S. et al. (2000c). Consequences of changing biodiversity. *Nature*, 405(6783), 234–242.

Chapin, F. S. et al. (2005). Role of land-surface changes in Arctic summer warming. *Science*, 310, 657–660.

Chapin, F. S. et al. (2006). Reconciling carbon-cycle concepts, terminology, and methods. *Ecosystems*, 9(7), 1041–1050.

Chapin, F. S. et al. (2010). Ecosystem stewardship: Sustainability strategies for a rapidly changing planet. *Trends in Ecology and Evolution*, 25, 241–249.

Chapman, V. (1975). Mangrove biogeography. Proceedings of the International Symposium on Biology and Management of Mangroves,ed. by Walsh, G., S. Snedaker and H. Teas, pp. 3–22.

Charles, D. (2001). Seeds of discontent. *Science*, 294(5543), 772–775.

Charley, J. L., and West, N. E. (1975). Plant-induced soil chemical patterns in some shrub dominated semi-desert ecosystems of Utah. *Journal of Ecology*, 63, 945–963.

Chen, H., Qualls, R. G., and Miller, G. C. (2002). Adaptive responses of Lepidium latifolium to soil flooding: Biomass allocation, adventitious rooting, aerenchyma formation and ethylene production. *Environmental and Experimental Botany*, 48, 119–128.

Chen, J., Franklin, J. F., and Spies, T. A. (1993). Contrasting microclimates among clearcut, edge, and interior of old-growth Douglas fir forest. *Agricultural and Forest Meteorology*, 63, 219–237.

Chen, T. C., Yoon, J., St. Croix, K. J., and Takle, E. (2001). Suppressing impacts of the Amazonian deforestation by global circulation change. *Bull. Am. Meteorol. Soc.*, 82, 2209–2216.

Chengrui, M., and Dregne, H. E. (2001). Review article: Silt and the future development of China's Yellow River. *The Geographical Journal*, 167(1), 7–22.

Chibnik, M. (1994). *Risky rivers: The economics and politics of floodplain farming in Amazonia*. University of Arizona Press, Tucson.

Chokkalingam, U, and De Jong, W. (2001). Secondary forest: A working definition and typology. *Int. For. Rev.*, 3, 19–26.

Chomitz, K. M., Alger, K., Thomas, T. S., Orlando, H., and Nova, P. V. (2005). Opportunity costs of conservation in a biodiversity hotspot: The case of southern Bahia. *Environment and Development Economics*, 10(03), 293–312.

Chomitz, K., Buys P., De Luca G., Thomas T. S., and Wertz-Kanounnikoff, S. (2006). At loggerheads? Agricultural expansion, poverty reduction and environment in the tropical forests. World Bank Publications.

Choné, T., Andreux, F., Correa, J. C., Volkoff, B., and Cerri, C. C. (1991). Changes in organic matter in an oxisol from the Central Amazonian forest during eight years as pasture, determined by 13C isotopic composition. In Berthelin, J. (ed.), *Diversity of Environmental Biogeochemistry*. Elsevier, Amsterdam, pp. 397–405.

Chopra, K. (1993). The value of non-timber forest products: An estimation for tropical deciduous forests in India. *Economic Botany*, 47(3), 251–257.

Chu, P. S., Yu, Z.-P., and Hastenrath, S. (1994). Detecting climate change concurrent with deforestation in the Amazon Basin: Which way has it gone? *Bull. Am. Meteorol. Soc.*, 75, 579–582.

Clarke, C. J., George, R. J., Bell, R. W., and Hatton, T. J. (2002). Dryland salinity in south-western Australia: Its origins, remedies and furute research directions. *Aust J Soil Res* 2002; 40, 93–113.

Claussen, M. (1997). Modeling bio-geophysical feedback in the African and Indian monsoon región. *Climate Dynamics*, 13(4), 247–257.

Claussen, M., Brovkin, V., Ganopolski, A., Kubatzki, C., and Petoukhov, V. (2003). Climate change in Northern Africa: The past is not the future. *Climatic Change*, 57, 99–118.

Clay, J. W., and Clement, C. R. (1993). Selected species and strategies to enhance income generation from Amazonian forests. FO:Misc/93/6 Working Paper. FAO, Rome.

Cleveland, C. C. et al. (1999). Global patterns of terrestrial biological nitrogen (N_2) fixation in natural ecosystems. *Global Biogeochem. Cycles*, 13(2), 623–645, doi:10.1029/1999GB900014.

Cleveland, C. C., and Liptzin, D. (2007). C: N: P stoichiometry in soil: Is there a "Redfield ratio" for the microbial biomass? *Biogeochemistry*, 85(3), 235–252.

Cleveland C. C., Townsend, A. R., and Schmidt, S. K. (2002). Phosphorus limitation of microbial processes in moist tropical forests. *Ecosystems*, 5, 680–691.

Cochrane, M. A. (2003). Fire science for rainforests. *Nature*, 421, 913–919.

Cochrane, M. A., and Barber, C. P. (2009). Climate change, human land use and future fires in the Amazon. *Global Chnage Biology*, 15, 601–612.

Cochrane, M. A. et al. (1999). Positive feedbacks in the fire dynamic of closed canopy tropical forests. *Science*, 284, 1832–1835.

Cohuet, A. et al. (2004). High malaria transmission intensity due to Anopheles funestus (Diptera: Culicidae) in a village of savannah-forest transition area in Cameroon. *J Med Entomol* 41(5), 901–905.

Coimbra, C. E. A. (1991). Environmental change and human disease: A view from Amazonia. *Journal of Human Ecology*, 2, 15–21.

Collatz, G. J. et al. (1991). Physiological and environmental regulations of stomatal conductance, photosynthesis and transpiration: A model that includes a laminar boundary-layer. *Agric. For. Meteorol.*, 54(2–4), 107–136, doi:10.1016/0168-1923(91) 90002–8.

Colpaert, J. V., van Laere, A., van Tichelen, K. K., and van Assche, J. A. (1997). The use of inositol hexaphosphate as a phosphorus source by mycorrhizal and non-mycorrhizal Scots pine (Pinus sylvestris). *Funct. Ecol.*, 11, 407–415.

Coluzzi, M. (1994). Malaria and the Afrotropical ecosystems: Impact of man-made environmental changes. *Parassitologia*, 36(1–2), 223–227.

Commonwealth of Australia. (1999). International forest conservation: Protected areas and beyond. A discussion paper for the intergovernmental forum on forests. Commonwealth of Australia, Canberra, Australia.

Conlin, T. S. S., and Lieffers, V. J. (1992). Root respiration of boreal forest conifers under anoxia and cool conditions. *Can. J. For. Res.*, 23, 767–771.

Convention on Biological Diversity. (2001). Report of the ad hoc technical expert group on forest biological diversity, available at: https://www.cbd.int/forest/definitions.shtml (accessed on June 25, 2015).

Cook, B. I., Anchuk, K. A., Kaplan, J., Puma, M. J., Kelley, M., and Gueyffier, D. (2012). Pre-Colombian deforestation as an amplifier of drought in Mesoamerica. *Geophysical Research Letters*, doi:10.1029/2012GL052565.

Cordell, D., Drangert, J. O., and White, S. (2009). The story of phosphorus: Global food security and food for thought. *Global Environmental Change*, doi:10.1016/j.gloenvcha.2008.10.09.

Corman, A., Crozat, Y., and Cleyet-Marel, J. C. (1987). Modeling of survival kinetics of some Bradyrhizobium japonicum strains in soils. *Biol. Fertil. Soils*, 4, 79–84.

Costa, M. H., and Foley, J. A. (1999). Trends in the hydrological cycle of the Amazon basin. *Journal of Geophysical Research – Atmospheres*, 104, 14189–14198.

Costa, M. H., and Foley, J. A. (2000). Combined effects of deforestation and doubled atmospheric CO_2 concentrations on the climate of Amazonia. *J. Clim.*, 13, 18–34.

Costa, M. H. and Foley, J. H. (1997). Water balance of the Amazon Basin: dependence on vegetation cover and canopy conductance. *J. Geophys. Res.*, 102, 23973–23989.

Costanza, R., and Folke, C. (1997). Valuing ecosystem services with efficiency, fairness, and sustainability as goals. In *Nature's Services: Societal Dependence on Natural Ecosystems*. Island Press, Washington, DC, pp. 49–70.

Costello, M. J., May, R. M., and Stork, N. E. (2013). Can we name Earth's species before they go extinct? *Science*, 339(6118), 413–416.

Costello, M. J., Wilson, S., and Houlding, B. (2012). Predicting total global species richness using rates of species description and estimates of taxonomic effort. *Systematic Biology*, 61(5), 871–883.

Cotula, L. 2011. *Land Deals in Africa: What is in the Contracts?* London: IIED. http://pubs.iied.org/pdfs/12568IIED.pdf

Covington, W. W., and Sackett, S. S. (1992). Soil mineral nitrogen changes following prescribed burning in ponderosa pine. *Forest Ecology and Management*, 54(1), 175–191.

Craft, A. B., and Simpson, R. D. (2001). The value of biodiversity in pharmaceutical research with differentiated products. *Environ. Resource Econ.* 18, 1–17.

Cramer, V. A., and Hobbs, R. J. (2002). Ecological consequences of altered hydrological regimes in fragmented ecosystems in southern Australia: Impacts and possible management responses. *Austral Ecol.*, 27, 546–564.

Crews, T. E. (1999). The presence of nitrogen fixing legumes in terrestrial communities: Evolutionary vs ecological considerations. *Biogeochemistry*, 46(1), 233–246.

Crews, T. E., Farrington, H., and Vitousek, P. M. (2000). Changes in asymbiotic, heterotrophic nitrogen fixation on leaf litter of metrosideros polymorpha with long-term ecosystem development in Hawaii. *Ecosystems*, 3(4), 386–395, doi:10.1007/s100210000034.

Cromack, K., Todd, R. L., and Monk, C. D. (1975). Patterns of basidiomycete nutrient accumulation in conifer and deciduous forest litter. *Soil Biology and Biochemistry* 7, 265–268.

Cropper, M., and Griffiths, C. (1994). The interaction of population growth and environmental quality. *American Economic Review*, 250–254.

Crozier, M. J. (2005). Multiple-occurrence regional landslide events in New Zealand: Hazard management issues. *Landslides*, 2(4), 247–256.

Culas, R. J. (2007). Deforestation and the environmental Kuznets curve: An institutional perspective. *Ecological Economics*, 61(2), 429–437.

Culf, A. D., Esteves, J. L., Marques Filho, A. O., and Rocha, H. R. (1996). Radiation, temperature and humidity over forest and pasture in Amazonia. In Gash, J. H. C., C. A. Nobre, J. Roberts, and R. L. Victoria (eds.), *Amazonian Deforestation and Climate*. Chichester, UK, John Wiley, pp. 175–191.

Da Silveira, L. and Sternberg, L. (2001). Savanna-forest hysteresis in the tropics. *Global Ecology and Biogeography*, 10(4), 369–378.

Daily, G. C., Ehrlich, P. R., and Sanchez-Azofeifa, G. A. (2001). Countryside biogeography: Use of human-dominated habitats by the avifauna of southern Costa Rica. *Ecological Applications*, 11(1), 1–13.

Dakora, F. D., and Phillips, D. A. (2002). Root exudates as mediators of mineral acquisition in low-nutrient. *Plant and Soil*, 245, 35–47.

Dakos, V., Scheffer, M,. van New, E. H., Brovkin, V., Petoukhov, V., and Held, H. (2008). Slowing down as an early warning signal for abrupt climate change. *PNAS*, 105(38), 14308–14312.

Dames and Moore. (1980). Report on the Rehabilitation of Rio Blanco Hydroelectric Project for Puerto Rico Electric Power Authority, Houston, TX, Job No. 11905-002-14.

D'Antonio, C. M., and Vitousek, P. M. (1992). Biological invasions by exotic grasses, the grass/fire cycle, and global change. *Annual Review of Ecology and Systematics*, 23, 63–87.

Darby, H. C. (1956). The clearing of the woodland in Europe. In Thomas, W. L. Jr. (ed.), *Man's Role in Changing the Face of the Earth*. University of Chicago Press, Chicago.

Das, R., Lawrence, D., D'Odorico, P., and DeLonge, M. (2011). Impact of land use change on atmospheric P inputs in a tropical dry forest. *J. Geophys. Res.*, doi:10.1029/2010JG001403.

Dasgupta P. (2013). The nature of economic development and the economic development of nature. *Economic and Political Weekly*, XLVIII, 38–51.

Daszak P., Cunningham, A. A., and Hyatt, A. D. (2001). Anthropogenic environmental change and the emergence of infectious diseases in wildlife. *Acta Tropica*, 78(2), 103–116.

Daubenmire, R. (1954) Alpine timberlines in the Americas and their interpretation. *Butler University Botanical Studies*, 11, 119–136.

Davidson, E. A. et al. (2004). Nitrogen and phosphorus limitation of biomass growth in a tropical secondary forest. *Ecological Applications*, 14(4), Supplement, S150–S163.

Davidson, E. A. et al. (2012). The Amazon basin in transition. *Nature*, 481, 321–328.

Davies, K. F., Margules, C. R., and Lawrence, J. F. (2004). A synergistic effect puts rare, specialized species at greater risk of extinction. *Ecology*, 85(1), 265–271.

Davies-Colley R. J., Payne, G. W., and van Elswijk, M. (2000). Microclimate gradients across a forest edge. *New Zealand Journal of Ecology*, 24(2), 111–121.

Davis, K. F., D'Odorico, P., and Rulli, M. C. (2014). Moderating diets to feed the future. *Earth's Future*, 2, doi:10.1002/2014EF000254.

Davis, K. F., Yu, K., Rulli, M. C., Pichdara, L., and D'Odorico, P. (2015). Accelerated deforestation by large-scale land acquisitions in Cambodia, *Nature Geoscience*, 8, 772-775, doi: 10.1038/NGEO2540.

Davison, E. M., and Tay, F. C. S. (1985). The effect of waterlogging on seedlings of Eucalyptus marginata. *New Phytologist*, 101(4), 743–753.

Dawson, T. E. (1998). Fog in the California redwood forest: Ecosystem inputs and use by plants. *Oecologia*, 117, 4, 476–485.

de Almeida, A. L. O., and Campari, J. S. (1995). *Sustainable settlement in the Brazilian Amazon*. Oxford University Press, Oxford.

DeBano, L. F. (1966). Formation of non-wettable soils involves heat transfer mechanism, USDA Forest Service Research Note PSW-132.

DeBano, L. F. (2000). The role of fire and soil heating on water repellency in wildland environments: A review. *J. Hydrol.*, 231, 195–206.

DeBano, L. F., Savage, S. M., and Hamilton, D. A. (1976). The transfer of heat and hydrophobic substances during burning. *Soil Science Society America Journal*, 40, 779–782.

de Castro, M. C., Monte-Mór, R. L., Sawyer, D. O., Singer, B. H. (2006). Malaria risk on the Amazon frontier. *Proc. Natl. Acad. Sci. USA*, 103(7), 2452–2457.

de Chantal, M., Holt Hanssen, K., Granhus, A., Bergsten, U., Ottosson Löfvenius, M., and Grip, H. (2007). Frost-heaving damage to one-year-old Picea abies seedlings increases with soil horizon depth and canopy gap size. *Canadian Journal of Forest Research*, 37, 1236–1243.

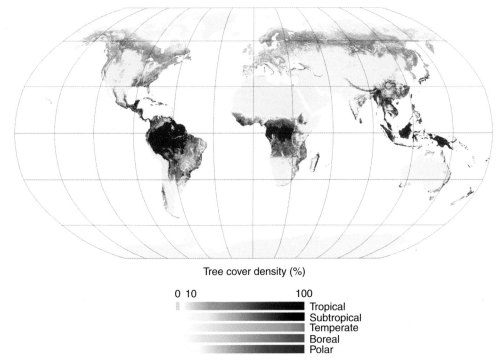

Figure 1.1. Distribution of the World's Forest by Climatic Domain. (FAO, 2010).
Source: Available at: http://www.fao.org/forestry/fra/80298/en/.

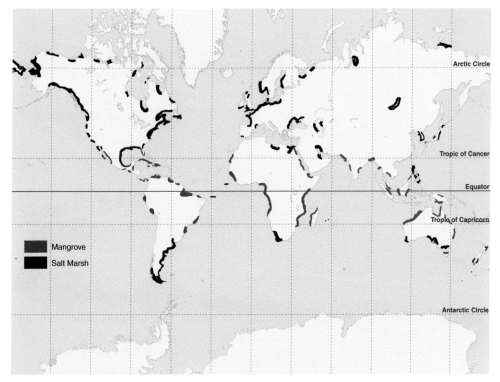

Figure 1.2. Global distribution of mangrove forests and salt marshes.
Source: D'Odorico et al., 2013. Redrawn after Chapman, 1975.

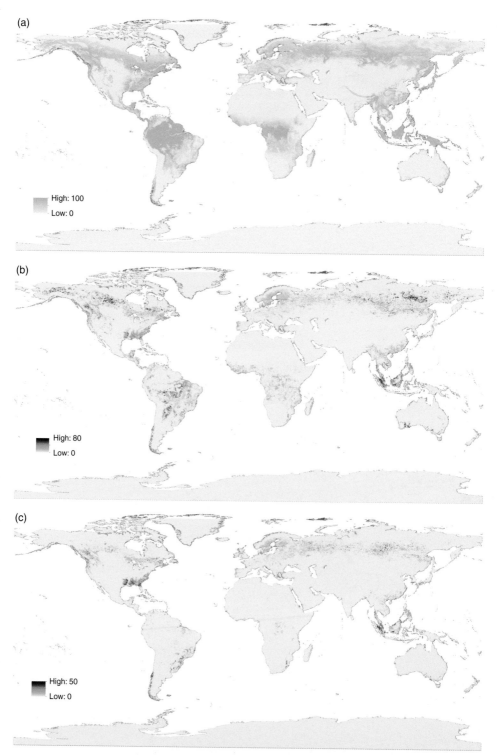

Figure 1.5. (a) Tree cover, (b) forest loss, and (c) forest gain.
Source: Redrawn using data sets from Hansen et al., 2013, for the period 2000–2012.

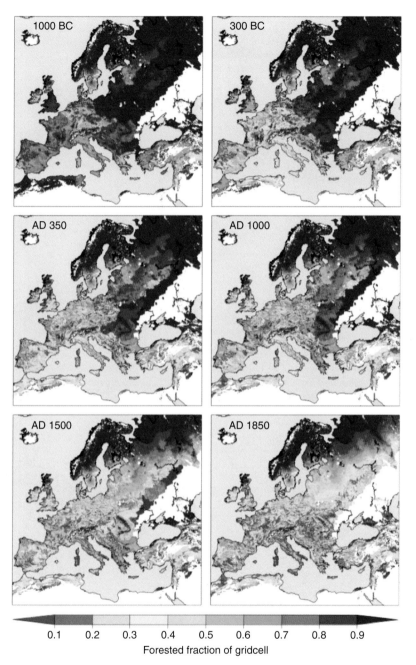

Figure 1.9. Historical forest clearance maps for 1000 BC, 300 BC, 350 AD, 1000 AD, 1500 AD, and 1850 AD.
Source: Kaplan et al., 2009.

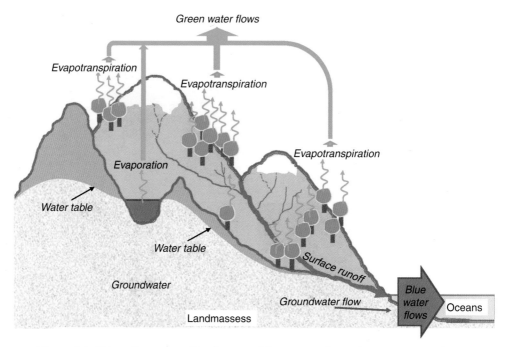

Figure 2.1. Water fluxes out of landmasses. Water vapor fluxes (or *green water fluxes*) are mostly mediated by vegetation (transpiration). Fluxes in the liquid phase (*blue water fluxes*) are due to surface runoff and (to a smaller extent) groundwater flow.

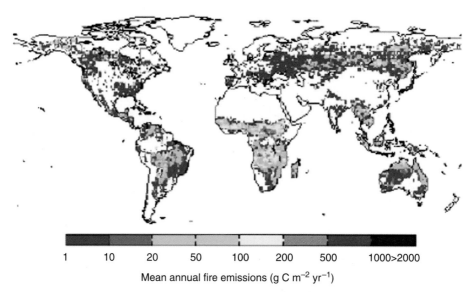

Figure 3.6. Mean annual fire emissions (g C m^{-2} yr^{-1}) averaged over the period 1997–2004.
Source: van der Werf et al., 2006.

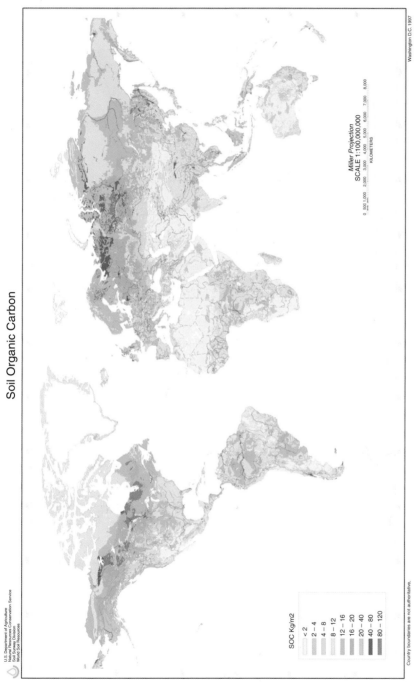

Figure 3.7. The global distribution of soil organic carbon to 1 meter depth.
Source: Reproduction is attributed to the U.S. Department of Agriculture and obtained from FAO-UNESCO, 2006.

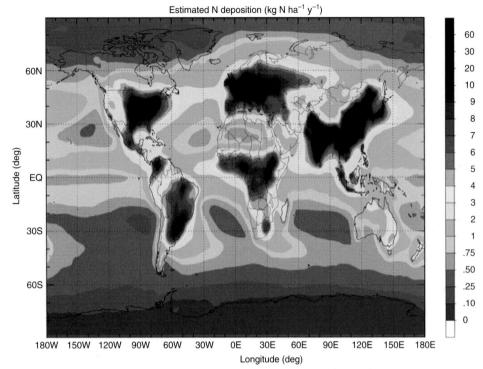

Figure 3.11. Estimated N deposition from global total N emissions, totaling 0.1 Pg N yr^{-1}.
Source: Galloway et al., 2008.

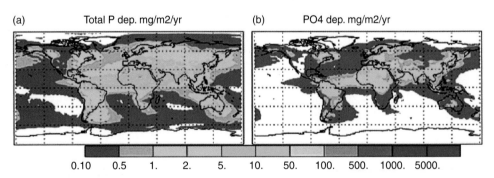

Figure 3.13. Model estimated deposition fluxes of (a) total phosphorus (TP) and (b) available P (PO$_4$).
Source: Mahowald et al., 2008.

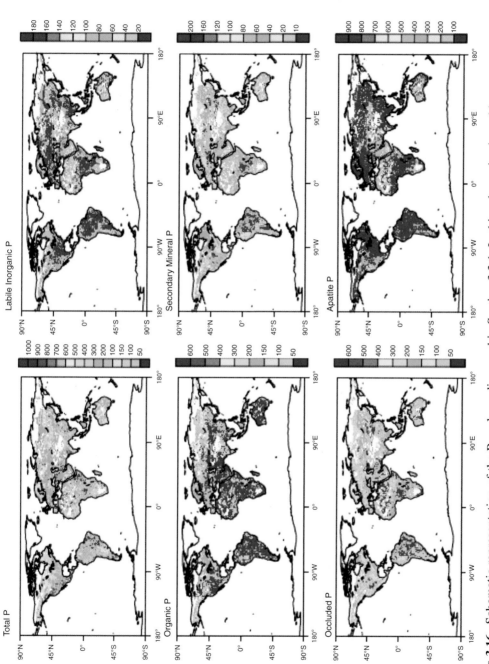

Figure 3.16. Schematic representation of the P cycle as discussed in Section 3.3.1. In this schematic, the red box represents organic pools while the blue box represents inorganic P pools.
Source: Yang et al., 2013.

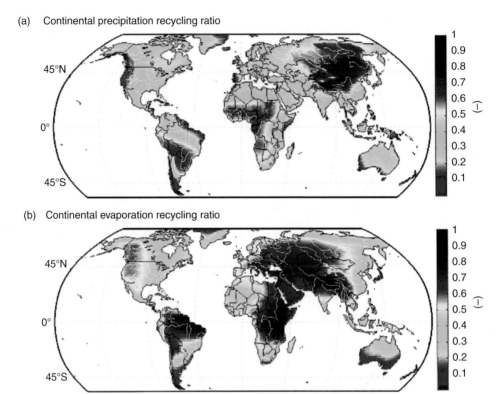

Figure 4.5. (a) Average global precipitation recycling (1999–2008), which is the part of precipitation falling in a region that originates from evaporation within that same region. This ratio describes the region's dependence on evaporation from within the region to sustain precipitation. (b) Average global evaporation recycling (1999–2008), which is the amount of evaporated water that returns as precipitation in the same region. High values of ε_c indicate locations from which the evaporated moisture will fall again as precipitation over continents.
Source: After van der Ent et al., 2010.

Figure 4.14. Stages of vegetation regrowth during shifting cultivation. A reduction in the duration of the fallow period (red arrow) reduces the recovery of the labile soil P pool and eventually leads to its depletion, thereby preventing forest regrowth.
Source: Photos were taken by the authors at field sites in southern Mexico.

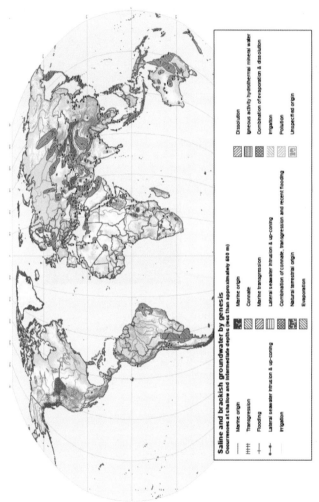

Figure 4.22. Global occurrence of saline groundwater at depths less than 500 m. Groundwater in these identified regions has a threshold limit of 1,000 mg/L TDS. Of the total land area on Earth 16% is affected by high salinity groundwater. The basins of West and Central Asia account for the largest area with high groundwater salinity contributing 14% to the total groundwater salinity area. The lowlands of South America and Europe, mountain belt of Central and Eastern Asia, and Eastern Australia all contribute individually at about 6%–7% to the total groundwater salinity area.
Source: Published by the International Groundwater Resources Assessment Centre (IGRAC).

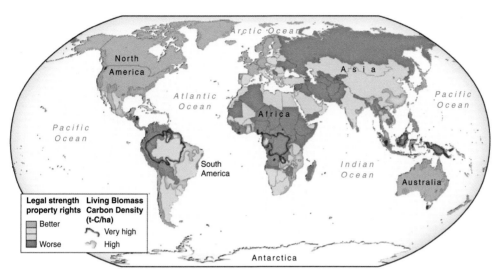

Figure 5.7. Global distribution of property rights security. Forests are located in areas with high live plant biomass and tend to be found in regions with poorly defined property rights.
Source: Bruce et al., 2010.

de Chantal, M., Rita, H., Bergsten, U., and Grip, H. (2009). Frost heaving of Picea abies seedlings as influenced by soil preparation, planting technique, and location along gap-shelterwood gradients. *Silva Fennica*, 43, 39–50.

de Moraes, J. F., Volkoff, B. C. C. C., Cerri, C. C., and Bernoux, M. (1996). *Soil properties under Amazon forest and changes due to pasture installation in Rondônia, Brazil.* Geoderma, 70(1), 63–81.

DeFries, R. S., Foley, J. A., and Asner, G. P. (2004). Land-use choices: Balancing human needs and ecosystem function. *Frontiers in Ecology and the Environment*, 2(5), 249–257.

DeFries, R. S., Houghton, R. A., Hansen, M. C., Field, C. B., Skole, D., & Townshend, J. (2002). Carbon emissions from tropical deforestation and regrowth based on satellite observations for the 1980s and 1990s. *Proceedings of the National Academy of Sciences*, 99(22), 14256–14261.

DeFries, R., and Rosenzweig, C. (2010). Toward a whole-landscape approach for sustainable land use in the tropics. *Proceedings of the National Academy of Sciences*, 107(46), 19627–19632.

DeFries, R. S., Rudel, T., Uriarte, M., and Hansen, M. (2010). Deforestation driven by urban population growth and agricultural trade in the twenty-first century. *Nature Geoscience*, 3(3), 178–181.

DeGraff, J. V. (1979). Initiation of shallow mass movement by vegetative-type conversion. *Geology*, 7, 426–429.

DeGraff, V. J., Bryce, R., Jibson, R. W., Mora, S., and Rogers, C. T. (1989). Landslides: Their Extent and Significance in the Caribbean. In Brabb, E. E., and B. L. Harrold (eds.), *Landslides: Extent and Ecological Significance*. Balkema, Rotterdam.

DeLonge, M. S. et al. (2006). Rain forest islands in the Chilean semiarid region: Fog-dependency. *Ecosystem* p4, 154–160.

DeLonge, M., D'Odorico, P., and Lawrence, D. (2008). Feedbacks between phosphorus deposition and canopy cover: The emergence of multiple stable states in tropical dry forests. *Global Change Biology*, 14(1), 154–160.

Del Pozo, A., Pérez, P., Gutiérrez, D., Alonso, A., Morcuende, R., and Martínez-Carrasco, R. (2007). Gas exchange acclimation to elevated CO_2 in upper-sunlit and lower-shaded canopy leaves in relation to nitrogen acquisition and partitioning in wheat grown in field chambers. *Environmental and Experimental Botany*, 59(3), 371–380.

DeLucia, E. V., Drake, J. E., Thomas, R. B., and Gonzalez-Meler, M. (2007). Forest carbon use efficiency: Is respiration a constant fraction of gross primary production? *Global Change Biology*, 13(6), 1157–1167.

del-Val, E. et al. (2006). Rain forest islands in the Chilean semiarid region: Fog-dependency, ecosystem persistence and tree regeneration. *Ecosystems*, 9(4), 598–608.

Dentener, F. et al. (2006). Nitrogen and sulfur deposition on regional and global scales: A multimodel evaluation. *Global Biogeochemical Cycles*, 20(4).

Deo, R. C. (2011). Links between native forest and climate in Australia. *Weather*, 66, 64–69.

DeVries, J., and Toenniessen, G. H. (2001). Securing the harvest: Biotechnology, breeding, and seed systems for African crops. CABI.

Diamond, J. (2005). *Collapse: How societies choose to fail or succeed*. Penguin, New York.

Diaz, R. J., and Rosenberg, R. (2008). Spreading dead zones and consequences for marine ecosystems. *Science*, 321(5891), 926–929.

Diaz-Ravina, M., Acea, M. J., and Carballas, T. (1995). Seasonal changes in microbial biomass and nutrient flush in forest soils. *Biology and Fertility of Soils*, 19, 220–226.

Dickinson, R. E., and Henderson-Sellers, A. (1988). Modelling tropical deforestation: A study of GCM land-surface parameterizations. *Q J Roy. Meteor. Soc.*, 114, 439–462.

Didia, D. O. (1997). Democracy, political instability and tropical deforestation. *Global Environmental Change*, 7(1), 63–76.

Dietrich, W. E., Reiss, R., Hsu, M. L., and Montgomery, D. (1995). A process-based model for colluvial soil depth and shallow landslides using digital elevation data. *Hydrol. Processes*, 9, 383–400.

Dirmeyer, P. A., and Brubaker, K. L. (2007). Characterization of the global hydrologic cycle from a back-trajectory analysis of atmospheric water vapor. *J. Hydrometeorol.*, 8(1), 20–37, doi:10.1175/JHM557.1.

Dirzo, R., and Raven, P. H. (2003). Global state of biodiversity and loss. *Annual Review of Environment and Resources*, 28(1), 137–167.

Dixon, R. K., Solomon, A. M., Brown, S., Houghton, R. A., Trexier, M. C., and Wisniewski, J. (1994). Carbon pools and flux of global forest ecosystems. *Science*, 263(5144), 185–190.

Doak, D. F., Bigger, D., Harding, E. K., Marvier, M. A., O'malley, R. E., and Thomson, D. (1998). The statistical inevitability of stability-diversity relationships in community ecology. *The American Naturalist*, 151(3), 264–276.

D'Odorico, P., Carr, J. A., Laio, F., Ridolfi, L., and Vandoni, S. (2014). Feeding humanity through global food trade. *Earth's Future*, 2, doi:10.1002/2014EF000250.

D'Odorico, P., Caylor, K., Okin, G. S., and Scanlon, T. M. (2007). On soil moisture–vegetation feedbacks and their possible effects on the dynamics of dryland ecosystems. *J. Geophys. Res.*, 112, G04010, doi:10.1029/2006JG000379.

D'Odorico, P., Engel, V., Carr, J. A., Oberbauer, S. F., Ross, M. S., and Sah, J. P. (2011). Tree-grass coexistence in the Everglades freshwater system. *Ecosystems*, 14, 298–310.

D'Odorico, P., and Fagherazzi, S. (2003). A probabilistic model of rainfall-triggered shallow landslides in hollows: A long-term analysis. *Water Resour. Res.*, 39(9), 1262.

D'Odorico, P., Fuentes, J. D., Pockman, W. T., Collins, S. L., He, Y., Medeiros, J. S., De Wekker, S., and Litvak, M. E. (2010a) Positive feedback between microclimate and shrub encroachment in the northern Chihuahuan desert. *Ecosphere*, 1, art17, doi:10.1890/ES10-00073.1.

D'Odorico, P., He, Y., Collins, S., De Wekker, S. F., Engel, V., and Fuentes, J. D. (2013). Vegetation–microclimate feedbacks in woodland-grassland ecotones. *Global Ecology and Biogeography*, 22(4), 364–379.

D'Odorico P., Laio, F., Porporato, A., Ridolfi, L., Rinaldo, A., and Rodriguez-Iturbe, I. (2010b). Ecohydrology of terrestrial ecosystems. *Bioscience*, 60(11), 898–907.

D'Odorico, P., Laio, F., and Ridolfi, L. (2006). A probabilistic analysis of fire-induced tree-grass coexistence in savannas. *American Naturalist*, 167(3), E79–E87.

D'Odorico, P., and Rulli, M. C. (2013). The fourth food revolution. *Nat. Geosci.*, 6(6), 417–418, doi:10.1038/ngeo1842.

D'Odorico P. and Rulli, M.C. (2014), The land and its people. *Nature Geoscience*, 4: 324–325.

Doerr, S. H., Shakesby, R. A., and Walsh, R. P. D. (2000). Soil water repellency: Its causes, characteristics and hydrogeomorphological significance. *Earth Sci. Rev.*, 51, 33–65.

Don, A., Schumacher, J., and Freibauer, A. (2011). Impact of tropical land-use change on soil organic carbon stocks–a meta-analysis. *Global Change Biology*, 17(4), 1658–1670.

Dornburg, V. et al. (2010). Bioenergy revisited: Key factors in global potentials of bioenergy. *Energy & Environmental Science*, 3(3), 258–267.

Dosskey, M. G., and Bertsch, P. M. (1997). Transport of dissolved organic matter through a sandy forest soil. *Soil Science Society of America Journal*, 61(3), 920–927.

Drury, W. H. (1956). Bog Flats and Physiographic Processes in the Upper Kuskokwim River Region, Alaska, Contributions from the Gray Herbarium of Harvard University, No. CLXXVIII, p. 130.

Dubayah, R. O., Drake, J. B. (2000). Lidar remote sensing for forestry. *Journal of Forestry*, 98, 44–46.

Dubé, S., and Plamondon, A. P. (1995). Watering up after clearcutting on forested wetlands of the St. Lawrence lowland. *Water Resour. Res.*, 31(7), 1741–1750.

Dublin, H. T., Sinclair, A. R. E., and McGlade, J. (1990). Elephants and fire as causes of multiple stable states in the Serengeti-Mara woodlands. *Journal of Animal Ecology*, 59, 1147–1164.

Dunham, A. E. (2008). Above and below-ground impacts of terrestrial mammals and birds in a tropical forest. *Oikos*, 117(4), 571–579.

Dunne, T. (1978). Field studies of hillsope flow processes. In Kirkby, M. J. (ed.), *Hillslope Hydrology*. J. Wiley & Sons, New York, pp. 227–293.

Dunne, T., and Black, R. D. (1970a). An experimental investigation of runoff prediction in permeable soils. *Water Resour. Res.*, 6(2), 478–490.

Dunne, T., and Black, R. D. (1970b). Partial area contributions to storm runoff in a small New England watershed. *Water Resour. Res.*, 6(5), 1296–1311.

Dunne, T., and Leopold, L. B. (1978). *Water in Environmental Planning*. W. H. Freeman and Co., New York.

Dutch, J., and Ineson, P. (1990). Denitrification of an upland forest site. *Forestry*, 63(4), 363–37

du Toit, J. T., Rogers, K. H., and Biggs, H. C. (2003). eds. *The Kruger experience: Ecology and management of savanna heterogeneity*. Island, Washington, DC.

Dutschke, M., and Wolf, R. (2007). Reducing emissions from deforestation and forest degradation in developing countries: The way forward. *Foresty, Risk and Climate Policy*, 125.

Dvořàk, K. A. (1992). Resource management by West African farmers and the economics of shifting cultivation. *American Journal of Agricultural Economics*, 74(3), 809–815.

Dy, G., and Payette, S. (2007). Frost hollows of the boreal forest as extreme environments for black spruce tree growth. *Canadian Journal of Forest Research*, 37, 492–504.

Dyrness, C. T. (1982). Control of depth to permafrost and soil temperature by the forest floor in black spruce/feathermoss communities, US For. Serv. Res. Note PNW-396;1–9 pp.

Easterling, D. R., Karl, T. R., Gallo, K. P., Robinson, D. A., Trenberth, K. E., and Dai, A. (2000). Observed climate variability and change of relevance to the biosphere. *J. Geophys. Res.*, 105(D15), 20, 101–120, 114, doi:10.1029/2000JD900166.

Easterling, D. R., Peterson, T. C., and Karl, T. R. (1996). On the development and use of homogenized climate datasets. *J. Clim.*, 9, 1429–1434.

Easterwood, G. W., and Sartain, J. B. (1990). Clover residue effectiveness in reducing orthophosphate sorption on ferric hydroxide coated soil. *Soil Science Society of America Journal*, 54(5), 1345–1350.

Eberbach, P. L. (2003). The eco-hydrology of partly cleared, native ecosystems in southern Australia: A review. *Plant and Soil*, 257(2), 357–369.

Eden, M. J., Furley, P. A., McGregor, D. F. M., Milliken, W., and Ratter, J. A., (1991). Effect of forest clearance and burning on soil properties in northern Roraima, Brazil. *For. Ecol. Manage.*, 38, 283–290.

Edmonds, R. L., and Lebo, D. S. (1998). Diversity, production, and nutrient dynamics of fungal sporocarps on logs in an oldgrowth temperate rain forest, Olympic National Park, Washington. *Canadian Journal of Forest Research*, 28, 665–673.

Egerton, J. J. G., Banks, J. C. G., Gibson, A., Cunningham, R. B., and Ball, Marilyn C. (2000). Facilitation of seedling establishment: Reduction in irradiance enchances winter growth of eucalyptus pauciflora. *Ecology*, 81, 1437–1449.

Ellison D., Futter, M. N., and Bishop, K. (2012). On the forest cover-water yield debate: From demand-to-supply-side thinking. *Global Change Biology*, 18, 806–820.

Elmqvist, T., Folke, C., Nyström, M., Peterson, G., Bengtsson, J., Walker, B., and Norberg, J. (2003). Response diversity, ecosystem change, and resilience. *Frontiers in Ecology and the Environment*, 1(9), 488–494.

Eltahir, E. A. B., and Bras, R. L. (1994). Precipitation recycling in the Amazon Basin. *Quarterly Journal of the Royal Meteorological Society*, 120, 861–880.

Eltahir, E. A. B., and Bras, R. L. (1996). Precipitation recycling. *Rev. Geophys.* 34(3), 367–379.

Elton, C. S. (1958). *The Ecology of Invasions by Plants and Animals*. Methuen, London, 18.

Engels C., and Marschner H. (1955). Plant uptake and utilization of nitrogen. In Bacon, P. E. (ed.), *Nitrogen Fertilization in the Environment*. Marcel Dekker, New York, pp. 41–81.

Ensminger, J. (1996). *Making a market: The institutional transformation of an African society*. Cambridge University Press, Cambridge.

Epstein, H. E. et al. (2004). The nature of spatial transitions in the Arctic. *Journal of Biogeography*, 31, 1917–1933.

Estes, J. A. et al. (2011). Trophic downgrading of planet Earth. *Science*, 333(6040), 301–306.

E. U. (2009). Directive 2009/28/EC of the European Parliament and of the Council of 23 April 2009 on the promotion of the use of energy from renewable sources and amending and subsequently repealing Directives 2001/77/EC and 2003/30/EC.

Eugster, W. et al. (2000). Land–atmosphere energy exchange in Arctic tundra and boreal forest: Available data and feedbacks to climate. *Global Change Biology*, 6, 84–115.

Eviner, V. T., and Chapin III, F. S. (2003). Functional matrix: A conceptual framework for predicting multiple plant effects on ecosystem processes. *Annu. Rev. Ecol. Evol. Syst.*, 34, 455–485.

Ewel, J., Berish, C., Brown, B., Price, N., and Raich, J. (1981). Slash and burn impacts on a Costa Rican wet forest site. *Ecology*, 816–829.

Ewers, R. M., Laurance, W. F., and Souza, C. M. (2008). Temporal fluctuations in Amazonian deforestation rates. *Environmental Conservation*, 35(04), 303–310.

Fairhead, J., Leach, M., and Scoones, I. (2012). Green grabbing: A new appropriation of nature? *Journal of Peasant Studies* 39(2), 237–261.

Fairman, J. G., Jr., Nair, U. S., Christopher, S. A., and Mölg, T. (2011). Land use change impacts on regional climate over Kilimanjaro. *J. Geophys. Res.*, 116, D03110, doi:10.1029/2010JD014712.

Falkenmark M. J., and Rockström, J. (2004). *Balancing Water for Humans and Nature: The New Approach in Ecohydrology*. Earthscan, London.

Falkenmark, M., and Rockstrom, J. (2006). The new blue and green water paradigm: Breaking new ground for water resources planning and management. *J. Water Resour. Plann. Manage.*, 132(3), 129–132.

FAO. (1982). Tropical forest resource, by J.P. Lanly. FAO Forestry Paper No. 30. Rome.

FAO. (1995). Forest Resources Assessment 1990. Global Synthesis. FAO Forestry Paper No. 124. FAO, Rome.

FAO. (1996). Forest resources assessment 1990: Survey of tropical forest cover and study of change processes. FAO Forestry Paper 130, Rome.

FAO. (2001). Global forest resources assessment 2000. FRA 2000 Main Report. FAO Forestry Paper 140, Rome.

FAO. (2005). FAOSTAT database. Rome, Italy: FAO.

FAO (2008). FAOSTAT – Production – ForesSTAT, available at: http://faostat.fao.org/DesktopDefault.aspx?PageID=381&lang=en.

FAO. (2009). State of the world's forests. Rome: FAO.

FAO. (2010). Global forest resource assessment 2010: Key findings, available at: http://www.fao.org/forestry/fra/fra2010/en/.

FAO. (2011a). State of the world's forests. Rome: FAO.

FAO. (2011b). Looking ahead in world food and agriculture: Perspectives to 2050. Edited by Piero Conforti. Agricultural Development Economics Division Economic and Social Development Department Food and Agriculture Organization of the United Nations 2011, Paris.

FAO. (2011c). Voluntary guidelines on the responsible governance of tenure of land and other natural resources (Zero Draft, Released 18 April 2011). Rome: Food and Agriculture Organization of the United Nations. www.fao.org/fileadmin/user_upload/nr/land_tenure/econsultation/english/Zero_Draft_VG_Final.pdf.

FAO. (2012). State of the world's forests. Rome: FAO.

FAO and JRC.(2012). Global forest land-use change 1990–2005, by E. J. Lindquist, R. D'Annunzio, A. Gerrand, K. MacDicken, F. Achard, R. Beuchle, A. Brink, H. D. Eva, P. Mayaux, J. San-Miguel-Ayanz & H-J. Stibig. FAO Forestry Paper No. 169. Food and Agriculture Organization of the United Nations and European Commission Joint Research Centre. Rome, FAO.

FAOSTAT. (2011). Rome, FAO, available at: http://www.faostat.fao.org.proxy1.library.jhu.edu/default.aspx.

FAO/UNEP. (1981). Tropical Forest Resources Assessment Project. FAO, Rome.

FAO-UNESCO. (2006). Soil Map of the World, digitized by ESRI. Soil climate map, USDA-NRCS, Soil Science Division, World Soil Resources, Washington DC. Soil Pedon database, USDA-NRCS National Soil Survey Center, Lincoln, NE.

Fargione, J., Hill, J., Tilman, D., Polasky, S., and Hawthorne, P. (2008). Land clearing and the biofuel carbon debt. *Science*, 319, 1235–1238.

Farzin, Y. H. (1984). The effect of the discount rate on depletion of exhaustible resources. *Journal of Political Economy*, 841–851.

Fearnside, P. M. (2005). Deforestation in Brazilian Amazonia: History, rates, and consequences. *Conservation Biology*, 19, 680–688, doi:10.1111/j.1523-1739.2005.00697.x.

Fearnside, P. M. (2007). Deforestation in Amazonia. In Cleveland, C. J. (ed.), *Encyclopedia of Earth*. Washington, DC: Environmental Information Coalition, National Council of Science and the Environment, available at: www.eoearth.org/article/Deforestation_in_Amazonia.

Fearnside, P. M., and Imbrozio Barbosa, R. (1998). Soil carbon changes from conversion of forest to pasture in Brazilian Amazonia. *Forest Ecology and Management*, 108(1), 147–166.

Feller, C., and Beare, M. H. (1997). Physical control of soil organic matter dynamics in the tropics. *Geoderma*, 79(1), 69–116.

Feller, I. C., Lovelock, C. E., Berger, U., McKee, K. L., Joye, S. B., and Ball, M. C. (2010). Biocomplexity in mangrove ecosystems. *Annual Review of Marine Science*, 2, 395–417.

Ferraz, S. F.d., Vettorazzi, C. A., Theobald, D. M., and Ballester, M. A. R. (2005). Landscape dynamics of Amazonian deforestation between 1984 and 2002 in central Rondonia, Brazil: Assessment and future scenarios. *Forest Ecology and Management*, 204, 67–83.

Field, C. B., van der Werf, G. R., and Shen, S. S. P. (2009). Human amplification of drought induced biomass burning in Indonesia since 1960. *Nature Geoscience*, 2, 185–188.

Finér, L., Mannerkoski, H., Piirainen, S., and Starr, M. (2003). Carbon and nitrogen pools in an old-growth, Norway spruce mixed forest in eastern Finland and changes associated with clear-cutting. *Forest Ecology and Management*, 174(1), 51–63.

Fisch, G. et al. (2004). The convective boundary layer over pasture and forest in Amazonia. *Theoretical and Applied Climatology*, 78(1–3), 47–59.

Fischer, J., Lindenmayer, D. B., and Manning, A. D. (2006). Biodiversity, ecosystem function, and resilience: Ten guiding principles for commodity production landscapes. *Frontiers in Ecology and the Environment*, 4(2), 80–86.

Fisher, B. (2010). African exception to drivers of deforestation. *Nature Geoscience*, 3(6), 375–376.

Fisher, J. B., Badgley, G., and Blyth, E. (2012). Global nutrient limitation in terrestrial vegetation. *Global Biogeochemical Cycles*, 26(3).

Fisher, R. A., Williams, M., Lola Da Costa, A., Malhi, Y., Da Costa, R. F., Almeida, S., and Meir, P. (2007). The response of an E. Amazonian rain forest to drought stress: Results and modelling analyses from a throughfall exclusion experiment. *Global Change Biol.*, 13, 2361–2378.

Fitzherbert, E. B. et al. (2008). How will oil palm expansion affect biodiversity? *Trends Ecol. Evol.*, 23, 538–545.

Flannery, T. (2002). *The future eaters: An ecological history of the Australasian lands and people.* Grove Press.

Flint, E. P. (1994). Changes in land use in South and Southeast Asia from 1880 to 1980: A data base prepared as part of a coordinated research program on carbon fluxes in the tropics. *Chemosphere*, 29(5), 1015–1062.

Foley, J. A., Kutzbach, J. E., Coe, M. T., and Levis, S. (1994). Feedbacks between climate and boreal forests during the Holocene epoch. *Nature*, 371, 52–55.

Foley, J. A., Prentice, I. C., Ramankutty, N., Levis, S., Polard, D, Sitch, S., and Haxeltine, A. (1996). An integrated biosphere model of land surface processes, terrestrial carbon balance and vegetation dynamics. *Global Biogeophysical Cycle*, 10, 603–629.

Foley, J. A. et al. (2007). Amazonia revealed: Forest degradation and loss of ecosystem goods and services in the Amazon Basin. *Frontiers in Ecology and the Environment*, 5(1), 25–32.

Foley, J. A. et al. (2011). Solutions for a cultivated planet. *Nature*, 478, 337–342, doi:10.1038/nature10452.

Folke, C., Carpenter, S., Walker, B., Scheffer, M., Elmqvist, T., Gunderson, L., and Holling, C. S. (2004). Regime shifts, resilience, and biodiversity in ecosystem management. *Annu. Rev. Ecol. Evol. Syst.*, 35, 557–581.

Folland, C. K. et al. (2001). Observed Climate Variability and Change. In Houghton, J. T. (ed.), *Climate Change 2001: The Scientific Basis.* Cambridge University Press, New York, pp. 99–181.

Folliott, P. F., and Thrud, D. B. (1977). Water resources and multiple-use forestry in the Southwest. *J. For.*, 75, 469–472.

Food and Agriculture Organization (FAO) (2002). *World Agriculture: Towards 2015/2030.* Food and Agriculture Organization of the United Nations, Rome.

Food and Agriculture Organization (FAO) (2007). *FAOSTAT Online Statistical Service.* United Nations Food and Agriculture Organization (FAO), Rome, available at: http://faostat.fao.org (accessed October 2007).

Food and Agriculture Organization (FAO). (2010). *Global Forest Resources Assessment 2010: Main Report.* United Nations Food and Agriculture Organization, Rome.

Foster, P. (2001). The potential negative impacts of global climate change on tropical montane cloud forests. *Earth-Science Reviews*, 55, 73–106.

Fraedrich, K., Kleidon, A., and Lunkeit, F. (1999). A green planet versus a desert world: Estimating the effect of vegetation extremes on the atmosphere. *J. Clim.* 12, 3156–3163.

Fredriksen, R. (1975). Nitrogen, phosphorus and particulate matter budgets of five coniferous forest ecosystems in the western Cascades Range, Oregon, Ph.D. dissertation, Oregon State University, Corvallis.

Freeze, R. A. (1972). Role of subsurface flow in generating surface runoff. 2. Upstream source areas. *Water Resour. Res.*, 8(5), 1272–1283, doi:10.1029/WR008i005p01272.

Froment, A. (2009). Biodiversity and health: The place of parasitic and infectious diseases. In Sala O. E., L. A. Meyerson, and C. Parmesan (eds.), *Biodiversity Change and Human Health.* Island Press, Washington, DC, pp. 211–227.

Fujiwara, K. (1970). A study on landslides by aerial photographs. *Res. Bull. Exp. Forests Hakkaido University*, 27(2), 297–345.

Fuller, D. O. (2006). Tropical forest monitoring and remote sensing: A new era of transparency in forest governance? *Singapore Journal of Tropical Geography*, 27(1), 15–29.

Galindo-Leal, C., and de Gusmao Camara, I. (2003). *The Atlantic Forest of South America: Biodiversity Status, Threats, and Outlook*. Island Press, Washington, DC.

Galloway, J. N., and Cowling, E. B. (2002). Reactive nitrogen and the world: 200 years of change. *AMBIO: A Journal of the Human Environment*, 31(2), 64–71.

Galloway, J. N., Schlesinger, W. H., Levy, H., Michaels, A., and Schnoor, J. L. (1995). Nitrogen fixation: Anthropogenic enhancement-environmental response. *Global Biogeochemical Cycles*, 9(2), 235–252.

Galloway, J. N. et al. (2004). Nitrogen cycles: Past, present, and future. *Biogeochemistry*, 70(2), 153–226.

Galloway, J. N. et al. (2008). Transformation of the nitrogen cycle: Recent trends, questions, and potential solutions. *Science*, 320(5878), 889–892.

Garcia-Montiel, D. C., Neill, C., Melillo, J., Thomas, S., Steudler, P. A., and Cerri, C. C. (2000). *Soil phosphorus transformations following forest clearing for pasture in the Brazilian Amazon.*

García-Oliva, F., Casar, I., Morales, P., and Maass, J. M. (1994). Forest-to-pasture conversion influences on soil organic carbon dynamics in a tropical deciduous forest. *Oecologia*, 99(3–4), 392–396.

Gash, J. H. C., and Nobre, C. A. (1997). Climatic effects of Amazonian deforestation: Some results from ABRACOS. *Bull. American Meteorol. Soc.*, 78(5), 823–830.

Gebremedhin, B., and Schwab, G. (1998). The economic importance of crop rotation systems: Evidence from the literature (No. 11690). Michigan State University, Department of Agricultural, Food, and Resource Economics, Lansing.

Geiger, R. (1965). *The Climate near the Ground*. Harvard University Press, Cambridge, MA.

Geist, H. J., and Lambin, E. F. (2002). Proximate causes and underlying driving forces of tropical deforestation: Tropical forests are disappearing as the result of many pressures, both local and regional, acting in various combinations in different geographical locations. *BioScience*, 52(2), 143–150.

Gelfand, I., Zenone, T., Jasrotia, P., Chen, J., Hamilton, S. K., and Robertson, G. P. (2011). Carbon debt of Conservation Reserve Program (CRP) grasslands converted to bioenergy production. *Proceedings of the National Academy of Sciences*, 108(33), 13864–13869.

Gerber, S., Hedin, L. O., Oppenheimer, M., Pacala, S. W., and Shevliakova, E. (2010). Nitrogen cycling and feedbacks in a global dynamic land model. *Global Biogeochemical Cycles*, 24(1).

Germino, M. J., and Smith, W. K. (1999). Sky exposure, crown architecture, and low-temperature photoinhibition in conifer seedlings at alpine treeline. *Plant, Cell and Environment*, 22, 407–415.

Germino, M. J., and Smith, W. K. (2000). Differences in microsite, plant form, and low-temperature photoinhibition in Alpine plants. *Arctic, Antarctic, and Alpine Research*, 32, 388–396.

Germino, M. J., Smith, W. K., and Resor, A. C. (2002). Conifer seedling distribution and survival in an alpine-treeline ecotone. *Plant Ecology*, 162, 157–168.

Ghassemi, F., Jakeman, A. J., and Nix, H. A. (1995). Salinization of Land and Water Resources: Human Causes, Extent, Management and Case Studies. CABI, Canberra.

Giardina, C. P., Sanford Jr., R. L., Dockersmith, I. C., and Jaramillo, V. J. (2000). The effects of slash and burning on ecosystem nutrients during the land preparation phase of shifting cultivation. *Plant and Soil*, 220, 247–260.

Gibbs, H. K., Brown, S., Niles, J. O., and Foley, J. A. (2007). Monitoring and estimating tropical forest carbon stocks: Making REDD a reality. *Environmental Research Letters*, 2(4), 045023.

Gibbs, H. K., Ruesch, A. S., Achard, F., Clayton, M. K., Holmgren, P., Ramankutty, N., and Foley, J. A. (2010). Tropical forests were the primary sources of new agricultural land in the 1980s and 1990s. *Proceedings of the National Academy of Sciences*, 107(38), 16732–16737.

Gibson, C. C., Williams, J. T., and Ostrom, E. (2005). Local enforcement and better forests. *World Development*, 33(2), 273–284.

Gibson, L. et al. (2011). Primary forests are irreplaceable for sustaining tropical biodiversity. *Nature*, 478(7369), 378–381.

Giller, K. E., Beare, M. H., Lavelle, P., Izac, A. M., and Swift, M. J. (1997). Agricultural intensification, soil biodiversity and agroecosystem function. *Applied Soil Ecology*, 6(1), 3–16.

Glade, T. (2003). Landslide occurrence as a response tol and use change: A review of evidence from New Zealand. *Catena*, 51, 297–314.

Global Forest Watch. (2000). *Canada's Forests at a Crossroads: An Assessment in the Year 2000*. World Resources Institute, Washington, DC.

Godfray, H. C. J. et al. (2010). Food security: The challenge of feeding 9 billion people. *Science*, 327, 812–818.

Godfray, H. C. (2011). Food for thoughts. *Proc. Natl. Acad. Sci. USA*, 108, 19845–6.

Godfray, H. C. J., and Garnett, T. (2014). Food security and sustainable intensification. *Philosophical Transactions of the Royal Society B: Biological Sciences*, 369(1639), 20120273.

Goeschl T., and Swanson T. (2002). The social value of biodiversity for RandD. *Environ. Resource Econ.* 22, 477–504.

Goetz, S., and Dubayah, R. (2011). Advances in remote sensing technology and implications for measuring and monitoring forest carbon stocks and change. *Carbon Management*, 2(3), 231–244.

Goetz, S. J. et al. (2009). Mapping and monitoring carbon stocks with satellite observations: A comparison of methods. *Carbon Balance and Management*, 4(1), 2.

GOFC-GOLD (2011). A sourcebook of methods and procedures for monitoring and reporting anthropogenic greenhouse gas emissions and removals caused by deforestation, gains and losses of carbon stocks in forests remaining forests, and forestation. GOFC-GOLD Report Version COP17-1. GOFC-GOLD Project Office, Natural Resources Canada, Alberta.

Goldewijk, K. K. (2001). Estimating global land use change over the past 300 years: The HYDE database. *Global Biogeochemical Cycles*, 15(2), 417–433.

Goldewijk, K., Beusen, A., Van Drecht, G., and De Vos, M. (2011). The HYDE 3.1 spatially explicit database of human-induced global land-use change over the past 12,000 years. *Global Ecology and Biogeography*, 20(1), 73–86.

Goll, D. S. et al. (2012). Nutrient limitation reduces land carbon uptake in simulations with a model of combined carbon, nitrogen and phosphorus cycling. *Biogeosciences Discussions*, 9(3), 3173–3232.

Gordon, L. J., Dunlop, M., and Foran, B. (2003). Land cover change and water vapour flows: Learning from Australia. *Philos. Trans. R. Soc. London, Ser. B*, 358, 1973–1984.

Gordon, L. J., Steffen, W., Jönsson, B. F., Folke, C., Falkenmark, M., and Johannessen, Å. (2005). Human modification of global water vapor flows from the land surface. *Proceedings of the National Academy of Sciences of the United States of America*, 102(21), 7612–7617.

Gosz, J. R. (1981). Nitrogen cycling in coniferous ecosystems. Ecological Bulletins, Sweden.

Grace, J. (1989). Tree lines. *Philosophical Transactions of the Royal Society of London. Series B, Biological Sciences*, 324, 233–245.

Grace, J., Allen, S. J., and Wilson, C. (1989). Climate and the meristem temperatures of plant communities near the tree-line. *Oecologia*, 79, 198–204.

Grau, H. R., Gasparri, N. I., and Aide, T. M. (2005). Agriculture expansion and deforestation in seasonally dry forests of north-west Argentina. *Environmental Conservation*, 32(02), 140–148.

Greeley, W. B. (1925). The relation of geography to timber supply. *Economic Geography*, 1, 1–11.

Greene, R. S. B. (1992). Soil physical-properties of three geomorphic zones in a semiarid Mulga woodland. *Aust. J. Soil Res.*, 30(1), 55–69.

Greene, R. S. B., Kinnell, P. I. A., and Wood, J. T. (1994). Role of plant cover and stock trampling on runoff and soil-erosion from semiarid wooded rangelands. *Aust. J. Soil Res.*, 32(5), 953–973.

Greene, R. S. B., Valentin, C., and Esteves, M. (2001). Runoff and erosion processes, in Banded Vegetation Patterning in Arid and Semiarid Environments: Ecological Processes and Consequences for Management. *Ecol. Stud.*, 149, 52–76.

Grey, M. J., Clarke, M. F., and Loyn, R. H. (1998). Influence of the Noisy Miner Manorina melanocephala on avian diversity and abundance in remnant Grey Box woodland. *Pacific Conservation Biology*, 4(1), 55.

Griffin, R. C., and McCarl, B. A. (1989). Brushland management for increased water yield in Texas. *Journal of the American Water Resources Association*, 25, 175–186. doi: 10.1111/j.1752-1688.1989.tb05679.x

Grimmond, C. S. B., Robeson, S. M., and Schoof, J. T. (2000). Spatial variability of micro-climatic conditions within a mid-latitude deciduous forest. *Climate Research*, 15, 137–149.

Groot, A., and Carlson, D. W. (1996). Influence of shelter on night temperatures, frost damage, and bud break of white spruce seedlings. *Canadian Journal of Forest Research*, 26, 1531–1538.

Gruber, N., and Galloway, J. N. (2008). An Earth-system perspective of the global nitrogen cycle. *Nature*, 451(7176), 293–296.

Guerra, C. A., Snow, R. W., and Hay, S. I. (2006). A global assessment of closed forests, deforestation and malaria risk. *Ann. Trop. Med. Parasitol.*, 100(3), 189–204.

Gutierrez, A. G., Barbosa, O., Christie, D. A., DEL-VAL, E. K., Ewing, H. A., Jones, C. G., ... & Armesto, J. J. (2008). Regeneration patterns and persistence of the fog-dependent Fray Jorge forest in semiarid Chile during the past two centuries. *Global Change Biology*, 14(1), 161–176.

Guo, Z. W., Xiao X. M., and Li, D. M. (2000). An assessment of ecosystem services: Water flow regulation and hydroelectric power production. *Ecol. Appl.*, 10, 925–936.

Gupta, R. K., and Abrol, I. P. (2000). Salinity build-up and changes in the rice-wheat system of the Indo-Gangetic Plains. *Exp. Agric.*, 36, 273–84.

Gurr, G. M., Wratten, S. D., and Luna, J. M. (2003). Multi-function agricultural biodiversity: Pest management and other benefits. *Basic and Applied Ecology*, 4(2), 107–116.

Guthrie, R. H. (2002). The effects of logging on frequency and distribution of landslides in three watersheds on Vancouver Island, British Columbia. *Geomorphology*, 43(3), 273–292.

Gutierrez, A. G., Barbosa, O., and Christie, D. A. (2008). DE Lequency and distribution of landslides in three watersheds on Vancouver Island, British Columbia. *Geomorphology*, 43(3), 273–292.

Haberl, H., Beringer, T., Bhattacharya, S. C., Erb, K. H., and Hoogwijk, M. (2010). The global technical potential of bio-energy in 2050 considering sustainability constraints. *Current Opinion in Environmental Sustainability*, 2(5), 394–403.

Haberl, H., Erb, K. H., Krausmann, F., Running, S., Searchinger, T. D., and Smith, W. K. (2013). Bioenergy: How much can we expect for 2050? *Environmental Research Letters*, 8(3), 031004.

Haigh, M. J., Rawat, J. S., Rawat, M. S., Bartarya, S. K., and Rai, S. P. (1995). Interactions between forest and landslide activity along new highways in the Kumaun Himalaya. *Forest Ecology and Management*, 78(1), 173–189.

Halladay, K., Malhi, Y., and New, M. (2012). Cloud frequency climatology at the Andes/Amazon transition. 2. Trends and variability. *J. Geophys. Res.*, 117, D23103, doi:10.1029/2012JD017789.

Hammond, D. S., and ter Steege, H. (1998). Propensity for fire in Guianan rainforests. *Conserv. Biol.* 12, 944–947.

Hannam, K. D., Quideau, S. A., and Kishchuk, B. E. (2007). The microbial communities of aspen and spruce forest floors are resistant to changes in litter inputs and microclimate. *Applied Soil Ecology* 35, 635–647.

Hansen, M. C., Stehman, S. V., Potapov, P. V., Arunarwati, B., Stolle, F., and Pittman, K. (2009). Quantifying changes in the rates of forest clearing in Indonesia from 1990 to 2005 using remotely sensed data sets. *Environmental Research Letters*, 4(3), 034001.

Hansen, M. C. et al. (2008). Humid tropical forest clearing from 2000 to 2005 quantified by using multitemporal and multiresolution remotely sensed data. *Proceedings of the National Academy of Sciences*, 105(27), 9439–9444.

Hansen, M. C. et al. (2013). High-resolution global maps of 21st-century forest cover change. *Science*, 342(6160), 850–853.

Hanson, J. D., Liebig, M. A., Merrill, S. D., Tanaka, D. L., Krupinsky, J. M., and Stott, D. E. (2007). Dynamic cropping systems. *Agronomy Journal*, 99(4), 939–943.

Hargrave, J., and Kis-Katos, K. (2013). Economic causes of deforestation in the Brazilian Amazon: A panel data analysis for the 2000s. *Environmental and Resource Economics*, 54(4), 471–494.

Harr, R. D., and McCorison, F. M. (1979). Initial effects of clearcut logging on size and timing of peak flows in a small watershed in western Oregon, Forest Service Res. Paper PNW-249 Pacific Northwest Forest and Rangeland Experimental Station.

Harrison, A. F. (1987), Mineralisation of organic phosphorus in relation to soil factors, determined using isotopic 32P labeling. In Rowland, A. P. (eds.), *Chemical Analysis in Environmental Research*. NERC/ITE, Abbotts Ripton, pp. 84–87.

Hart, S. C. (1999). Nitrogen transformations in fallen tree boles and mineral soil of an old-growth forest. *Ecology*, 80(4), 1385–1394.

Hartwick, J. M. (1978). Substitution among exhaustible resources and intergenerational equity. *The Review of Economic Studies*, 45(2), 347–354.

Harvey, C. A. et al. (2008). Integrating agricultural landscapes with biodiversity conservation in the Mesoamerican hotspot. *Conserv. Biol.*, 22, 8–15.

Hasler, N., Werth, D., and Avissar, R. (2009). Effects of tropical deforestation on global hydroclimate: A multimodel ensemble analysis. *J. Clim.* 22, 1124–1141.

Hauggaard-Nielsen, H., Ambus, P., and Jensen, E. S. (2001). Interspecific competition, N use and interference with weeds in pea–barley intercropping. *Field Crops Research*, 70(2), 101–109.

Hauggaard-Nielsen, H., and Jensen, E. S. (2005). Facilitative root interactions in intercrops. *Plant and Soil*, 274(1–2), 237–250.

Hayes, T. M. (2006). Parks, people, and forest protection: An institutional assessment of the effectiveness of protected areas. *World Development*, 34(12), 2064–2075.

Hazell, P., and Wood, S. (2008). Drivers of change in global agriculture. *Philosophical Transactions of the Royal Society B: Biological Sciences*, 363(1491), 495–515.

He, Y., De Wekker, S. F., Fuentes, J. D., and D'Odorico, P. (2011). Coupled land-atmosphere modeling of the effects of shrub encroachment on nighttime temperatures. *Agric. For. Meteorol.*, 151, 1690–1697, doi:10.1016/j.agrformet.2011.07.005.

He, Y., D'Odorico, P., DeWekker, S. (2015).The role of vegetation-microclimate feedback in promoting shrub encroachment in the northern Chihuahuan desert. *Global Change Biology*, doi:10.1111/gcb.12856.

Heath, J. A., and Huebert, B. J. (1999). Cloudwater deposition as a source of fixed nitrogen in a Hawaiian montane forest. *Biogeochemistry*, 44, 119–134.

Hecht, S. B. (2005). Soybeans, development and conservation on the Amazon frontier. *Development and Change*, 36(2), 375–404.

Henderson-Sellers, A., and Gornitz, V. (1984). Possible climatic impacts of land cover transformations, with particular emphasis on tropical deforestation. *Clim. Change*, 6, 231–257.

Henrot, J., and Robertson, G. P. (1994). Vegetation removal in two soils of the humid tropics: Effect on microbial biomass. *Soil Biol. Biochem.*, 26(1), 111–116.

Hermele K. (2014). *The Appropriation of Ecological Space. Agrofuels, unequal exchange and environmental load displacements*, Rutledge, New York, pp. 158.

Hese, S. et al. (2005). Global biomass mapping for an improved understanding of the CO_2 balance – the Earth observation mission Carbon-3D. *Remote Sensing of Environment*, 94(1), 94–104.

Hewlett, J. D. (1969). *Principles of Forest Hydrology*. University of Georgia University Press, Athens.

Hietz, P., Turner, B. L., Wanek, W., Richter, A., Nock, C. A., and Wright, S. J. (2011). Long-term change in the nitrogen cycle of tropical forests. *Science*, 334(6056), 664–666.

Hillel, D. (2000). *Salinity Management for Sustainable Irrigation: Integrating Science, Environment and Economics*. The World Bank, Washington, DC.

Hirota, M., Holmgren, M., Van Nes, E. H., and Scheffer, M. (2011). Global resilience of tropical forest and savanna to critical transitions. *Science*, 334(6053), 232–235.

Hobbie, E. A., Weber, N. S., and Trappe, J. M. (2001). Mycorrhizal vs saprotrophic status of fungi: The isotopic evidence. *New Phytologist*, 150(3), 601–610.

Hobbie, S. E. (1992). Effects of plant species on nutrient cycling. *Tree*, 7(10), 336–339.

Hobbie, S. E., and Vitousek, P. M. (2000). Nutrient limitation of decomposition in Hawaiian forests. *Ecology*, 81(7), 1867–1877.

Hobson, K. A., Bayne, E. M., & Van Wilgenburg, S. L. (2002). Large-scale conversion of forest to agriculture in the boreal plains of Saskatchewan. *Conservation Biology*, 16(6), 1530–1541.

Hoekman, D. H. (1997). Radar monitoring system for sustainable forest management in Indonesia IGARSS '97: 1997 Int. Geoscience and Remote Sensing Symposium (Singapore, 3–8 August 1997) pp. 1731–1733, doi:10.1109/IGARSS.1997.609048.

Hoffman, W. A., Schroeder, W., and Jackson, R. B. (2003). Regional feedbacks among fire, climate and tropical deforestation. *Journal of Geophysical Research*, 108(D23), 4721, doi:10.1029/2003JD003494.

Holden, S. T. (1993). Peasant household modelling: Farming systems evolution and sustainability in northern Zambia. *Agricultural Economics*, 9(3), 241–267.

Holder, C. D. (2003). Fog precipitation in the Sierra de las Minas Biosphere Reserve, Guatemala. *Hydrol. Process.*, 17, 2001–2010.

Holder, C. D. (2006). The hydrological significance of cloud forests in the Sierra de las Minas Biosphere Reserve, Guatemala. *Geoforum*, 37(1), 82–93.

Holling, C. S. (1973). Resilience and stability of ecological systems. *Annual Review of Ecology and Systematics*, 4, 1–23.

Hong, Y., Adler, R., and Huffman, G. (2006). Evaluation of the potential of NASA multi-satellite precipitation analysis in global landslide hazard assessment. *Geophysical Research Letters*, 33, L22402, doi:10.1029/2006GL028010.

Hooper, D. U. et al. (2012). A global synthesis reveals biodiversity loss as a major driver of ecosystem change. *Nature*, 486(7401), 105–108.

Hopkins, A. D. (1938), Bioclimatics: A science of life and climate relations. U.S. Dept. of Agriculture.

Horsthemke, W., and Léfèver, R. (1984). *Noise-Induced Transitions: Theory and Applications in Physics, Chemistry, and Biology*. Springer, Berlin.

Hosonuma, N. et al. (2012). An assessment of deforestation and forest degradation drivers in developing countries. *Environmental Research Letters*, 7(4), 044009.

Houghton R. A. (1996). Land-use change and terrestrial carbon: The temporal record. In Apps M. J., and D. T. Price (eds.), *Forest Ecosystems, Forest Management and the Global Carbon Cycle*. NATO ASI Series I, Vol. 40, Springer-Verlag, Berlin, Heidelberg, pp. 117–134.

Houghton, R. A. (1999). The annual net flux of carbon to the atmosphere from changes in land use 1850–1990. *Tellus B*, 51(2), 298–313.

Houghton, R. A. (2002). Temporal patterns of land-use change and carbon storage in China and tropical Asia. *Science in China Series C Life Sciences-English Edition*, 45(Suppl), 10–17.

Houghton, R. A. (2003). Revised estimates of the annual net flux of carbon to the atmosphere from changes in land use and land management 1850–2000. *Tellus B*, 55, 378–390, doi:10.1034/j.1600-0889.2003.01450.x.

Houghton, R. A. (2005). Aboveground forest biomass and the global carbon balance. *Global Change Biology*, 11(6), 945–958.

Houghton, R. A., Lefkowitz, D. S., and Skole, D. L. (1991). Changes in the landscape of Latin America between 1850 and 1985. I. Progressive loss of forests. *For. Ecol. Manag.*, 38, 143–72.

Hudak, A. T. et al. (2012). Quantifying aboveground forest carbon pools and fluxes from repeat LiDAR surveys. *Remote Sensing of Environment*, 123, 25–40.

Huenneke, L. F., Hamburg, S. P., Koide, R., Mooney, H. A., and Vitousek, P. M. (1990). Effects of soil resources on plant invasion and community structure in Californian serpentine grassland. *Ecology*, 71, 478–491.

Hughes, M. A., and Dunn, M. A. (1996). The molecular biology of plant acclimation to low temperature. *Journal of Experimental Botany*, 47, 291–305.

Hulm, S. C., and Killham, K. (1988). Gaseous nitrogen losses from soil under Sitka spruce following the application of fertilizer 15N urea. *Journal of Soil Science*, 39(3), 417–424.

Hurtt, G. C. et al. (2006). The underpinnings of land-use history: Three centuries of global gridded land-use transitions, wood-harvest activity, and resulting secondary lands. *Global Change Biology*, 12(7), 1208–1229.

Hyde, W. F., Belcher, B. M., and Xu, J. (eds.). (2003). *China's forests: Global lessons from market reforms. Resources for the Future*. and CIFOR, Washington, DC.

IEA. (2009). *Bioenergy: A Sustainable and Reliable Energy Source*. Main Report. Paris: International Energy Agency.

IEA. (2011). *Biofuels for Transport*. International Energy Agency, Paris.

IEA. (2012). *World Energy Outlook*. The International Energy Agency, Paris.

Ikerd, J. E. (1991). A decision support system for sustainable farming. *Northeastern J. Agricult. Resource Economics*, 20, 109–113.

Ingebo, P. A. (1971). Suppression of channel-side chaparral cover increases streamflow. *J. Soil Water Conserv.*, 26, 79–81.

INPE (2014). Annual Amazonian deforestation rates, available at: http://www.obt.inpe.br/prodes/index.php (accessed on March 23, 2015).

International Groundwater Resources Assessment Centre (IGARC). (2015). Saline and brackish groundwater by genesis, Available at: http://www.un-igrac.org/publications/344 (accessed on June 25, 2015).

IPCC. (2000). Land-use, land-use change and forestry: Summary for policymakers, available at: http://www.ipcc.ch/ipccreports/sres/land_use/index.php?idp=1 (accessed on June 25, 2015).

IPCC.(2014). Summary for Policymakers. In Field, C. B. (eds.), *Climate Change 2014: Impacts, Adaptation, and Vulnerability. Part A: Global and Sectoral Aspects*. Contribution of

Working Group II to the Fifth Assessment Report of the Intergovernmental Panel on Climate Change. Cambridge University Press, Cambridge and New York, pp. 1–32.

IUCN. (2014). The IUCN red list of threatened species, available at: http://www.iucnredlist.org. (Accessed on June 1, 2015).

Jactel, H., and Brockerhoff, E. G. (2007). Tree diversity reduces herbivory by forest insects. *Ecology Letters*, 10(9), 835–848.

Jakob, M. (2000). The impacts of logging on landslide activity at Clayoquot Sound, British Columbia. *Catena*, 38(4), 279–300.

Jakobsen, I., Chen, B. D., Munkvold, L., Lundsgaard, T., and Zhu, Y. G. (2005). Contrasting phosphate acquisition of mycorrhizal fungi with that of root hairs using root hairless barley mutant. *Plant Cell Environ.*, 28, 928–938.

Janos, D. P. (1988). Mycorrhiza applications in tropical forestry: Are temperate-zone approaches appropriate? In Ng, S. P. (ed.), *Trees and Mycorrhiza*. Forest Research Institute, Kuala Lumpur, Malaysia, pp. 133–188

Janzen, D. H. (1988). *Tropical Dry Forests: The Most Endangered Major Tropical Ecosystem.* in *Biodiversity*, EO Wilson (ed.), Washington, D.C., National Academy Press, pp 130-137.

Jenkins, M. (2003). Prospects for biodiversity. *Science*, 302(5648), 1175–1177.

Jetz, W., Wilcove, D. S., and Dobson, A. P. (2007). Projected impacts of climate and land-use change on the global diversity of birds. *PLoS Biology*, 5(6), e157.

Jipp, P. H., Nepstad, D. C., Cassel, D. K., and Reis De Carvalho, C. (1998). Deep soil moisture storage and transpiration in forests and pastures of seasonally-dry Amazonia. *Climatic Change*, 39, 2–3, 395–412.

Joffre, R., and Rambal, S. (1988). Soil-water improvement by trees in the rangelands of Southern Spain. *Acta Oecologica – Oecologia Plantarum*, 9, 405–422.

Johnson, D. W., and Henderson, P. (1995). Effects of forest management and elevated carbon dioxide on soil carbon storage. *Soil Management and the Greenhouse Effects*, 137–145.

Jones, J. A. (2000). Hydrologic processes and peak discharge response to forest removal, regrowth, and roads in 10 small experimental basins, western Cascades, Oregon. *Water Resour. Res.*, 36(9), 2621–2642.

Jones, D. L., and Oburger, E. (2011). Solubilization of phosphorus by soil microorganisms. In Buenemann, E. K., A. Oberson, and E. Frossard (eds.), *Phosphorus in Action: Biological Processes in Phosphorus Cycling, Soil Biology. 26.* Springer, Heidelberg, 169–198.

Jones, D. T., Sah, J. P., Ross, M. S., Oberbauer, S. F., Hwang, B., and Jayachandran, K. (2006). Growth and physiological responses of twelve tree species common in Everglades tree islands to simulated hydrologic regimes. *Wetlands*, 26, 830–844.

Jorgenson, M. T., Racine, C. H., Walters, J. C., and Osterkamp, T. E. (2001). Permafrost degradation and ecological changes associated with a warming climate in Central Alaska. *Climatic Change*, 48, 551–579.

Ju, X. T. et al. (2009). Reducing environmental risk by improving N management in intensive Chinese agricultural systems. *Proceedings of the National Academy of Sciences*, 106(9), 3041–3046.

Juang, J.-Y. et al. (2007). Hydrologic and atmospheric controls on initiation of convective precipitation events. *Water Resour. Res.*, 43, W03421, doi:10.1029/2006WR004954.

Jull, C. (ed.) (2007). Recent trends in the law and policy of bioenergy production, promotion and use (No. 95). Food & Agriculture Organization, Rome.

Kaimowitz, D., Mertens, B., Wunder, S., and Pacheco, P. (2004). Hamburger connection fuels Amazon destruction. Centre for International Forestry Research, Bogor, Indonesia.

Kaiser, B., and Roumasset, J. (2002). Valuing indirect ecosystem services: the case of tropical watersheds. *Environment and Development Economics*, 7(04), 701–714.

Kallio, A., and Rieger, S. (1969). Recession of permafrost in a cultivated soil of interior Alaska. *Soil Sci. Soc. Am. Proc.* 33, 430–432.

Kaplan, J. O., Krumhardt, K. M., & Zimmermann, N. (2009). The prehistoric and preindustrial deforestation of Europe. *Quaternary Science Reviews*, 28(27), 3016–3034.

Karlen, D. L., Varvel, G. E., Bullock, D. G., and Cruse, R. M. (1994). Crop rotations for the 21st century. *Advances in Agronomy*, 53, 1–45.

Karp, A., and Shield, I. (2008). Bioenergy from plants and the sustainable yield challenge. *New Phytologist*, 179(1), 15–32.

Kasischke, E. S., Melack, J. M., Dobson, M. C. (1997). The use of imaging radars for ecological applications – a review. *Remote Sensing of Environment*, 59, 141–156.

Kastner, T., Rivas, M. J. I., Koch, W., and Nonhebel, S. (2012). Global changes in diets and the consequences for land requirements for food. *Proc Natl Acad Sci USA*, 109(18): 6868–6872.

Katul, G. G., and Novick, K. A. (2009). Evapotranspiration. In Likens, G. E. (ed.), *Encyclopedia of Inland Waters*. Elsevier, Oxford, UK, pp. 661–667. doi:10.1016/B978-012370626-3.00012-0.

Katul, G. G., Oren, R., Manzoni, S., Higgins, C., and Parlange, M. B. (2012). Evapotranspiration: A process driving mass transport and energy exchange in the soil-plant-atmosphere-climate system. *Rev. Geophys.*, 50(3), RG3002, doi:10.1029/2011RG000366.

Kauffman, J. B., Cummings, D. L., and Ward, D. E. (1994). Relationships of fire, biomass and nutrient dynamics along a vegetation gradient in the Brazilian cerrado. *Journal of Ecology*, 82(3), 519–531.

Kauffman, J. B., Sanford Jr., R. L., Cummings, D. L., Salcedo, I. H., and Sampaio, E. V. (1993). Biomass and nutrient dynamics associated with slash fires in neotropical dry forests. *Ecology*, 74(1), 140–151.

Kazianga, H., and Masters, W. A. (2006). Property rights, production technology, and deforestation: Cocoa in Cameroon. *Agricultural Economics*, 35(1), 19–26.

Kellman, M. (1979). Soil enrichment by Neotropical savanna trees. *J. Ecol.*, 67, 565-577

Kendall C., and McDonnell, J. J (eds.). (1998). *Isotope Tracers in Catchment Hydrology*. Elsevier Science B.V., Amsterdam.

Kennedy, G., Nantel, G., and Shetty, P. (2004). Globalization of food systems in developing countries: Impacts on food security and nutrition. FAO Food and Nutrition Paper, 83, 1–26.

Khanna, P. K., Raison, R. J., and Falkiner, R. A. (1994). Chemical properties of ash derived from Eucalyptus litter and its effects on forest soils. *Forest Ecology Management*, 66, 107–125.

Kinzig, A. P., Perrings, C., Chapin, F. S., Polasky, S., Smith, V. K., Tilman, D., and Turner, B. L. (2011). Paying for ecosystem services promise and peril. *Science*, 334(6056), 603–604.

Kirschbaum, M. U. (1995). The temperature dependence of soil organic matter decomposition, and the effect of global warming on soil organic C storage. *Soil Biology and Biochemistry*, 27(6), 753–760.

Kleidon, A., Fraedrich, K., and Heimann, M. (2000). A green planet versus a desert world: Estimating the maximum effect of vegetation on the land surface climate. *Climat. Change*, 44, 471–493.

Kleidon, A., and Heimann, M. (1999). Deep-rooted vegetation, Amazonian deforestation, and climate: Results from a modeling study. *Global Ecol. Biogeogr.*, 8, 397–405.

Kleinen, T., Held, H., and Petschel-Held, G. (2003). The potential role of spectral properties in detecting thresholds in the Earth system: Application to the thermohaline circulation. *Ocean Dynam.*, 53, 53–63.

Kleinman, P., Pimentel, D., and Bryant, R. (1996), The ecological sustainability of slash-and-burn agriculture. *Agric. Ecosyst. Environ.*, 52, 235–249.

Klepeis, P. (2003). Development policies and tropical deforestation in the southern Yucatan peninsula: Centralized and decentralized approaches. *Land Degradation and Development*, 14(6), 541–561.

Knight, R. (2015). Balancing the numbers: Using grassroots land valuation to empower communities in land investment negotiations, 2015 Land and Poverty Conference. World Bank, Washington, DC.

Knowles, R. (1982). Denitrification. *Microbiological Reviews*, 46(1), 43.

Koh, L. P. (2008). Birds defend oil palms from herbivorous insects. *Ecol. Appl.*, 18, 821–825.

Koh, L. P., Miettinen, J., Liew, S. C., and Ghazoul, J. (2011). Remotely sensed evidence of tropical peatland conversion to oil palm. *Proceedings of the National Academy of Sciences*, 108(12), 5127–5132.

Koh, L. P., and Wilcove, D. S. (2008). Is oil palm agriculture really destroying tropical biodiversity? *Conservation Letters*, 1(2), 60–64.

Koop, G., and Tole, L. (1999). Is there an environmental Kuznets curve for deforestation? *Journal of Development Economics*, 58(1), 231–244.

Körner, C. (1998). A re-assessment of high elevation treeline positions and their explanation. *Oecologia*, 115, 445–459.

Körner, C. (2006). Plant CO_2 responses: An issue of definition, time and resource supply. *New Phytologist*, 172(3), 393–411.

Körner, C., and Paulsen, J. (2004). A world-wide study of high altitude treeline temperatures. *Journal of Biogeography*, 31, 713–732.

Kozlowski, T. T. (1965). Responses of woody plants to flooding. In Kozlowski, T. T. (ed.), *Flooding and Plant Growth*. Elsevier, New York, pp. 129–163.

Krammers, J. S., and DeBano, L. F. (1965). Soil wettability: A neglected factor in watershed management. *Water Resources Research*, 1, 283–286.

Krause, G. H. (1994). *Photoinhibition Induced by Low Temperatures. Photoinhibition of Photosynthesis: From Molecular Mechanisms to the Field*, edited by N. R. Baker and J. R. Bowyer. BIOS Scientific, Oxford, pp. 331–348.

Krauss, K. W., Lovelock, Catherine E., McKee, Karen L., López-Hoffman, L., Ewe, S. M. L., and Sousa, W. P. (2008). Environmental drivers in mangrove establishment and early development: A review. *Aquatic Botany*, 89, 105–127.

Kremen, C. et al. (2000). Economic incentives for rain forest conservation across scales. *Science*, 288(5472), 1828–1832.

Kress, M. R., Graves, M. R., and Bourne, S. G. (1996). Loss of bottomland hardwood forests and forested wetlands in the Cache River basin, Arkansas. *Wetlands* 16, 258–263.

Krupinsky, J. M., Bailey, K. L., McMullen, M. P., Gossen, B. D., and Turkington, T. K. (2002). Managing plant disease risk in diversified cropping systems. *Agronomy Journal*, 94(2), 198–209.

Kullman, L. (1988). Holocene history of the forest – Alpine tundra ecotone in the Scandes Mountains (central Sweden). *New Phytologist*, 108, 101–110.

Kumari, K. (1994). Sustainable forest management in Peninsular Malaysia: Towards a total economic valuation, Doctoral dissertation, University of East Anglia.

Kummu, M., de Moel, H., Porkka, M., Siebert, S., Varis, O., and Ward, P. J. (2012). Lost food, waste resources: Global food supply chain losses and their impacts on freshwater, cropland, and fertilizer use. *Sci. Total Environ.*, 438, 477–489.

Kuriakose, S. L., van Beek, L. P. H, and van Westen, C. J. (2009). Parameterizing a physically based shallow landslide model in a data poor region. *Earth Surface Processes and Landforms*, 34(6), 867–881.

Kuznetsov, Y. A. (1995). *Elements of applied bifurcation theory*. Springer-Verlag, New York.

Lade, S. J., Tavoni, A., Levin, S. A., and Schlüter, M. (2013). Regime shifts in a social-ecological system. *Theoretical Ecology*, 6(3), 359–372.

Lamarque, J. F. et al. (2005). Assessing future nitrogen deposition and carbon cycle feedback using a multimodel approach: Analysis of nitrogen deposition. *Journal of Geophysical Research: Atmospheres* (1984–2012), 110(D19).

Lambers, H., Chapin III, F. S., and Pons, T. L. (2008b). Interactions among Plants. In *Plant Physiological Ecology*. Springer, New York, pp. 505–531.

Lambers, H., Raven, J. A., Shaver, G. R., and Smith, S. E. (2008a). Plant nutrient-acquisition strategies change with soil age. *Trends in Ecology & Evolution*, 23(2), 95–103.

Lambin, E. F. (1994). Modeling deforestation processes: A review. TREES (Tropical Ecosystem Environment Observation by Satellites). Research Report 1. European Commission Joint Research Centre/European Space Agency, Brussels.

Lambin, E. F., and Geist, H. J. (2003). Regional differences in tropical deforestation. *Environment: Science and Policy for Sustainable Development*, 45(6), 22–36.

Lambin, E. F., Geist, H. J., and Lepers, E. (2003). Dynamics of land-use and land-cover change in tropical regions. *Annu. Rev. Environ. Resour.*, 28, 205–241.

Lambin, E. F., Rounsevell, M. D. A., and Geist, H. J. (2000). Are agricultural land-use models able to predict changes in land-use intensity? *Agriculture, Ecosystems and Environment*, 82(1), 321–331.

Lane, J. (2013). IEA says cellulstic biofuels capacity has tripled since 2010, International Energy Agency, New Task 39 Global report, Biofuels Digest, available at: www.biofuelsdigest.com.

Langford, A. O., Fehsenfeld, F. C., Zachariassen, J., and Schimel, D. S. (1992). Gaseous ammonia fluxes and background concentrations in terrestrial ecosystems of the United States. *Global Biogeochemical Cycles*, 6(4), 459–483.

Langvall, O., and Örlander, G. (2001). Effects of pine shelterwoods on microclimate and frost damage to Norway spruce seedlings. *Canadian Journal of Forest Research*, 31, 155–164.

Langvall, O., and Ottosson Löfvenius, M. (2002). Effect of shelterwood density on nocturnal near-ground temperature, frost injury risk and budburst date of Norway spruce. *Forest Ecology and Management*, 168, 149–161.

Lapola, D. M., Schaldach, R., Alcamo, J., Bondeau, A., Koch, J., Koelking, C., and Priess, J. A. (2010). Indirect land-use changes can overcome carbon savings from biofuels in Brazil. *Proceedings of The national Academy of Sciences*, 107(8), 3388–3393.

Larcher, W. (1995). *Physiological Plant Ecology*, 3rd ed. Springer, Berlin.

Larsen, J. A. (1980). *The Boreal Ecosystem*. Academic Press, New York.

Laurance, W. F., Albernaz, A. K., Schroth, G., Fearnside, P. M., Bergen, S., Venticinque, E. M., & Da Costa, C. (2002). Predictors of deforestation in the Brazilian Amazon. *Journal of Biogeography*, 29(5–6), 737–748.

Laurance, W. F., Cochrane, M. A., Bergen, S., Fearnside, P. M., Delamônica, P., Barber, C., ... & Fernandes, T. (2001). The future of the Brazilian Amazon. *Science (Washington)*, 291(5503), 438–439.

Laurance, W. F. et al. (2012). Averting biodiversity collapse in tropical forest protected areas. *Nature*, 489(7415), 290–294.

Lawrence, D., D'Odorico, P., Diekmann, L., DeLonge, M., Das, R., and Eaton, J. (2007). Ecological feedbacks following deforestation create the potential for a catastrophic ecosystem shift in tropical dry forest. *Proc. Natnl Acad. Sci, USA, PNAS*, 104(52), 20696–20701.

Lawrence, D., and Schlesinger, W. H. (2001). Changes in soil phosphorus during 200 years of shifting cultivation in Indonesia. *Ecology*, 82, 2769–2780.

Lawrence, D., and Vandecar, K. (2015). Effects of tropical deforestation on climate and agriculture. *Nature Climate Change*, 5(1), 27–36.

Lawton, J. H., and Brown, V. K. (1993). Redundancy in ecosystems. In Schulze, E. D. and H. A. Mooney (eds.), *Biodiversity and Ecosystem Function*. pp. 255–270, Springer, Berlin.

Lawton, R. O., Nair, U. S., Pielke Sr., R. A., and Welch, R. M. (2001). Climatic impact of tropical lowland deforestation on nearby montane cloud forests. *Science*, 294, 584–587.

Le Maitre, D. C., Scott, D. F., and Colvin, C. (1999). A review of information on interactions between vegetation and groundwater. – *Water SA*, 25(2), 137–152.

Le Toan, T. et al. (2011). The BIOMASS mission: Mapping global forest biomass to better understand the terrestrial carbon cycle. *Remote Sensing of Environment*, 115(11), 2850–2860.

Lean, J., and Rowntree, P. R. (1999). Correction note on "Understanding the sensitivity of a GCM simulation of Amazonian deforestation to the specification of vegetation and soil characteristics." *J. Clim.*, 12, 1549–1551.

Lean, J., and Warrilow, D. A. (1989). Simulation of the regional climatic impact of Amazon deforestation. *Nature*, 342, 411–413.

LeBauer, D. S., and Treseder, K. K. (2008). Nitrogen limitation of net primary productivity in terrestrial ecosystems is globally distributed. *Ecology*, 89(2), 371–379.

Lee, J. E., Lintner, B. R., Boyce, C. K., and Lawrence, P. J. (2011). Land use change exacerbates tropical South American drought by sea surface temperature variability. *Geophys. Res. Lett.*, 38, L19706, doi:10.1029/2011GL049066.

Lee, J. E. et al. (2012). Reduction of tropical land region precipitation variability via transpiration. *Geophys. Res. Lett.*, 39, L19704, doi:10.1029/2012GL053417.

Lee, R. (1978). *Forest Microclimatology*. Columbia University Press, New York.

Lee, R. (1980). *Forest Hydrology*. Columbia University Press, New York.

Lee, R. (2011). The outlook for population growth. *Science*, 333, 569–573.

Leighton, M., and Wirawan, N. (1986). Catastrophic drought and fire in Borneo tropical rain forest associated with the 1982–83 El Nino southern oscillation event, 75–102. In Prance, G. T. (ed.) Tropical rain forests and the world atmosphere. AAAS Selected Symposium 101. West-view, Boulder, CO.

Lemus, R., and Lal, R. (2005). Bioenergy crops and carbon sequestration. *Critical Reviews in Plant Sciences*, 24(1), 1–21.

Leopoldo, P. R., Chaves, J. G., and Franken, W. K. (1993). Solar energy budgets in central Amazonian ecosystems: A comparison between natural forest and bare soil areas. *Forest Ecology and Management*, 59(3–4), 313–328.

Letey, J. (2001). Causes and consequences of fire-induced soil water repellency. *Hydrol. Process.*, 15, 2867–2875.

Lewis, S. L. et al. (2013). Above-ground biomass and structure of 260 African tropical forests. *Philosophical Transactions of the Royal Society B: Biological Sciences*, 368(1625), 20120295.

Li, Y., Zhao, M., Motesharrei, S., Mu, Q., Kalnay E., and Li S. (2015). Local cooling and warming effects of forests based on satellite observations. *Nature Communications*, 2015; 6: 6603 DOI: 10.1038/ncomms7603.

Liebman, M., and Altieri, M. A. (1986). *Insect, weed and plant disease management in multiple cropping systems*. MacMillan, New York.

Lieffers, V. J., and Rothwell, R. L. (1986). Effects of depth of water table and substrate temperature on root and top growth of Picea mariana and Larix lancina seedlings. *Canadian Journal of Forest Research*, 16(6), 1201–1206.

Lieth, H., and Werger, M. J. A. (1989). *Ecosystems of the world. 14B. Tropical Rain Forest Ecosystems: Biogeographical and Ecological Studies*. Elsevier, Amsterdam.

Likens, G. E., Bormann, F. H., and Johnson, N. M. (1969). Nitrification: Importance to nutrient losses from a cutover forested ecosystem. *Science*, 163(3872), 1205–1206.

Likens, G. E., Bormann, F. H., Johnson, N. M., Fisher, D. W., and Pierce, R. S. (1970). Effects of forest cutting and herbicide treatment on nutrient budgets in the Hubbard Brook watershed-ecosystem. *Ecological Monographs*, 40(1), 23–47.

Lima, M., Skutsch, M., and de Medeiros Costa, G. (2011). Deforestation and the social impacts of soy for biodiesel: Perspectives of farmers in the south Brazilian Amazon. *Ecology and Society*, 16(4), 04, available at: http://dx.doi.org/10.5751/ES-04366-160404.

Linde, M., Galbe, M., and Zacchi, G. (2008). Bioethanol production from non-starch carbohydrate residues in process streams from a dry-mill ethanol plant. *Bioresource Technology*, 99(14), 6505–6511.

Lindenmayer, D. B., and Franklin, J. F. (2002). *Conserving Forest Biodiversity: A Comprehensive Multiscaled Approach*. Island Press, Washington, DC.

Lindenmayer, D. B., Franklin, J. F., and Fischer, J. (2006). General management principles and a checklist of strategies to guide forest biodiversity conservation. *Biological conservation*, 131(3), 433–445.

Linn, D. M., and Doran, J. W. (1984). Effect of water-filled pore space on carbon dioxide and nitrous oxide production in tilled and nontilled soils. *Soil Science Society of America Journal*, 48(6), 1267–1272.

Liscow, Z. D. (2013). Do property rights promote investment but cause deforestation? Quasi-experimental evidence from Nicaragua. *Journal of Environmental Economics and Management*, 65(2), 241–261.

Liski, J., Ilvesniemi, H., Mäkelä, A., and Starr, M. (1998). Model analysis of the effects of soil age, fires and harvesting on the carbon storage of boreal forest soils. *European Journal of Soil Science*, 49(3), 407–416.

Liu, W., Meng, F., Zhang, Y., Liu, Y., and Li, H. (2004). Water input from fog drip in the tropical seasonal rain forest of Xishuangbanna, South-West China. *Journal of Tropical Ecology*, 20, 517–524.

Lloyd, A. H., Yoshikawa, K., Fastie, C. L., Hinzman, L., and Fraver, M. (2003). Effects of permafrost degradation on woody vegetation at Arctic treeline on the Seaward Peninsula. *Permafrost and Periglacial Processes*, 14, 93–101.

Loarie, S. R., Lobell, D. B., Asner, G. P., and Field, C. B. (2011). Land-cover and surface water change drive large albedo increases in South America. *Earth Interact*, 15, 1–16, doi:10.1175/2010EI342.1.

Loik, M. E., and Nobel, P. S. (1993). Freezing tolerance and water relations of Opuntia fragilis from Canada and the United States. *Ecology*, 74, 1722–1732.

Lopez, R. (1994). The environment as a factor of production: The effects of economic growth and trade liberalization. *Journal of Environmental Economics and Management*, 27(2), 163–184.

Lovett, G. M. (1994). Atmospheric deposition of nutrients and pollutants in North America: An ecological perspective. *Ecological Applications*, 4(4), 629–650.

Lovett, G. M., and Kinsman, J. D. (1990). Atmospheric pollutant deposition to high-elevation. *Atmos. Environ.* 24(11), 2767–2786.

Lovett, G. M., and Reiners, W. A. (1986). Canopy structure and cloud water deposition in subalpine coniferous forests. *Tellus TELLAL,* 38B (5), 319–327.

Lu, D., Batistella, M., and Moran, E. (2005). Satellite estimation of aboveground biomass and impacts of forest stand structure. *Photogrammetric Engineering & Remote Sensing*, 71(8), 967–974.

Luck, G. W., Daily, G. C., and Ehrlich, P. R. (2003). Population diversity and ecosystem services. *Trends in Ecology and Evolution*, 18(7), 331–336.

Lundmark, T., and Hällgren, J.-E. (1987). Effects of frost on shaded and exposed spruce and pine seedlings planted in the field. *Canadian Journal of Forest Research*, 17, 1197–1201.

Luo, Y. et al. (2004). Progressive nitrogen limitation of ecosystem responses to rising atmospheric carbon dioxide. *Bioscience*, 54(8), 731–739.

Luyssaert, S. et al. (2007). CO_2 balance of boreal, temperate, and tropical forests derived from a global database. *Global Change Biology*, 13(12), 2509–2537.

MacArthur, R. H., and Wilson, E. O. (1967). *The Theory of Island Biogeography*. Princeton University Press, Princeton, NJ.

MacDonald, G. E. (2004). Cogongrass (Imperata cylindrica) – biology, ecology, and management. *Critic. Rev. Plant Sci.* 23, 367–380.

Machimura, T., Kobayashi, Y., Iwahana, G., Hirano, T., Lopez, L., Fukuda, M., and Fedorov, A. N. (2005). Change of carbon dioxide budget during three years after deforestation in Eastern Siberian Larch Forest. *J. Agric. Meteorol.*, 60(5), 653–656.

Madi, M. A. C. (2004). Financial liberalization and macroeconomic policy options: Brazil, 1994–2003. Campinas: Instituto de Economia da UNICAMP, p. 32, available at: www.eco.unicamp.br/Downloads/Publicacoes/TextosDiscussao/texto117.pdf.

Maertens, M., Zeller, M., and Birner, R. (2006). Sustainable agricultural intensification in forest frontier areas. *Agricultural Economics*, 34(2), 197–206.

Magnusson, T. (1994). Studies of the soil atmosphere and related physical characteristics in peat forest soils. *For. Ecol. Manage.*, 67, 203–224.

Mahar, D. J. (1979). *Frontier Development Policy in Brazil: A Study of Amazonia*. Praeger, New York.

Maher, E. L., and Germino, Matthew J. (2006). Microsite differentiation among conifer species during seedling establishment at alpine treeline. *Ecoscience*, 13, 334–341.

Maher, E. L., Germino, Matthew J., and Hasselquist, N. J. (2005). Interactive effects of tree and herb cover on survivorship, physiology, and microclimate of conifer seedlings at the alpine tree-line ecotone. *Canadian Journal of Forest Research*, 35, 567–574.

Maherali, H., Pockman, W. T., and Jackson, R. B. (2004). Adaptive variation in the vulnerability of woody plants to xylem cavitation. *Ecology*, 85, 2184–2199.

Mahowald, N. et al. (2008). The global distribution of atmospheric phosphorus deposition and anthropogenic impacts. *Global Biogeochem. Cycles*, 22, GB4026, doi:10.1029/2008GB003240.

Malézieux, E. et al. (2009). Mixing plant species in cropping systems: Concepts, tools and models: A review. *Agronomy for Sustainable Development*, 29(1), 43–62.

Malhi, Y., Adu-Bredu, S., Asare, R. A., Lewis, S. L., and Mayaux, P. (2013). African rainforests: Past, present and future. *Philosophical Transactions of the Royal Society B: Biological Sciences*, 368(1625), 20120312.

Malhi, Y., Baldocchi, D. D., and Jarvis, P. G. (1999). The carbon balance of tropical, temperate and boreal forests. *Plant, Cell & Environment*, 22(6), 715–740.

Malhi, Y., and Grace, J. (2000). Tropical forests and atmospheric carbon dioxide. *Trends in Ecology & Evolution*, 15(8), 332–337.

Malhi, Y., Roberts, J. T., Betts, R. A., Killeen, T. J., Li, W., and Nobre, C. A. (2008). Climate change, deforestation and the fate of the Amazon. *Science*, 319, 169–172.

Malhi, Y. et al. (2009). Exploring the likelihood and mechanism of a climate-change-induced dieback of the Amazon rainforest. *PNAS*, 106(49), 20610–20615.

Malingreau, J. P., Stephens, G., and Fellows, L. (1985). Remote sensing of forest fires: Kalimantan and North Borneo in 1982–83, *Ambio*, 14(6), 314–321.

Man, R., and Lieffers, V. J. (1999). Effects of shelterwood and site preparation on microclimate and establishment of white spruce seedlings in a boreal mixedwood forest. *The Forestry Chronicle*, 75, 837–844.

Manzoni, S., Trofymow, J. A., Jackson, R. B., and Porporato, A. (2010). Stoichiometric controls on carbon, nitrogen, and phosphorus dynamics in decomposing litter. *Ecological Monographs*, 80(1), 89–106.

Markewitz, D., Davidson, E. A., Figueiredo, R. D. O., Victoria, R. L., and Krusche, A. V. (2001). Control of cation concentrations in stream waters by surface soil processes in an Amazonian watershed. *Nature*, 410(6830), 802–805.

Marshall, E., Schreckenberg, K., and Newton, A. C. (eds.) (2006). *Commercialization of Non-Timber Forest Products: Factors Influencing Success: Lessons Learned from*

Mexico and Bolivia and Policy Implications for Decision-Makers. UNEP World Conservation Monitoring Centre, Cambridge.

Martinez-Meze, E., and Whitford, W. G. (1996). Stemflow, throughfall and channelization of stemflow by roots in the three Chihuahuan desert shrubs. *J. Arid Environ.*, 32, 271–287.

Martínez-Yrízar, A., Bullock, S. H., Mooney, H. A., and Medina, E. (1995). *Seasonally Dry Tropical Forests.* Cambridge University Press, Cambridge.

Mas, J. F., Sorani, V., and Alvarez, R. (1996). Elaboración de un modelo de simulación del proceso de deforestación., *Investigaciones Geográficas*, 5, 43–57.

Masera, O. R., Díaz-Jiménez, R., and Berrueta, V. (2005). From cookstoves to cooking systems: The integrated program on sustainable household energy use in Mexico. In *Energy for Sustainable Development*, vol. IX(1). IEI, India.

Mather, A. S. (1992). The forest transition. *Area*, 367–379.

Mather, A. S., Needle, C. L., and Fairbairn, J. (1999). Environmental Kuznets curves and forest trends. *Geography*, 55–65.

Matson, P. (1990). Plant-soil interactions in primary succession at Hawaii Volcanoes National Park. *Oecologia*, 85, 241–246.

Matson, P. A., Billow, C., Hall, S., and Zachariassen, J. (1996). Fertilization practices and soil variations control nitrogen oxide emissions from tropical sugar cane. *Journal of Geophysical Research: Atmospheres (1984–2012)*, 101(D13), 18533–18545.

Matson, P. A., McDowell, W. H., Townsend, A. R., and Vitousek, P. M. (1999). The globalization of N deposition: Ecosystem consequences in tropical environments. *Biogeochemistry*, 46(1–3), 67–83.

Matson, P. A., Naylor, R., and Ortiz-Monasterio, I. (1998). Integration of environmental, agronomic, and economic aspects of fertilizer management. *Science*, 280 (5360), 112–115.

Matson, P. A., and Vitousek, P. M. (1981). Nitrogen mineralization and nitrification potentials following clearcutting in the Hoosier National Forest, Indiana. *Forest Science*, 27(4), 781–791.

Matson, P. A., and Vitousek, P. M. (2006). Agricultural intensification: Will land spared from farming be land spared for nature? *Conservation Biology*, 20(3), 709–710.

Matthews, E. (1983). Global vegetation and land use: New high-resolution data bases for climate studies. *Journal of Climate and Applied Meteorology*, 22(3), 474–487.

May, R.M. (1973). *Stability and Complexity in Model Ecosystems.* Princeton University Press, Princeton, New Jersey.

May, R. M. (1977). Thresholds and breakpoints in ecosystems with a multiplicity of stable states. *Nature*, 269, 471–477.

Mayaux, P. et al. (2013). State and evolution of the African rainforests between 1990 and 2010. *Philosophical Transactions of the Royal Society B: Biological Sciences*, 368(1625), 20120300.

McBirney, A. R. (1993). *Igneous Petrology*, 2nd ed. Jones & Bartlett, Boston.

McCann, K. S. (2000). The diversity–stability debate. *Nature*, 405(6783), 228–233.

McDowell, N. et al. (2008). Mechanisms of plant survival and mortality during drought: Why do some plants survive while others succumb to drought? *New Phytologist*, 178(4), 719–739.

McGill, W. B., and Cole, C. V. (1981). Comparative aspects of cycling of organic C, N, S and P through soil organic matter. *Geoderma*, 26, 267–286.

McGrady-Steed, J., and Morin, P. J. (2000). Biodiversity, density compensation, and the dynamics of populations and functional groups. *Ecology*, 81(2), 361–373.

McGroddy, M. E., Silver, W. L., and de Oliveira, R. C. (2004). The effect of phosphorus availability on decomposition dynamics in a seasonal lowland Amazonian forest. *Ecosystems*, 7, 172–179.

McGuffie, K., Henderson-Sellers, A., Zhang, H., Durbridge, T. B., and Pitman, A. J. (1995). Global climate sensitivity to tropical deforestation. *Global Planet. Change*, 10, 97–128.

McGuire, A. D. et al. (2001). Carbon balance of the terrestrial biosphere in the twentieth century: Analyses of CO_2, climate and land use effects with four process-based ecosystem models. *Global Biogeochemical Cycles*, 15(1), 183–206.

McJannet, D., Wallace, J., and Reddell, P. (2007). Precipitation interception in Australian tropical rainforests. II. Altitudinal gradients of cloud interception, stemflow, throughfall and interception. *Hydrological Processes*, 21(13), 1703–1718.

McKee, K. L., and Rooth, J. E. (2008). Where temperate meets tropical: Multi-factorial effects of elevated CO_2, nitrogen enrichment, and competition on a mangrove-salt marsh community. *Global Change Biology*, 14, 971–984.

McNaughton, S. J. (1977). Diversity and stability of ecological communities: A comment on the role of empiricism in ecology. *American Naturalist*, 515–525.

McRae, L., Freeman, R., and Deinet, S. (2014). The Living Planet Index. In McLellan, R., L. Iyengar, B. Jeffries, and N. Oerlemans (eds.), *Living Planet Report 2014: Species and Spaces, People and Places*. WWF, Gland, Switzerland.

Medeiros, J. S., and Pockman, W. T. (2011). Drought increases freezing tolerance of both leaves and xylem of Larrea tridentata. *Plant, Cell and Environment*, 34, 43–51.

Medvigy, D., Walko, R. L., and Avissar, R. (2011). Effects of deforestation on spatiotemporal distributions of precipitation in South America. *J. Clim.*, 24, 2147–2163, doi:10.1175/2010JCLI3882.1.

Megevand, C. (2013). Deforestation trends in the Congo Basin: Reconciling economic growth and forest protection. World Bank Publications.

Meggers, B. J. (1994). Archeological evidence for the impact of Mega-Nino events on Amazonia during the past two millennia. *Clim. Change*, 28, 321–338.

Melillo, J. M. et al. (2009). Indirect emissions from biofuels: How important? *Science*, 326(5958), 1397–1399.

Melo, F. P. L, Pinto, S. R. R., Brancalion, P. H. S., Castro, P. S., Rodrigues, R. R., Aronson, J., and Tabarelli, M. (2013). Priority setting for scaling up tropical forest restoration projects: Early lessons from Atlantic Forest Restoration Pact. *Environmental Science and Policy*, 33, 395–404.

Mendelsohn, R. (1994). Property rights and tropical deforestation. *Oxford Economic Papers*, 750–756.

Mendez, M., and Popkin, B. (2004). Globalization, urbanization and nutritional change in the developing world: Globalization of food systems in developing countries. *Impact on Food Security and Nutrition*, 55–80.

Merriam, R. A. (1973). Fog drip from artificial leaves in a fog wind tunnel. *Water Resources Research*, 9, 1591–1598.

Merry, F. D., Hildebrand, P. E., Pattie, P., & Carter, D. R. (2002). An analysis of land conversion from sustainable forestry to pasture: a case study in the Bolivian Lowlands. *Land Use Policy*, 19(3), 207–215.

Meusburger, K., and Alewell, C. (2008). Impacts of anthropogenic and environmental factors on the occurrence of shallow landslides in an alpine catchment (Urseren Valley, Switzerland). *Natural Hazards and Earth System Science*, 8(3), 509–520.

Meyer, W. B. (1995). Past and present land use and land cover in the USA. *Consequences*, 1(1), 25–33.

Meyfroidt, P., and Lambin, E. F. (2011). Global forest transition: Prospects for an end to deforestation. *Annual Review of Environment and Resources*, 36(1), 343–371.

Meyfroidt, P., Lambin, E. F., Erb, K.-H., and Hertel, T. W. (2013). Globalization of land use: Distant drivers of land change and geographic displacement of land use. *Curr. Opin. Environ. Sustain*, 5, 438–444.

Meyfroidt, P., Rudel, T. K., and Lambin, E. F. (2010). Forest transitions, trade, and the global displacement of land use. *Proc. Natl. Acad. Sci. USA* 107, 20917–20922.

Miettinen, J., Shi, C., and Liew, S. C. (2011). Deforestation rates in insular Southeast Asia between 2000 and 2010. *Global Change Biology*, 17(7), 2261–2270.

Minetti, J. L., and Lamelas, C. M. (1997). Respuesta regional de la soja en Tucuman a la variabilidad clim´atica. *Revista Industrial y Agr´ıcola de Tucuman*, 72, 63–68.

Minetti, J. L., and Vargas, W. M. (1997). Trends and jumps in the annual rainfall in South America, south of 15° S. *Atmosfera*, 11, 205–221.

Minh, D. H. T., Le Toan, T., Rocca, F., Tebaldini, S., d'Alessandro, M. M., and Villard, L. (2014). Relating P-band synthetic aperture radar tomography to tropical forest biomass. *Geoscience and Remote Sensing, IEEE Transactions on*, 52(2), 967–979.

Mitchard, E. T. et al. (2011). Measuring biomass changes due to woody encroachment and deforestation/degradation in a forest–savanna boundary region of central Africa using multi-temporal L-band radar backscatter. *Remote Sensing of Environment*, 115(11), 2861–2873.

Mitchard, E. T. et al. (2013). A novel application of satellite radar data: Measuring carbon sequestration and detecting degradation in a community forestry project in Mozambique. *Plant Ecology & Diversity*, 6(1), 159–170.

Mitsch, W. J., and Gosselink, J. (2000). *Wetlands*, 3rd ed. John Wiley, New York.

Mittermeier, R. A., Myers, N., Thomsen, J. B., Da Fonseca, G. A., and Olivieri, S. (1998). Biodiversity hotspots and major tropical wilderness areas: Approaches to setting conservation priorities. *Conservation Biology*, 12(3), 516–520.

Miyake, S., Renouf, M., Peterson, A., McAlpine, C., and Smith, C. (2012). Land-use and environmental pressures resulting from current and future bioenergy crop expansion: A review. *Journal of Rural Studies*, 28(4), 650–658.

Molotch, N. P., Blanken, P. D., and Link, T. E. (2011). Snow: Hydrological and ecological feedbacks in forests. In Levia, Delphis F., Darryl Carlyle-Moses, and Tadashi Tanaka (eds.), *Forest Hydrology and Biogeochemistry*. Synthesis of Past Research and Future Directions Series: Ecological Studies, Vol. 216. Springer, New York.

Monela, G. C. (1995). Tropical rainforest deforestation, biodiversity benefits and sustainable landuse: Analysis of economic and ecological aspects related to the Nguru Mountains, Tanzania. Doctor Scientiarum Theses (Norway), Agricultural University of Norway.

Montgomery, D. R. (2008). *Dirt: The erosion of civilizations*. University of California Press, Berkeley.

Montgomery, D. R., and Dietrich, W. E. (1994). A physically-based model for the topographic control on shallow landsliding. *Water Resour. Res.*, 30, 1153–1171.

Mora, C., Tittensor, D. P., Adl, S., Simpson, A. G., and Worm, B. (2011). How many species are there on Earth and in the ocean? *PLoS Biology*, 9(8), e1001127.

Morton, D. C. et al. (2006). Cropland expansion changes deforestation dynamics in the southern Brazilian Amazon. *Proceedings of the National Academy of Sciences*, 103(39), 14637–14641.

Mueller, N. D., Gerber, J. S., Johnston, M., Ray, D. K., Ramankutty, N., and Foley, J. A. (2012). Closing yield gaps through nutrient and water management. *Nature*, 490, 254–257, doi:10.1038/nature11420.

Muller, S. W. (1945). Permafrost or perennially frozen ground and related engineering problems, 2nd ed., U.S. Geol. Surv. Spec. Rep. Strategic Eng., Study No. 62.

Muller, S. W. (1947). *Permafrost or permanently frozen ground and related engineering problems*. Edwards, Ann Arbor, MI.

Müller, M. M. L., Guimaraes, M. F., Desjardins, T., and Mitja, D. (2004). The relationship between pasture degradation and soil properties in the Brazilian Amazon: A case study. *Agriculture, Ecosystems and Environment*, 103(2), 279–288.

Mulvaney, T. J. (1851). On the use of self-registering rain and flood gauges in making observations of the relations of rainfall and of flood discharges in a given catchment. Proceedings of the Institute of Civil Engineers of Ireland, 4, pp. 18–31.

Munns, R., and Tester, M. (2008). Mechanisms of salinity tolerance. *Annu. Rev. Plant Biol.*, 59, 651–81.

Murdiyarso, D., Hergoualc'h, K., and Verchot, L. V. (2010). Opportunities for reducing greenhouse gas emissions in tropical peatlands. *Proceedings of the National Academy of Sciences*, 107(46), 19655–19660.

Murphy, P. G., and Lugo, A. E. (1986). Ecology of tropical dry forest. *Annu. Rev. Ecol. Syst.*, 17, 67–88, doi:10.1146/annurev.es.17.110186.000435.

Murray, M. B., Smith, R. I., Friend, A., and Jarvis, P. G. (2000). Effect of elevated [CO_2] and varying nutrient application rates on physiology and biomass accumulation of Sitka spruce (Picea sitchensis). *Tree Physiology*, 20(7), 421–434.

Murty, D., Kirschbaum, M. U., Mcmurtrie, R. E., and Mcgilvray, H. (2002). Does conversion of forest to agricultural land change soil carbon and nitrogen? A review of the literature. *Global Change Biology*, 8(2), 105–123.

Myers, N. (1980). Role of forest farmers in conversion of tropical moist forests. A report prepared for the committee on research priorities in tropical Biology of the National Research Council. National Academy of Sciences, Washington, DC

Myers, N. (1988). Tropical forests: Much more than stocks of wood. *Journal of Tropical Ecology*, 4, 209–21.

Myers, N., Mittermeier, R. A., Mittermeier, C. G., Da Fonseca, G. A., and Kent, J. (2000). Biodiversity hotspots for conservation priorities. *Nature*, 403(6772), 853–858.

Myers, N., and Tucker, R. (1987). Deforestation in Central America: Spanish legacy and North American consumers. *Environmental Review*: 11, 55–71.

Myers, S. S. et al. (2013). Human health impacts of ecosystem alteration. *Proceedings of the National Academy of Science, USA, PNAS*, 110(47), 18753–18760, doi:10.1073/pnas.1218656110.

Myrold, D. D., Matson, P. A., and Peterson, D. L. (1989). Relationships between soil microbial properties and aboveground stand characteristics of conifer forests in Oregon. *Biogeochemistry*, 8(3), 265–281.

Naeem, S. (1998). Species redundancy and ecosystem reliability. *Conservation Biology*, 12(1), 39–45.

Naeem, S., and Li, S. (1997). Biodiversity enhances ecosystem reliability. *Nature*, 390(6659), 507–509.

Naiman, R. J., Decamps, H., and McClain, M. E. (2010). *Riparia: Ecology, Conservation, and Management of Streamside Communities*. Academic Press, New York.

Nair, U. S., Lawton, R. O., Welch, R. M., and Pielke Sr., R. A. (2003). Impact of land use on Costa Rican tropical montane cloud forests: Sensitivity of cumulus cloud field characteristics to lowland deforestation. *J. Geophys. Res.*, 108(D7), 4206, doi:10.1029/2001JD001135.

Nannipieri, P., Giagnoni, L., Landi, L., and Renella, G. (2011). Role of phosphatase enzymes in soil. In Buenemann, E. K., A. Oberson, and E. Frossard (eds.), *Phosphorus in Action: Biological Processes in Phosphorus Cycling*. Soil Biology, 26. Springer, Heidelberg, 215–241.

Nassar, A. M., Alvesde, B. F. T., R., and D., Aguiar, M., Bacchi, R. P., and Adami, M. (2008). Prospects of the sugarcane expansion in Brazil: Impacts on direct and indirect land use changes. In Zuurbier, P., and J. van de Vooren (eds.), *Sugarcane Ethanol: Contributions to Climate Change Mitigation and the Environment*. Wageningen Academic Publishers, Wageningen.

National Research Council (2008). *Hydrologic effects of a changing forest landscape.* National Academies Press, Washington, DC.

Naumburg, E., Mata-Gonzales, R., Hunter, R. G., McLendon, T., and Martin, D. W. (2005). Phreatophytic vegetation and groundwater fluctuations: A review of current research and application of ecosystem response modeling with an emphasis on Great Basin vegetation. *Environ. Manag.*, 35, 726–740.

Návar, J., and Bryan, R. (1990). Interception loss and rainfall redistribution by three semi-arid growing shrubs in northeastern Mexico. *Journal of Hydrology*, 115(1), 51–63.

Naylor, R. L. (1996). Energy and resource constraints on intensive agricultural production. *Annual Review of Energy and the Environment*, 21(1), 99–123.

Naylor, R. (2011). Expanding the boundaries of agricultural development. *Food Sec.*, 3, 233–251.

Negri, A. J., Adler, R. F., Xu, L., and Surratt, J. (2004). The impact of Amazonian deforestation on dry season rainfall. *J. Clim.* 17, 1306–1319.

Nelson, P. N. et al. (2014). Oil palm and deforestation in Papua New Guinea. *Conservation Letters*, 7(3), 188–195.

Nepstad, D. C., Stickler, C. M., and Almeida, O. T. (2006). Globalization of the Amazon soy and beef industries: Opportunities for conservation. *Conserv. Biol.*, 20, 1595–1602.

Nepstad, D. C., Stickler, C. M., Soares-Filho, B., and Merry, F. (2008). Interactions among Amazon land use, forests and climate: Prospects for a near-term forest tipping point. *Phil. Trans. R. Soc. B.*, 363, 1737–1746.

Nepstad, D. C., Tohver, I. M., Ray, D., Moutinho, P., and Cardinot, G. (2007). Mortality of large trees and lianas following experimental drought in an amazon forest. *Ecology*, 88, 2259–2269.

Nepstad, D. C., Uhl, C., and Serrão, E. A. S. (1991). Recuperation of a degraded Amazonian landscape: Forest recovery and agricultural restoration. *Ambio*, 20, 248–255.

Nepstad, D. C. et al. (1994). The role of deep roots in the hydrological and carbon cycles of Amazonian forests and pastures. *Nature*, 372, 666–669.

Nepstad, D. et al. (2009). The end of deforestation in the Brazilian Amazon. *Science*, 326(5958), 1350–1351.

Neumann, R. P., and Hirsch, E. (2000). *Commercialisation of Non-Timber Forest Products: Review and Analysis of Research.* CIFOR, Bogor.

Newmark, W. D. (2001). Tanzanian forest edge microclimatic gradients: Dynamic patterns. *Biotropica*, 33, 2–11.

Nickel, J. L. (1973). Pest situation in changing agricultural systems – a review. *Bulletin of the ESA*, 19(3), 136–142.

Niklaus, P. A., Spinnler, D., and Körner, C. (1998). Soil moisture dynamics of calcareous grassland under elevated CO_2. *Oecologia*, 117(1–2), 201–208.

Nobre, C. A., Sellers, P. J., and Shukla, J. (1991). Amazonian deforestation and regional climate change. *J. Climate*, 4, 957–988.

Norton, T. W. (1996). Conserving biological diversity in Australia's temperate eucalypt forests. *For. Ecol. Manage.*, 85, 21–33.

Noy-Meir, I. (1973). Desert ecosystems: Environment and producers. *Ann. Rev. Ecol. Systemat.*, 4, 25–51.

Noy-Meir, I. (1975). Stability of grazing systems: Application of predator-prey graphs. *J. Ecol.*, 63, 459–481.

Nunez, D., Nahuelhual, L., and Oyarzun, C. (2006). Forests and water: The value of native temperate forests in supplying water for human consumption. *Ecol. Econ.*, 58, 606–616.

Nusslein, K., and Tiedje, J. M. (1999). Soil bacterial community shift correlated with change from forest to pasture vegetation in a tropical soil. *Appl. Environ. Microbiol.*, 65, 3622–3626.

Nye, P. H., and Greenland, D. J. (1964). Changes in the soil after clearing tropical forest. *Plant and Soil*, 21(1), 101–112.

O'Loughlin, C. L., and Pearce, A. J. (1976). Influence of Cenozoic geology on mass movement and sediment yield response to forest removal, North Westland, New Zealand. *Bull. Int. Assoc. Eng. Geol.*, 14, 41–46.

Oberson, A., and Joner, E. J. (2005). Microbial turnover of phosphorus in the soil. In Turner, B. L., E. Frossard, and D. S. Baldwin, (eds.), *Organic Phosphorus in the Environment*. CABIL, Wallingford, pp. 133–164.

Oerke, E. C. (2006). Crop losses to pests. *Journal of Agricultural Science*, 144(01), 31–43.

Oliveira, G. D. T. (2013). Land regularization in Brazil and the global land grab. *Development and Change*, 44(2), 261–283.

Oliveira-Filho, A. T., and Fontes, M. A. L. (2000). Patterns of foristic differentiation among atlantic forests in southeastern Brazil and the infuence of climate. *Biotropica*, 32(4b), 793–810.

Olschewski, R., and Benitez, P. C. (2005). Secondary forests as temporary carbon sinks? The economic impact of accounting methods on reforestation projects in the tropics. *Ecological Economics*, 55(3), 380–394.

Örlander, G. (1993). Shading reduces both vsible and invisible frost damage to Norway spruce seedlings in the field. *Forestry*, 66, 27–36.

Oslisly, R., White, L., Bentaleb, I., Favier, C., Fontugne, M., Gillet, J. F., and Sebag, D. (2013). Climatic and cultural changes in the west Congo Basin forests over the past 5000 years. *Philosophical Transactions of the Royal Society of London B: Biological Sciences*, 368(1625), 20120304.

Osterkamp, T. E. et al. (2000). Observations of Thermokarst and its impact on Boreal Forests in Alaska, U.S.A. *Arct. Antart. Alp. Res.*, 32, 303–315.

Ostrom, E. (1990). *Governing the Commons: The Evolution of Institutions for Collective Action*. Cambridge University Press, Cambridge.

Ouellet, F. (2007). Cold Acclimation and Freezing Tolerance in Plants. *Encyclopedia of Life Sciences*, 1-6, doi: 0.1002/9780470015902.a0020093

Ovington J. D (1983). *Ecosystems of the world. 10. Temperate Broad-Leaved Evergreen Forests*. Elsevier, Amsterdam.

Oyama, M. D., and Nobre, C. A. (2003). A new climate-vegetation equilibrium state for Tropical South America. *Geophysical Research Letters*, 30, 2199, doi:10.1029/2003GL018600.

Palmer M. A., Lee, J., Matthews, J. H., and D'Odorico, P. (2015). Building water security: The role of natural systems. *Science*, 349(6248), 584-585.

Pan, Y. et al. (2011). A large and persistent carbon sink in the world's forests. *Science*, 333(6045), 988–993.

Paquette, A., Bouchard, A., and Cogliastro, A. (2006). Survival and growth of under-planted trees: A meta-analysis across four biomes. *Ecological Applications*, 16, 1575–1589.

Pastor, J., and Post, W. M. (1988). Response of northern forests to CO_2-induced climatic change. *Nature*, 334, 55–58.

Patenaude, G., Hill, R. A., Milne, R., Gaveau, D. L. A., Briggs, B. B. J., and Dawson, T. P. (2004). Quantifying forest above ground carbon content using LiDAR remote sensing. *Remote Sensing of Environment*, 93(3), 368–380.

Pattanayak, S. K., and Kramer, R. A. (2001). Pricing ecological services: Willingness to pay for drought mitigation from watershed protection in eastern Indonesia. *Water Resources Research*, 37(3), 771–778.

Pattanayak, S. K., and Sills, E. O. (2001). Do tropical forests provide natural insurance? The microeconomics of non-timber forest product collection in the Brazilian Amazon. *Land Economics*, 77(4), 595–612.

Paustian, K. A. O. J. H. et al. (1997). Agricultural soils as a sink to mitigate CO_2 emissions. *Soil Use and Management*, 13(s4), 230–244.

Pearce, D., and Brown, K. (eds.) (1994). *The Causes of Tropical Deforestation*. University of British Columbia Press, Vancouver, pp. 2–26

Peck, A. J., and Hurle, D. H. (1973). Chloride balance of some farmed and forested catchments in Southwestern Australia. *Water Resour. Res.*, 9(3), 648–657.

Peck, A. J., and Williamson, D. R. (1987). Effects of forest clearings on groundwater. *J. Hydrol.*, 94, 47–65.

Peng, S., Garcia, F. V., Laza, R. C., Sanico, A. L., Visperas, R. M., and Cassman, K. G. (1996). Increased N-use efficiency using a chlorophyll meter on high-yielding irrigated rice. *Field Crops Research*, 47(2), 243–252.

Pennington R. T., Lavin, M., and Oilveira-Filho, A. (2009). Woody plant diversity, evolution, and ecology in the tropics: Perspectives from seasonally dry tropical forests. *Annu. Rev. Ecol. Evol. Syst.*, 40, 437–457.

Peñuelas, J. et al. (2013). Human-induced nitrogen–phosphorus imbalances alter natural and managed ecosystems across the globe. *Nature Communications*, 4.

Pereira, H. M. et al. (2010). Scenarios for global biodiversity in the 21st century. *Science*, 330(6010), 1496–1501.

Perez, D., and Gonzáles-Lelong, A. (2003). Estimacíon de rendimiento y produccíon de soja y maíz en Tucuman. Campaña 2001–2002. *Avance Agroindustrial* 23, 19–23.

Perez, D., Gonzales Lelong, A. & Devani, M. (2002) Evolucion de algunos aspectos economico-productivos de la producci´on de soja en la ultima decada en la provincia de Tucuman. *Avance Agroindustrial*, 22, 31–34.

Perfecto, I., Mas, A., Dietsch, T., and Vandermeer, J. (2003). Conservation of biodiversity in coffee agroecosystems: A tri-taxa comparison in southern Mexico. *Biodivers. Conserv.*, 12, 1239–1252.

Perfecto, I., and Vandermeer, J. (2008). Biodiversity conservation in tropical agroecosystems. *Ann. N.Y. Acad. Sci.*, 1134, 173–200.

Perrin, R. M. (1977). The role of environmental diversity in crop protection. *Prot. Ecology*, 2, 77–114.

Perry, D. A., Amaranthus, M. P., Borchers, J. G., Borchers, S. L., and Brainerd, R. E. (1989). Bootstrapping in ecosystems. *Bioscience*, 39(4), 230–237.

Perry D. A., Orem, R., and Hart, S. C. (2008). *Forest Ecosystems*. J. Hopkins University Press, Baltimore.

Perry, D. A., Molina, R., and Amaranthus, M. P. (1987). Mycorrhizae, mycorrhizospheres, and reforestation: Current knowledge and research needs. *Can. J. For. Res.*, 17, 929–940.

Persha, L., Agrawal, A., and Chhatre, A. (2011). Social and ecological synergy: Local rulemaking, forest livelihoods, and biodiversity conservation. *Science*, 331(6024), 1606–1608.

Petheram, C., Zhang, L., Walker, G. R., and Grayson, R. (2000). Towards a framework for predicting impacts of land-use on recharge. 1. A review of recharge studies in Australia, *Aust. J. Soil Res.*, 40, 397–417.

Petraitis, P. (2013). *Multiple Stable States in Natural Ecosystems*. Oxford University Press, Oxford.

Pfaff, A., and Walker, R. (2010). Regional interdependence and forest "transitions": Substitute deforestation limits the relevance of local reversals. *Land Use Policy*, 27(2), 119–129.

Pfaff, A. S. (1996). What drives deforestation in the Brazilian Amazon? MIT Joint Program on the Science and Policy of Global Change, pp. 16.

Phelps, J., Carrasco, L. R., Webb, E. L., Koh, L. P., and Pascual, U. (2013). Agricultural intensification escalates future conservation costs. *Proceedings of the National Academy of Sciences*, 110(19), 7601–7606.

Phillips, O. L. et al. (1998). Changes in the carbon balance of tropical forests: Evidence from long-term plots. *Science*, 282, 439–442.

Pielke R. A. et al. (2011). Land use/land cover changes and climate: Modelling analysis and observational evidence, *WIREs Clim. Change*, doi:10.1002/wcc.144.

Pimentel, D., and Kounang, N. (1998). Ecology of soil erosion in ecosystems. *Ecosystems*, 1: 416–426.

Pimm, S. L., and Brooks, T. M. (2000). The sixth extinction: How large, where, and when. In *Nature and Human Society: The Quest for a Sustainable World*. National Academy Press, Washington DC, pp. 46–62.

Pimm, S. L., and Raven, P. (2000). Biodiversity: Extinction by numbers. *Nature*, 403(6772), 843–845.

Pockman, W. T., and Sperry, J. S. (1997). Freezing-induced xylem cavitation and the northern limit of Larrea tridentata. *Oecologia*, 109, 19–27.

Poeplau, C., Don, A., Vesterdal, L., Leifeld, J., Van Wesemael, B. A. S., Schumacher, J., and Gensior, A. (2011). Temporal dynamics of soil organic carbon after land-use change in the temperate zone-carbon response functions as a model approach. *Global Change Biology*, 17(7), 2415–2427.

Poffenberger, M. (1989). The deforestation of South Asia: History of management conflicts. In *Environmental Education and Sustainable Development*. Indian Environmental Soc., New Delhi, pp. 155–172.

Pointing, C. (1991). *A New Green History of the World: The Environment and the Collapse of Great Civilizations*. Penguin Books, New York.

Pommerening, A., and Murphy, S. T. (2004). A review of the history, definitions and methods of continuous cover forestry with special attention to afforestation and restocking. *Forestry*, 77, 27–44.

Pongratz, J., Reick, C., Raddatz, T., and Claussen, M. (2008). A reconstruction of global agricultural areas and land cover for the last millennium. *Global Biogeochemical Cycles*, 22(3).

Popp, J., Lakner, Z., Harangi-Rakos, M., and Fari, M. (2014). The effect of bioenergy expansion: Food, energy, and environment. *Renewable and Sustainable Energy Reviews*, 32, 559–578.

Porporato, A., D'Odorico, P., Laio, F., and Rodriguez-Iturbe, I. (2003). Hydrologic controls on soil carbon and nitrogen cycles. I. Modeling scheme. *Advances in Water Resources*, 26(1), 45–58.

Porter-Bolland, L., Ellis, E. A., Guariguata, M. R., Ruiz-Mallén, I., Negrete-Yankelevich, S., and Reyes-García, V. (2012). Community managed forests and forest protected areas: An assessment of their conservation effectiveness across the tropics. *Forest Ecology and Management*, 268, 6–17.

Poschl, U. et al. (2010). Rainforest Aerosols as Biogenic Nuclei of Clouds and Precipitation in the Amazon. *Science*, 329, 1513–1516.

Potapov, P., Hansen, M. C., Stehman, S. V., Loveland, T. R., and Pittman, K. (2008). Combining MODIS and Landsat imagery to estimate and map boreal forest cover loss. *Remote Sensing of Environment*, 112(9), 3708–3719.

Potter, C. S., Matson, P. A., Vitousek, P. M., and Davidson, E. A. (1996). Process modeling of controls on nitrogen trace gas emissions from soils worldwide. *Journal of Geophysical Research: Atmospheres (1984–2012)*, 101(D1), 1361–1377.

Pounds, J. A., Fogden, M. P. L., and Campbell, J. H. (1999). Biological response to climate change on a tropical mountain. *Nature*, 398, 611–614.

Prandini, L., Guidiini, G., Bottura, J. A., Pancano, W. L., and Santos, A. R. (1977). Behavior of the vegetation in slope stability: A critical review. *Bulletin of Engineering Geology and the Environment*, 16(1), 51–55.

Prather, M., Derwent, R., Ehhalt, D., Fraser, P., Sanhueza, E., and Zhou, X. (1994). Other trace gases and atmospheric chemistry. *Climate Change*, 94, 77–126.

Pressland, A. J. (1976). Soil moisture distribution as affected by throughfall and stemflow in an arid zone shrub community. *Aust. J. Bot.*, 24, 641–649.

Pretty, J. (2008). Agricultural sustainability: Concepts, principles and evidence. *Philosophical Transactions of the Royal Society of London B: Biological Sciences*, 363(1491), 447–465.

Price, T. D. (2000). *Europe's First Farmers*. Cambridge University Press, Cambridge.

Pritchard, S. G., and Rogers, H. H. (2000). Spatial and temporal deployment of crop roots in CO_2-enriched environments. *New Phytologist*, 147(1), 55–71.

Putz, F. E., Redford, K. H. (2010). The importance of defining "forest": Tropical forest degradation, deforestation, long-term phase shifts, and further transitions. *Biotropica*, 42, 10–20.

Qadir, M., Ghafoor, A., and Murtaza, G. (2000). Amelioration strategies for saline soils: A review. *Land. Degrad. Dev.*, 11, 501–21.

Qadir, M., Oster, J. D., Schubert, S., Noble, A. D., and Sahrawat, K. L. (2007). Phytoremediation of Sodic and Saline-Sodic Soils. *Advances in Agronomy*, 96, 197–247.

Raison, R. J., Khanna, P. K., and Woods, P. (1985). Mechanisms of element transfer to the atmosphere during vegetation burning. *Can. J. For. Res.*, 15, 132–140.

Ramankutty, N., and Foley, J. A. (1999). Estimating historical changes in global land cover: Croplands from 1700 to 1992. *Global Biogeochemical Cycles*, 13(4), 997–1027.

Ramankutty, N., Gibbs, H. K., Achard, F., Defries, R., Foley, J. A., and Houghton, R. A. (2007). Challenges to estimating carbon emissions from tropical deforestation. *Global Change Biology*, 13(1), 51–66.

Randerson, J. T., Chapin III, F. S., Harden, J. W., Neff, J. C., and Harmon, M. E. (2002). Net ecosystem production: A comprehensive measure of net carbon accumulation by ecosystems. *Ecological Applications*, 12(4), 937–947.

Rasmussen, M. et al. (2011). An Aboriginal Australian genome reveals separate human dispersals into Asia. *Science*, 334, 94–98.

Rausser, G. C., and Small, A. A. (2000). Valuing research leads: Bioprospecting and the conservation of genetic resource. *J. Polit. Economy*, 108, 173–206.

Ray, D. K., Mueller, N. D., West, P. C., and Foley, J. A. (2013). Yield trends are insufficient to double global crop production by 2050. *PLoS ONE*, 8(6), e66428, doi:10.1371/journal.pone.0066428.

Ray, D. K., Nair, U. S., Lawton, R. O., Welch, R. M., and Pielke Sr., R. A. (2006). Impact of land use on Costa Rican tropical montane cloud forests: Sensitivity of orographic cloud formation to deforestation in the plains. *J. Geophys. Res.*, 111, D02108, doi:10.1029/2005JD006096.

Ray, D. K., Ramankutty, N., Mueller, N. D., West, P. C., and Foley, J. A. (2012). Recent patterns of crop yield growth and stagnation. *Nat. Comm.*, 3, 1293, doi:10.1038/ncomms2296.

Raynor, G. S. (1971). Wind and temperature structure in a coniferous forest and a contiguous field. *Forest Science*, 17, 351–363.

Redfield, G. W. (2002). Atmospheric deposition of phosphorus to the Everglades: Concepts, constraints and published deposition rates for ecosystem management. *Scientific World Journal*, 2, 1843–1873.

Redo, D. J., Grau, H. R., Aide, T. M., and Clark, M. L. (2012). Asymmetric forest transition driven by the interaction of socioeconomic development and environmental heterogeneity in Central America. *Proceedings of the National Academy of Sciences*, 109(23), 8839–8844.

Reed, S. C., Townsend, A. R., Taylor, P. G., and Cleveland, C. C. (2011). Phosphorus cycling in tropical forests growing on highly weathered soils. In Buenemann, E. K.,

A. Oberson, and E. Frossard (eds.), *Phosphorus in Action: Biological Processes in Phosphorus Cycling, Soil Biology*. 26, Springer, Heidelberg, pp. 339–369.

Regan, H. M., Lupia, R., Drinnan, A. N., and Burgman, M. A. (2001). The currency and tempo of extinction. *American Naturalist*, 157(1), 1–10.

REN21. (2008). Renewable 2007 Global Status Report. REN21 Secretariat, Paris and Washington, DC.

REN21. (2013). Renewables 2013 Global Status Report. REN21 Secretariat, Paris, p. 177.

Renaud, V., and Rebetez, M. (2009). Comparison between open-site and below-canopy climatic conditions in Switzerland during the exceptionally hot summer of 2003. *Agricultural and Forest Meteorology*, 149, 873–880.

Rengasamy, P. (2006). World salinization with emphasis on Australia. *J. Exp. Bot.*, 57(5), 1017–1023.

Rengasamy, P., Chittleborough, D., and Helyar, K. (2003). Root-zone constraints and plant-based solutions for dryland salinity. *Plant and Soil*, 257, 249–260.

Renssen, H., Goosse, H., and Fichefet, T. (2003). On the non-linear response of the ocean thermohaline circulation to global deforestation. *Geophysical Research Letters*, 30(2).

Resende, J. C. F., Markewitz, D., Klink, C. A., Bustamante, M. M., and Davidson, E. A. (2010). Phosphorus cycling in a small watershed in the Brazilian Cerrado: Impacts of frequent burning. *Biogeochemistry*, doi:10.1007/s10533-010-9531-5.

RFA (Renewable Fuels Association). (2010). Statistics 2010, available at: http://www.ethanolrfa.org/pages/statistics.

Rice, R. M. (1977). Forest management to minimize landslide risk. *Guidelines for Watershed Management*, 271–287.

Ricker, M., Mendelsohn, R. O., Daly, D. C., and Angeles, G. (1999). Enriching the rainforest with native fruit trees: An ecological and economic analysis in Los Tuxtlas (Veracruz, Mexico). *Ecological Economics*, 31(3), 439–448.

Ricklefs, R. (2008). *The Economy of Nature*. 6th ed., W. H. Freeman, pp. 700

Ridolfi, L., D'Odorico, P., and Laio, F. (2006). Effect of vegetation-water table feedbacks on the stability and resilience of plant ecosystems. *Water Resources Research*, 42, W01201, doi:10.1029/2005WR004444.

Ridolfi, L, D'Odorico, P., and Laio, F. (2007). Vegetation dynamics induced by phreatophyte-water table interactions. *J. Theor. Biol.*, 248, 301–310.

Ridolfi, L., D'Odorico, P., and Laio, F. (2011). *Noise-Induced Phenomena in the Environmental Sciences*. Cambridge University Press, New York.

Ridolfi, L., D'Odorico, P., Laio, F., Tamea, F. S., and Rodriguez-Iturbe, I. (2008). Coupled stochastic dynamics of water table and soil moisture in bare soil conditions. *Water Resour. Res.*, 44, W01435, doi:10.1029/2007WR006707.

Riekerk, H. (1989). Influence of silvicultural practices on the hydrology of pine flatwoods in Florida. *Water Resour. Res.*, 25(4), 713–719.

Rietkerk, M., and van de Koppel, J. (1997). Alternate stable states and threshold effects in semiarid grazing systems. *Oikos*, 79, 69–76.

Robinson, B. E., Holland, M. B., and Naughton-Treves, L. (2013). Does secure land tenure save forests? A meta-analysis of the relationship between land tenure and tropical deforestation. *Global Environmental Change*. 29, 281-293.

Robertson, F. A., Myers, R. J., and Sagna, P. G. (1993). Carbon and nitrogen mineralisation in cultivated and grassland soils in subtropical grassland. *Australian Journal of Soil Research*, 31, 611–619.

Robichaud, P. R., and Hungerford, R. D. (2000). Water repellency by laboratory burning of four northern Rocky Mountain forest soils. *Journal of Hydrology*, 231–232, 207–219.

Rodrigues, A. S., Ewers, R. M., Parry, L., Souza, C., Veríssimo, A., and Balmford, A. (2009). Boom-and-bust development patterns across the Amazon deforestation frontier. *Science*, 324(5933), 1435–1437.

Rodrigues, R. R., Gandolfi, S., Nave, A. G., Aronson, J., Barreto, T. E., Vidal, C. Y., and Brancalion, P. H. S. (2011). Large-scale ecological restoration of high-diversity tropical forests in SE Brazil. *For. Ecol. Manage.*, 261, 1605–1613.

Rodriguez-Iturbe, I., and Rinaldo, A. (1997) *Fractal River Basins: Chance and Self-Organization.* Cambridge Univ. Press, New York.

Roering, J. J., Schmidt, K. M., Stock, J. D., Dietrich, W. E., and Montgomery, D. R. (2003). Shallow landsliding, root reinforcement, and the spatial distribution of trees in the Oregon Coast Range. *Canadian Geotechnical Journal*, 40, 237–253.

Röhrig, E., and Ulrich, B. (1991) *Ecosystems of the world. 7. Temperate Deciduous Forests.* Elsevier, Amsterdam.

Rojstaczer, S., Sterling, S. M., Moore, N. J. (2001). Human appropriation of photosynthesis products. *Science*, 294, 2549–2552.

Rolett, B., and Diamond, J. (2004). Environmental predictors of pre-European deforestation on Pacific islands. *Nature*, 431(7007), 443–446.

Ronco, F. (1970). Influence of high light intensity on survival of planted Engelmann spruce. *Forest Science*, 16, 331–339.

Rosado, B. H. P., and Holder, C. (2012). The significance of leaf water repellency in ecohydrological research: A review. *Ecohydrology*, doi:10.1002/eco.1340.

Rosenfeld, D. (1999). TRMM observed first direct evidence of smoke from forest fires inhibiting rainfall. *Geophysical Res. Lett.* 26(20), 3105–3108.

Rosenqvist, A., Shimada, M., Igarashi, T., Watanabe, M., Tadono, T., and Yamamoto, H. (2003, July). Support to multi-national environmental conventions and terrestrial carbon cycle science by ALOS and ADEOS-II-the Kyoto and Carbon Initiative. In International Geoscience and Remote Sensing Symposium, Vol. 3, pp. III-1471.

Ross, D. J., Kelliher, F. M., and Tate, K. R. (1999). Microbial processes in relation to carbon, nitrogen and temperature regimes in litter and a sandy mineral soil from a central Siberian *Pinus sylvestris* L. forest. *Soil Biology and Biochemistry*, 31(5), 757–767.

Roy, V., Ruel, J. C., and Plamondon, A. C. (2000). Establishment, growth and survival of natural regeneration after clearcutting and drainage on forested wetlands. *Forest Ecology and Management*, 129, 253–267.

Royama, T. (1984). Population dynamics of the spruce budworm Choristoneura fumiferana. *Ecological Monographs*, 429–462.

Royama, T., MacKinnon, W. E., Kettela, E. G., Carter, N. E., and Hartling, L. K. (2005). Analysis of spruce budworm outbreak cycles in New Brunswick, Canada, since 1952. *Ecology*, 86(5), 1212–1224.

Royer, P. D., Cobb, N. S., Clifford, M. J., Huang, C., Breshears, D. D., Adams, H. D., and Villegas, J. C. (2011). Extreme climatic event-triggered overstorey vegetation loss increases understorey solar input regionally: Primary and secondary ecological implications. *Journal of Ecology*, 99, 714–723.

Rozema, J., and Flowers, T. J. (2008). Crops for a salinized world. *Science*, 322(5907), 1478–1480.

Ruben, R., Kruseman, G., and Hengsdijk, H. (1994). Farm household modelling for estimating the effectiviness of price instruments on sustainable land use in the Atlantic Zone of Costa Rica. DLV report No. 4. AB-DLO, Wageningen.

Rudel, T. K. (1998). Is there a forest transition? Deforestation, reforestation, and development. *Rural Sociology*, 63(4), 533–552.

Rudel, T. K. (2005). *Tropical Forests – Regional Paths of Destruction and Regeneration in the Late Twentieth Century.* New York: Columbia University Press, p. 237.

Rudel, T. K. (2012). The human ecology of regrowth in the tropics. *Journal of Sustainable Forestry*, 31, 4–5, 340–354.

Rudel, T. K. (2013). The national determinants of deforestation in sub-Saharan Africa. *Philosophical Transactions of the Royal Society of London B: Biological Sciences*, 368(1625), 20120405.

Rudel, T. K., Coomes, O. T., Moran, E., Achard, F., Angelsen, A., Xu, J., and Lambin, E. (2005). Forest transitions: Towards a global understanding of land use change. *Global Environmental Change*, 15(1), 23–31.

Rudel, T. K., Defries, R., Asner, G. P., and Laurance, W. F. (2009a). Changing drivers of deforestation and new opportunities for conservation. *Conserv. Biol.*, 23, 1396–1405.

Rudel, T. K. et al. (2009b). Agricultural intensification and changes in cultivated areas, 1970–2005. *Proceedings of the National Academy of Sciences*, 106(49), 20675–20680.

Runge-Metzger, A. (1995). Closing the cycle: Obstacles to efficient P management for improved global food security. In Tiessen, H. (ed.), *Phosphorus in the Global Environment: Transfers, Cycles and Management*. Wiley, New York, pp. 27–42.

Runyan, C. W., and D'Odorico, P. (2010). Ecohydrological feedbacks between salt accumulation and vegetation dynamics: The role of vegetation-groundwater interactions. *Water Resources Research*, 46, W11561, doi:10.1029/2010WR009464.

Runyan, C. W., and D'Odorico, P. (2012). Hydrologic controls on phosphorus dynamics: A modeling framework. *Advances in Water Resources*, 35, 94–109.

Runyan, C. W., and D'Odorico, P. (2013). Positive feedbacks and bistability associated with phosphorus-vegetation-microbial interactions. *Advances in Water Resources*, 52, 151–164.

Runyan, C. W., and D'Odorico, P. (2014). Bistable dynamics between vegetation disturbance and landslide occurrence. *Water Resources Research*, 50(2), 1112-1130, doi: 10.1002/2013WR014819.

Runyan, C. W., D'Odorico, P., and Lawrence, D. (2012a). Physical and biological feedbacks of deforestation. *Reviews of Geophysics*, 50(4).

Runyan, C. W., D'Odorico, P., and Lawrence, D. L. (2012b). The effect of repeated deforestation on vegetation dynamics for phosphorus limited ecosystems. *Journal of Geophysical Research*, 117, G01008, doi:10.1029/2011JG001841.

Runyan, C. W., D'Odorico. P., and Shobe, W. (2015). The economic impacts of positive feedbacks associated with deforestation, *Ecological Economics*, 120, 93-99, doi:10.1016/j.ecolecon.2015.10.007.

Runyan, C. W., and D'Odorico, P., Vandecar, K. L., Das, R., Schmook, B., and Lawrence, D. (2013). Positive feedbacks between phosphorus deposition and forest canopy trapping, evidence from Southern Mexico. *Journal of Geophysical Research-Biogeosciences*, 118(4), 1521–1531.

Ruprecht, J. K., and Schofield, N. J. (1989). Analysis of streamflow generation following deforestation in southwest western Australia. *J. Hydrol.*, 105, 1–17.

Saa, A., Trasar-Cepeda, M. C., Gil-Sotres, F., and Carballas, T. (1993). Changes in soil phosphorus and acid phosphatase activity immediately following forest fires. *Soil Biology and Biochemistry*, 25(9), 1223–1230.

Saad, S. I., da Rocha, H. R., Silva Dias, M. A. F., and Rosolem, R. (2010). Can the deforestation breeze change the rainfall in Amazonia? A case study for the BR-163 highway region. *Earth Interact.*, 14, 18.

Saatchi, S. S. et al. (2011). Benchmark map of forest carbon stocks in tropical regions across three continents. *Proceedings of the National Academy of Sciences*, 108(24), 9899–9904.

Sahin, V., and Hall, M. J. (1996). The effects of afforestation and deforestation on water yields. *Journal of Hydrology*, 178(1), 293–309.

Sala, O. E., Meyerson, L. A., and Parmesan, C. (2009). *Biodiversity Change and Human Health*. Island Press, Washington, DC.

Sala, O. E. et al. (2000). Global biodiversity scenarios for the year 2100. *Science*, 287(5459), 1770–1774.

Salati, E., Dall'Olio, A., Matsui, E., and Gat, J. R. (1979). Recycling of water in the Amazon Basin: An isotopic study. *Water Resour. Res.*, 15(5), 1250–1258, doi:10.1029/WR015i005p01250.

Samarakoon, A. B., and Gifford, R. M. (1995). Soil water content under plants at high CO_2 concentration and interactions with the direct CO_2 effects: A species comparison. *Journal of Biogeography*, 193–202.

Sample, E. C., Soper, R. J., and Racz, G. J. (1980). Reaction of phosphate fertilizers in soils. In Khasawneh F. E., A. C. Sample, and E. J. Kamprath, (eds.),*The Role of Phosphorus in Agriculture*. American Society of Agronomy Crop Science Society of America, Soil Science Society of America. Madison, WI, pp. 263–310.

Santiago, L. S. (2007). Extending the leaf economics spectrum to decomposition: Evidence from a tropical forest. *Ecology*, 88(5), 1126–1131.

Sardans, J., and Peñuelas, J. (2012). The role of plants in the effects of global change on nutrient availability and stoichiometry in the plant-soil system. *Plant Physiology*, 160(4), 1741–1761.

Sarraf, M. (2004). Assessing the costs of environmental degradation in the Middle East and North Africa countries. Environment Strategy Notes 9. Environment Department, World Bank, Washington DC.

Savage, S. M., Osborn, J., Letey, J., Heaton, C. (1972). Substances contributing to fire-induced water repellency in soils. *Soil Science Society America Proceedings*, 36, 674–678.

Scheffer, M., Carpenter, S., Foley, J. A., Folke, C., and Walker, B. H. (2001). Catastrophic shifts in ecosystems. *Nature*, 413, 591–596.

Scheffer, M., Holgren, M., Brovkin, V., and Claussen, M. (2005). Synergy between small- and large-scale feedbacks of vegetation on the water cycle. *Global Change Biology*, 11, 1003–1012, doi:10.1111/j.1365-2486.2005.00962.x.

Scheffer, M. et al. (2009). Early-warning signals for critical transitions. *Nature*, 461(7260), 53–59.

Schimel, J. P., Gulledge, J. M., Clein-Curley, J. S., Lindstrom, J. E., and Braddock, J. F. (1999). Moisture effects on microbial activity and community structure in decomposing birch litter in the Alaskan taiga. *Soil Biology and Biochemistry*, 31(6), 831–838.

Schlesinger, W. (1997). *Biogeochemistry: An Analysis of Global Change*. Academic Press, San Diego.

Schlesinger, W. H., and Hartley, A. E. (1992). A global budget for atmospheric NH_3. *Biogeochemistry*, 15(3), 191–211.

Schlesinger, W. H, and Pilmanis, A. M. (1998). Plant-soil interactions in deserts. *Biogeochemistry*, 42, 169–187.

Schlesinger, W. H. et al. (1990). Biological feedbacks in global desertification. *Science*, 247, 1043–1048.

Schmitt, C. B. et al. (2008). Global ecological forest classification and forest protected area gap analysis. Analyses and recommendations in view of the 10% target for forest protection under the Convention on Biological Diversity (CBD). University of Freiburg, Freiburg.

Scholes, R. J., and Archer, S. R. (1997). Tree-grass interactions in savannas. *Annu. Rev. Ecol. Syst.*, 28, 517–544.

Schoups, G., Hopmans, J. W., Young, C. A., Vrugt, J. A., Wallender, W. W., Tanji, K. K., and Panday, S. (2005). Sustainability of irrigated agriculture in the San Joaquin Valley, California. *Proceedings of the National Academy of Sciences*, 102(43), 15352–15356.

Schuur, E. A. G. et al. (2008). Vulnerability of permafrost carbon to climate change: Implications for the global carbon cycle. *BioScience*, 58(8), 701–714.

Searchinger, T. D. et al. (2009). Fixing a critical climate accounting error. *Science*, 326(5952), 527.

Serrão, E. A. S., and Falesi, I. C. (1977). *Pastagens do Trópico Úmido Brasileiro*. Empresa Brasileira de Pesquisa Agropecuária-Centro de Pesquisas Agro-Pecuárias do Trópico Úmido (EMBRAPA- CPATU), Belém, Pará, Brazil.

Seto, K. C., Güneralp, B., and Hutyra, L. R. (2012). Global forecasts of urban expansion to 2030 and direct impacts on biodiversity and carbon pools. *Proceedings of the National Academy of Sciences*, 109(40), 16083–16088.

Seubert, C., Sanchez, P., and Valverde, C. (1977). Effects of land clearing methods on soil properties of an ultisol and crop performance in the Amazon jungle of Peru. *Tropical Agriculture (Trinidad)*, 54, 307–321.

Sharratt, B. S. (1998), Radiative exchange, near-surface temperature and soil water of forest cropland in interior Alaska. *Agricultural and Forest Meteorology*, 89, 269–280.

Shimabukuro, Y. et al. (2007). Near real time detection of deforestation in the Brazilian Amazon using MODIS imagery. *Ambiente & Água-An Interdisciplinary Journal of Applied Science*, 1(1), 37–47.

Shimada, M., Itoh, T., Motooka, T., Watanabe, M., Shiraishi, T., Thapa, R., and Lucas, R. (2014). New global forest/non-forest maps from ALOS PALSAR data (2007–2010). *Remote Sensing of Environment*, 155, 13–31.

Shimada, M., Rosenqvist, A., Watanabe, M., and Tadono, T. (2005). The polarimetric and interferometric potential of ALOS PALSAR. In ESA Special Publication, Vol. 586, p. 41.

Shively, G. E., and Pagiola, S. (2004). Agricultural Intensification, local labor markets, and deforestation in the Philippines. *Environment and Development Economics*, 9(02), 241–266.

Shukla, J., Nobre, C., and Sellers, P. (1990). Amazon deforestation and climate change. *Science*, 247, 1322–1325.

Shuttleworth, W. J. (1977). The exchange of wind-driven fog and mist between vegetation and the atmosphere. *Boundary-Layer Meteorology*, 12, 463–489.

Sidle, R. C., Pearce, A. J., and O'Loughlin, C. L. (1985). Hillslope Stability and Land Use. *Water Resour. Monogr. Ser.*, 11.

Sidle, R. C., and Swanston, D. N. (1982). Analysis of a small debris slide in coastal Alaska. *Can. Geotech. J.*, 19, 167–174.

Sidle, R. C., Ziegler, A. D., Negishi, J. N., Nik, A. R., Siew, R., and Turkelboom, F. (2006). Erosion processes in steep terrain – truths, myths, and uncertainties related to forest management in Southeast Asia. *Forest Ecology and Management*, 224(1), 199–225.

Sills, E., and Pattanayak, S. (2006). Tropical trade-offs: An economic perspective on tropical forests. In S. Spray and K. McGlothlin (eds.), *Tropical Deforestation*. Rowman & Littlefield, New York.

Silver, W. (1994). Is nutrient availability related to plant nutrient use in humid tropical forests? *Oecologia (Berl)*, 98, 336–343.

Silver, W. L., and Miya, R. K. (2001). Global patterns in root decomposition: Comparisons of climate and litter quality effects. *Oecologia*, 129(3), 407–419.

Simpson, R. D., Sedjo, R. A., and Reid, J. W. (1996). Valuing biodiversity for use in pharmaceutical research. *J. Polit. Economy* 104, 163–185.

Singer, B., de Castro, M. C. (2006). Enhancement and suppression of malaria in the Amazon. *Am. J. Trop. Med. Hyg.*, 74(1), 1–2.

Skiba, U. et al. (2012). UK emissions of the greenhouse gas nitrous oxide. *Philosophical Transactions of the Royal Society B: Biological Sciences*, 367(1593), 1175–1185.

Skidmore, T. E., and Smith, P. H. (1984). *Modern Latin America*. Oxford University Press, New York and Oxford, 39–56.

Skole, D. L., Chomentowski, W. H., Salas, W. A., and Nobre, A. D. (1994). Physical and human dimensions of deforestation in Amazonia. *BioScience*, 314–322.

Skopp, J., Jawson, M. D., and Doran, J. W. (1990). Steady-state aerobic microbial activity as a function of soil water content. *Soil Sci. Soc. Am. J.*, 54, 1619–1625.

Smil, V. (1999). Nitrogen in crop production: An account of global flows. *Global Biogeochemical Cycles*, 13(2), 647–662.

Smil, V. (2000). Phosphorus in the environment: Natural flows and human interferences. *Annual Review of Energy and the Environment*, 25(1), 53–88.

Smith, S. E., Smith, F. A., and Jakobsen, I. (2004). Functional diversity in arbuscular mycorrhizal (am) symbioses: The contribution of the mycorrhizal P uptake pathway is not correlated with mycorrhizal responses in growth or total P uptake. *New Phytol.*, 162, 511–524.

Smith, W. K., Germino, M. J., Hancock, T. E., and Johnson, D. M. (2003). Another perspective on altitudinal limits of alpine timberlines. *Tree Physiology*, 23, 1101–1112

Soares-Filho, B. S. et al. (2006). Modelling conservation in the Amazon basin. *Nature*, 440(7083), 520–523.

Sollins, P., and McCorison, F. M. (1981). Nitrogen and carbon solution chemistry of an old growth coniferous forest watershed before and after cutting. *Water Resources Research*, 17(5), 1409–1418.

Solow, R. M. (1974). Intergenerational equity and exhaustible resources: Review of economic studies. symposium of the economics of exhaustible resources, 29–46.

Sorenson, S. K., Dileanis, P. D., and Branson, F. A. (1991). Soil water and vegetation responses to precipitation and changes in depth to ground water in Owens Valley, California. USGS, Denver 54.

Southgate, D. (1990). The causes of land degradation along "spontaneously" expanding agricultural frontiers in the Third World. *Land Economics*, 93–101.

Souza, E. P., Renno N. O., Silva Dias M. A. F. (2000). Convective circulations induced by surface heterogeneities. *J. Atmos. Sci.*, 57, 2915–2922.

Spracklen, D. V., Arnold, S. R., and Taylor, C. M. (2012). Observations of increased tropical rainfall preceded by air passage over forests. *Nature*, 489, 282–286.

Stark, N. (1972). Nutrient cycling pathways and litter fungi. *Bioscience*, 355–360.

Staver, A. C., Archibald, S., and Levin, S. A. (2011). The global extent and determinants of savanna and forest as alternative stable states. *Science*, 334, 230–232.

Steadman, D. W. (1995). Prehistoric extinctions of Pacific island birds: Biodiversity meets zooarchaeology. *Science*, 267(5201), 1123–1131.

Stednick, J. D. (1996). Monitoring the effects of timber harvest on annual water yield. *Journal of Hydrology*, 176(1/4), 79–95.

Steininger, M. K., Tucker, C. J., Townshend, J. R. G., Killeen, T. J., Desch, A., Bell, V., and Ersts, P. (2001). Tropical deforestation in the Bolivian Amazon. *Environmental Conservation*, 28, 127–134.

Stevens, G. C., and Fox, J. F. (1991). The causes of treeline. *Annual Review of Ecology and Systematics*, 22, 177–191.

Stevenson, C.M., et al. (2015). Variation in Rapa Nui (Easter Island) land use indicates production and population peaks prior to European contact. *PNAS*, 112(4), 1025–1030; doi: 10.1073/pnas.1420712112.

Stokes, A. et al. (2008). How vegetation reinforces soil on slopes. In Norris, J. E., A. Stokes, S. B. Mickovski, E. Cammeraat, L. P. H. van Beek, B. Nicoll, and A. Achim (eds.), *Slope Stability and Erosion Control: Ecotechnological Solutions*. Springer, Dordrecht, pp. 65–118.

Stoorvogel, J. J., Smaling, E. M., and Janssen, B. H. (1993). Calculating soil nutrient balances in Africa at different scales I. *Supra-National Scale, Fertilizer Reseach*, 35, 227–235.

Stork, N. E. (2010). Re-assessing current extinction rates. *Biodiversity and Conservation*, 19(2), 357–371.

Strogatz, S. H. (1994). *Nonlinear Dynamics and Chaos*. Westview Press, Cambridge, MA.

Stuart, S. A., Choat, B., Martin, K. C., Holbrook, N. M., and Ball, M. C. (2007). The role of freezing in setting the latitudinal limits of mangrove forests. *New Phytologist*, 173, 576–583.

Sturm, M., Douglas, T., Racine, C., and Liston, G. E. (2005). Changing snow and shrub conditions affect albedo with global implications. *Journal of Geophysical Research*, 110, G01004, doi:10.1029/2005JG000013.

Sturm, M., Racine, C., and Tape, K. (2001). Increasing shrub abundance in the Arctic. *Nature*, 411, 546–547.

Suding, K. N., Collins, S. L., Gough, L., Clark, C., Cleland, E. E., Gross, K. L., Milchunas, D. G., and Pennings. S. (2005). Functional- and abundance-based mechanisms explain diversity loss due to N fertilization. *Proceedings of the National Academy of Sciences (USA)*, 102, 4387–4392.

Sun, G. et al. (2001). Effects of forest management on the hydrology of wetland forests in the southern United States. *For. Ecol. Manag.*, 143, 227–236.

Sutton, M. A., Pitcairn, C. E., and Fowler, D. (1993). *The Exchange of Ammonia between the Atmosphere and Plant Communities*. Academic Press, San Diego.

Suweis S., and D'Odorico, P. (2014). Early warning signs in social-ecological networks. *PLoS-One*, 9(7), e101851, doi:10.1371/journal.pone.0101851.

Sveinbjörnsson, B. (2000). North American and European treelines: External forces and internal processes controlling position. *AMBIO: A Journal of the Human Environment*, 29, 388–395.

Swank, W. T., and Johnson, C. E. (1994). Small catchment research in the evaluation and development of forest management practices. In Moldan, B., and J. Cerny (eds.), *Biogeochemistry of Small Catchments: A Tool for Environmental Research*. Wiley, Chichester, pp. 383–408.

Swanson, F. J., and Dyrness, C. T. (1975). Impact of clear-cutting and road construction on soil erosion by landslides in the western Cascade Range, Oregon. *Geology*, 3(7), 393–396.

Swanston, D. N. (1988). Timber harvest and progressive deformation of slopes in southwestern Oregon. *Bulletin of the Association of Engineering Geologists*, 25, 371–381.

Swift, M. J., Izac, A. M., and van Noordwijk, M. (2004). Biodiversity and ecosystem services in agricultural landscapes – are we asking the right questions? *Agriculture, Ecosystems and Environment*, 104(1), 113–134.

Tarafdar, J. C., and Marschner, H. (1994). Phosphatase activity in the rhizosphere and hyphosphere of VA mycorrhizal wheat supplied with inorganic and organic phosphorus. *Soil Biol. Biochem.*, 26, 387–395.

Tatem, A. J., Goetz, S. J., Hay, S. I. (2008). Fifty years of earth observation satellites. *American Scientist*, 96, 390–398.

Tavoni, A., Schlueter, M., and Levin, S. (2012). The survival of the conformist: Social pressure and renewable resource management. *J. Theor. Biol.*, 299, 152–161.

Terborgh, J. et al. (2001). Ecological meltdown in predator-free forest fragments. *Science*, 294(5548), 1923–1926.

Terman, G. L. (1980). Volatilization losses of nitrogen as ammonia from surface-applied fertilizers, organic amendments, and crop residues. *Advances in Agronomy*, 31, 189–223.

Terwilliger, V. J., and Waldron, L. J. (1991). Effects of root reinforcement on soil-slip patterns in the Transverse Ranges of southern California. *Geological Society of America Bulletin*, 103, 775–785.

Thompson, I. D. et al. (2011). Forest biodiversity and the delivery of ecosystem goods and services: Translating science into policy. *BioScience*, 61(12), 972–981.

Tilman, D. (1987). Secondary succession and the pattern of plant dominance along experimental nitrogen gradients. *Ecological Monographs*, 57, 189–214.

Tilman, D. (1999). Global environmental impacts of agricultural expansion: The need for sustainable and efficient practices. *Proceedings of the National Academy of Sciences*, 96(11), 5995–6000.

Tilman, D., Lehman, C. L., and Bristow, C. E. (1998). Diversity-stability relationships: Statistical inevitability or ecological consequence? *American Naturalist*, 151(3), 277–282.

Tilman, D. et al. (2001). Forecasting agriculturally driven global environmental change. *Science*, 292(5515), 281–284.

Tilman, D., Cassman, K. G., Matson, P. A., Naylor, R., and Polasky, S. (2002). Agricultural sustainability and intensive production practices. *Nature*, 418(6898), 671–677.

Tilman, D., Hill, J., and Lehman, C. (2006). Carbon-negative biofuels from low-input high-diversity grassland biomass. *Science*, 314(5805), 1598–1600.

Tilman, D., Isbell, F., and Cowles, J. M. (2014). Biodiversity and Ecosystem Functioning. *Annual Review of Ecology, Evolution, and Systematics*, 45(1), 471.

Torres, R., Dietrich, W. E., Montgomery, D. R., Anderson, S. P., and Loague, K. (1998). Unsaturated zone processes and the hydrologic response of a steep, unchanneled catchment. *Water Resour. Res.*, 34, 1865–1879

Townsend, A. R., Vitousek, P. M., and Holland, E. A. (1992). Tropical soils could dominate the short-term carbon cycle feedbacks to increased global temperatures. *Climatic Change*, 22(4), 293–303.

Toy, ADF. (1973). *The Chemistry of Phosphorus*. Pergamon, Oxford.

Tranquillini, W. (1979). *Physiological ecology of the alpine timberline: Tree existence at high altitudes with special reference to the European Alps*, Springer, Berlin.

Trappe, J. M. (1987). Phylogenetic and ecologic aspects of mycotrophy in the angiosperms from an evolutionary standpoint. In Safir, G. R. (ed.), *Ecophysiology of VA Mycorrhizal Plants*. CRC, Boca Raton, FL, pp. 5–25.

Trenbath, B. R. (1993). Intercropping for the management of pests and diseases. *Field Crops Research*, 34(3), 381–405.

Trenberth, K. E. (1999). Atmospheric moisture recycling: Role of advection and local evaporation. *J. Climate*, 12, 1368–1381.

Troendle, C. A. (1983). The potential for water yield augmentation from forest management in the Rocky Mountain region. *Water Resour. Bull.*, 19(3), 359–373.

Trumbore, S. E., Davidson, E. A., Camargo, P. B., Nepstad, D. C., and Martinelli, L. A. (1995). Below-ground cycling of carbon in forests and pastures of eastern Amazonia. *Global Biogeochem. Cycles*, 9(4), 515–528.

Trustrum, N. A., and De Rose, R. C. (1988). Soil depth-age relationship of landslides on deforested hillslopes. Taranaki, New Zeland. *Geomorphology*, 1, 143–160.

Trustrum, N. A., Lambert, M. G., and Thomas, V. J. (1983). The impact of soil slip erosion on hill country pasture production in New Zealand. Proceedings of the Second International Conference on Soil Erosion and Conservation, Honolulu, January 1983.

Tscharntke, T., Klein, A. M., Kruess, A., Steffan-Dewenter, I., and Thies, C. (2005). Landscape perspectives on agricultural intensification and biodiversity–ecosystem service management. *Ecology Letters*, 8(8), 857–874.

Turazza, D. (1880). *Trattato di idrometria o di idraulica pratica*. Padova, Italy: F. Sacchetto.

Turner, B. L., Lambin, E. F., and Reenberg, A. (2007b). The emergence of land change science for global environmental change and sustainability. *Proceedings of the National Academy of Sciences*, 104(52), 20666–20671.

Turner, B. L., and Robbins, P. (2008). Land-change science and political ecology: Similarities, differences, and implications for sustainability science. *Annual review of environment and resources*, 33, 295–316.

Turner, B.L., and Sabloff, J.A. (2012). Classic Period collapse of the Central Maya Lowlands: Insights about human–environment relationships for sustainability, *Proc. Natnl. Acad. Sci., USA, PNAS*, 109(35), 13908–13914.

Turner, I. M., Tan, H. T. W., Wee, Y. C., Ibrahim, A. B., Chew, P. T., and Corlett, R. T. (1994). A study of plant species extinction in Singapore: Lessons for the conservation of tropical biodiversity. *Conservation Biology*, 8(3), 705–712.

Turner, M. G., Smithwick, E. A., Metzger, K. L., Tinker, D. B., and Romme, W. H. (2007a). Inorganic nitrogen availability after severe stand-replacing fire in the Greater Yellowstone ecosystem. *Proceedings of the National Academy of Sciences*, 104(12), 4782–4789.

Uhl, C., and Kauffman, H. J. B. (1990). Deforestation, fire susceptibility and potential tree responses to fire in the eastern Amazon. *Ecology*, 71(2), 437–449.

UN Comtrade. (2014). Available at: http://comtrade.un.org/ (accessed on June 25, 2014).

UNEP. (2007). Global environment outlook 4. United Nations Environment Programme, Nairobi, Kenya.

UNEP/GRID Arendal. (2002). Arctic environmental atlas, available at: http://maps.grida.no/arctic/.

UNPD. UN Population Division, New York (2005). Long-range world population projections: Based on the 1998 revision.

Urli, M., Porté, A. J., Cochard, H., Guengant, Y., Burlett, R., and Delzon, S. (2013). Xylem embolism threshold for catastrophic hydraulic failure in angiosperm trees. *Tree Physiology*, 33(7), 672–683.

USDA-FAS (2010). Oilseeds: World markets and trade. Circular Series FOP 8–10 August 2010. US Depart of Agric-Foreign Agric Serv, Washington, DC.

US Forest Service. (2008). North American forest outlook study: US country report. Unpublished SOFO 2009 contribution.

Valdes, C. (2006). Brazil's booming agriculture faces obstacles, Amber Waves Economic Research Service/USDA, 4, 28–35, available at: www.ers.usda.gov/AmberWaves/November06/.

van de Koppel, Rietkerk, J., M., and Weissing, F. J. (1997). Catastrophic vegetation shifts and soil degradation in terrestrial grazing systems. *Trends in Ecology and Evolution*, 12, 352–356.

van der Ent, R. J., Savenije, H. H., Schaefli, B., and Steele-Dunne, S. C. (2010). Origin and fate of atmospheric moisture over continents. *Water Resour. Res.*, 46, W09525, doi:10.1029/2010WR009127.

van der Werf, G. R., Randerson, J. T., Giglio, L., Collatz, G. J., Kasibhatla, P. S., and Arellano Jr., A. F. (2006). Interannual variability in global biomass burning emissions from 1997 to 2004. *Atmos. Chem. Phys.*, 6, 3423–3441, doi:10.5194/acp-6-3423-2006.

van der Werf, G. R. et al. (2009). CO_2 emissions from forest loss. *Nature Geoscience*, 2(11), 737–738.

Van Laerhoven, F. (2010). Governing community forests and the challenge of solving two-level collective action dilemmas—a large-N perspective. *Global Environmental Change*, 20(3), 539–546.

van Langevelde, F., C. et al. (2003). Effects of fire and herbivory on the stability of savanna ecosystems. *Ecology*, 84, 337–350.

van Meeteren, M. M., Tietema, A., van Loon, E. E., and Verstraten, J. M. (2008). Microbial dynamics and litter decomposition under a changed climate in a Dutch heathland. *Applivan meeed Soil Ecology* 38, 119–127.

van Nes, E. H., and Scheffer, M. (2007). Slow recovery from perturbations as a generic indicator of a nearby catastrophic shift. *Am. Nat.*, 169, 738–747.

van Noordwijk, M., Cerri, C., Woomer, P. L., Nugroho, K., Bernoux, M. (1997). Soil carbon dynamics in the humid tropical forest zone. *Geoderma*, 79, 187–225.

van Vuuren, D. P., Sala, O. E., Pereira, H. M. (2006). The future of vascular plant diversity under four global scenarios. *Ecol. Soc.*, 11(2), 25.

van Wilgen, B. W., Cowling, R. M., and Burgers, C. J. (1996). Valuation of ecosystem services. *BioScience*, 184–189.

van Wilgen, B. W., Trollope, W. S. W, Biggs, H. C., Potgieter, A. L. F., and Brockett, B. H. (2003). Fire as a driver of ecosystem variability. In Du Toit, J. T., K. H. Rogers, and H. C. Biggs (eds.), *The Kruger Experience: Ecology and Management of Savanna Heterogeneity*. Island Press, Washington, DC, pp. 149–170.

Vanacker, V., Vanderschaeghe, M., Govers, G., Willems, E., Poesen, J., Deckers, J., and De Bievre, B. (2003). Linking hydrological, infinite slope stability and land-use change models through GIS for assessing the impact of deforestation on slope stability in high Andean watersheds. *Geomorphology*, 52(3), 299–315.

Vance, C. P., Uhde-Stone, C., and Allan, D. L. (2003). Phosphorus acquisition and use: Critical adaptations by plants for securing a nonrenewable resource. *New Phytologist*, 157, 423–447.

Vandermeer, J. H. (1989). *The Ecology of Intercropping*. Cambridge University Press, Cambridge.

Verhoeven, J. T. A., and Schmitz, M. B. (1991). Control of plant growth by nitrogen and phosphorus in mesotrophic fens. *Biogeochemistry* 12, 135–148.

Vermeer, E. B. (1998). Population and ecology along the frontier in Qing China, in Elvin, M., and T. -J. Liu (eds.), *Sediments of Time, Environment and Society in Chinese History*. Cambridge University Press, Cambridge.

Vetaas, O. R. (1992). Micro-site effects of trees and shrubs in dry savannas. *J. Veg. Sci.*, 3(3), 337–344.

Viana, V. M., Tabanez, A. A. (1996). Biology and conservation of forest fragments in the Brazilian Atlantic moist forest. In Schelhas, J., and R. Greenberg (eds.), *Forest Patches in Tropical Landscapes*. Island Press, Washington, DC, pp. 151–167.

Villegas, J. C., Breshears, D. D., Zou, C. B., and Royer, P. D. (2010). Seasonally pulsed heterogeneity in microclimate: Phenology and cover effects along deciduous grassland–forest continuum. *Vadose Zone Journal*, 9, 537–547.

Villoria, N. B., Byerlee, D., and Stevenson, J. (2014). The effects of agricultural technological progress on deforestation: What do we really know? *Applied Economic Perspectives and Policy*, 36(2), 211–237.

Vitousek, P. M. (1984). Litterfall, nutrient cycling, and nutrient limitation in tropical ecosystems. *Ecology*, 65, 285–298.

Vitousek, P. M., and Hobbie, S. (2000). Heterotrophic nitrogen fixation in decomposing litter: Patterns and regulation. *Ecology*, 81(9), 2366–2376.

Vitousek, P. M., and Howarth, R. W. (1991). Nitrogen limitation on land and in the sea: How can it occur? *Biogeochemistry*, 13(2), 87–115.

Vitousek, P. M., and Matson, P. A. (1985). Disturbance, nitrogen availability, and nitrogen losses in an intensively managed loblolly pine plantation. *Ecology*, 1360–1376.

Vitousek, P. M., Mooney, H. A., Lubchenco, J., and Melillo, J. M. (1997b). Human domination of Earth's ecosystems. *Science*, 277(5325), 494–499.

Vitousek, P. M., Porder, S., Houlton, B. Z., and Chadwick, O. A. (2010). Terrestrial phosphorus limitation: Mechanisms, implications and nitrogen-phosphorus interactions. *Ecological Applications*, 20(1), 5–15.

Vitousek, P. M., Turner, D. R., Parton, W. J., and Sanford, R. L. (1994). Litter decomposition on the Mauna Loa Environmental Matrix, Hawai'i: Patterns, mechanisms and models. *Ecology*, 75(2), 418–429.

Vitousek, P. M. et al. (1997a). Human alteration of the global nitrogen cycle: Sources and consequences. *Ecological Applications*, 7(3), 737–750.

Vitt, D. H., Halsey, L. A., and Zoltai, S. C. (1994). The bog landforms of continental Western Canada in relation to climate and permafrost patterns. *Arct. Alp. Res.*, 26, 1–13.

Vitt, D. H., Halsey, L. A., and Zoltai, S. C. (1999). The changing landscape of Canada's western boreal forest: The current dynamics of permafrost. *Can. J. For. Res.*, 30, 283–287.

Vittor, A. Y. et al. (2006). The effect of deforestation on the human-biting rate of Anopheles darlingi, the primary vector of falciparum malaria in the Peruvian Amazon. *American Journal of Tropical Medicine and Hygiene*, 74(1), 3–11.

Voicu, M. F., and Comeau, P. G. (2006). Microclimatic and spruce growth gradients adjacent to young aspen stands. *Forest Ecology and Management*, 221, 13–26.

von Randow, C. et al. (2004). Comparative measurements and seasonal variations in energy and carbon exchange over forest and pasture in South West Amazonia. *Theor. Appl. Climatol.*, 78, 5–26.

Von Uexkull, H. R., and Mutert, E. (1995). Global extent, development and economic impact of acid soils. *Plant and Soil*, 171, 1–15.

Von Uexkull, H. R., and Mutert, E. (1998). Global extent, development and economic impact of acid soils. In Date, R. A., N. J. Grundon, G. E. Rayment, and M. E. Probert (eds.), *Plant-Soil Interactions at Low pH: Principles and Management*. Dordrecht: Kluwer, 5–19.

Vosti, S. A., Witcover, J., and Carpentier, C. L. (2002). Agricultural intensification by smallholders in the western Brazilian Amazon: From deforestation to sustainable land use. *Intl Food Policy Res Inst.*, 130.

Wagener, S. M., and Schimel, J. P. (1998). Stratification of soil ecological processes: A study of the birch forest floor in the Alaskan taiga. *Oikos*, 63–74.

Wakker, E. (2006). The Kalimantan border oil palm mega-project. 38(19), International Atomic Energy Agency.

Waldrop, M. P., Balser, T. C., and Firestone, M. K. (2000). Linking microbial community composition to function in a tropical soil. *Soil. Biol. Biochem.*, 24, 317–323.

Walker, B., and Salt, D. (2006). *Resilience Thinking*. Island Press, Wahington, DC.

Walker, B. H., Ludwig, D., Holling, C. S., and Peterman, R. M. (1981). Stability of semiarid savanna grazing systems. *J. Ecol.*, 69, 473–498.

Walker, J., Bullen, F., and Williams, B. (1993). Ecohydrological changes in the Murray-Darling Basin. I. The number of trees cleared over the last two centuries. *J. Appl. Ecol.*, 30(2), 265–273, doi:10.2307/2404628.

Walker, R., (1993). Deforestation and economic development. *Canadian Journal of Regional Science*, 16, 481–497.

Walker, T. W., and Syers, J. K. (1976). The fate of phosphorus during pedogenesis. *Geoderma*, 15, 1–19.

Wan, S., Hui, D., and Luo, Y. (2001). Fire effects on nitrogen pools and dynamics in terrestrial ecosystems: A meta-analysis. *Ecological Applications*, 11(5), 1349–1365.

Wang, G. L., Eltahir, E. A. B. (2000a). Biosphere-atmosphere interactions over West Africa. Part 1: Development and validation of a coupled dynamic model. *Q J R Meteorol. Soc.*, 126, 1239–1260.

Wang, G. L., Eltahir, E. A. B. (2000b). Biosphere-atmosphere interactions over West Africa. Part 2: Multiple climate equilibria. *Q J R Meteorol. Soc.*, 126, 1261–1280.

Wang, G. L. and E. A. B. Eltahir (2000C), Ecosystem dynamics and the Sahel drought. *Geophys. Res. Lett.*, 27, 795–798.

Wang, J. et al. (2009). Impact of deforestation in the Amazon basin on cloud climatology. *Proc. Natl. Acad. Sci.*, 106, 3670–3674, doi:10.1073/pnas.0810156106.

Wanner, H. J. et al. (2008). Mid- to Late Holocene climate change: An overview. *Quaternary Science Reviews*, 27, 1791–1828.

Wardle, D. A. (1998). Controls of temporal variability of the soil microbial biomass: A global-scale synthesis. *Soil Biol. Biochem.*, 30(13), 1627–1637.

Warneck, P. (1988). Chemistry of the Natural Atmosphere. *International* Geophysics, 41, San Diego, California, 757.

Wasilewska, L. (1995). Differences in development of soil nematode communities in single-and multi-species grass experimental treatments. *Applied Soil Ecology*, 2(1), 53–64.

Wasowski, J. (1998). Understanding rainfall-landslide relationships in man-modified environments: A case-history from Caramanico Terme, Italy. *Environmental Geology*, 35(2–3), 197–209.

Wassenaar, T., Gerber, P., Verburg, P. H., Rosales, M., Ibrahim, M., and Steinfeld, H. (2007). Projecting land use changes in the Neotropics: The geography of pasture expansion into forest. *Global Environmental Change*, 17(1), 86–104.

Weathers, K. C., Cadenasso, M. L., and Pickett, S. T. (2001). Forest edges as nutrient and pollutant concentrators: Potential synergisms between fragmentation, forest canopies, and the atmosphere. *Conservation Biology*, 15, 1506–1514.

Weathers, K. C., and Likens, G. E. (1997). Clouds in Southern Chile: An important source of nitrogen to nitrogen-limited ecosystems? *Environ. Sci. Technol.*, 31, 210–213.

Webb, E. L., Jachowski, N. R. A. Phelps, J., Friess, D. A., Than, M. M., and Ziegler, A. D. (2014). Deforestation in the Ayeyarwady Delta and the conservation implications of an internationally-engaged Myanmar. *Global Environmental Change - Human and Policy Dimensions*, 24, 321–333.

Webb, T., Bartlein, P. J., and Kutzbach, J. E. (1987). Climatic Change in Eastern Noth America during the past 18,000 years: Comparisons of pollen data with model results *in North America and Adjacent Oceans during the last Glaciation*. (eds. Ruddiman, W. F., and H. E. Wright), 447-462, Geological Soc. Am., Boulder, Co.

Webb, T. J., Woodward, F. I., Hannah, L., and Gaston, K. J. (2005). Forest cover–rainfall relationships in a biodiversity hotspot: The Atlantic forest of Brazil. *Ecological Applications*, 15(6), 2005, 1968–1983.

Weitzman, M. L. (1994). On the "environmental" discount rate. *Journal of Environmental Economics and Management*, 26(2), 200–209.

Werth, D., and R. Avissar (2002). The local and global effects of Amazon deforestation. *J. Geophys. Res.*, 107(D20), 8087, doi:10.1029/2001JD000717.

West, P. C., Gibbs, H. K., Monfreda, C., Wagner, J., Barford, C. C., Carpenter, S. R., and Foley, J. A. (2010). Trading carbon for food: Global comparison of carbon stocks vs. crop yields on agricultural land. *Proceedings of the National Academy of Sciences*, 107(46), 19645–19648.

Westfall, J. (2004). 2003 Summary of forest health conditions in British Columbia. Ministry of Forests, Forest Practices Branch.

Westheimer, F. H. (1987). Why nature chose phosphates. *Science*, 235(4793), 1173–1178.

Wetzel, P. R. et al. (2005). Maintaining tree islands in the Florida Everglades: Nutrient redistribution is the key. *Front. Ecol. Environ.*, 3, 370–376.

Whitford, W. G., Anderson, J., and Rice, P. M. (1997). Stemflow contribution to the "fertile island" effect in cresotebush, Larrea-tridentata. *J. Arid Environ.*, 35, 451–457.

Wieczorek, G. F., Morgan, B. A., and Campbell, R. H. (2000). Debris-flow hazards in the Blue Ridge of central Virginia. *Environ. Eng. Geol.*, 6, 3–23.

Wieder, W. R., Cleveland, C. C., and Townsend, A. R. (2009). Controls over leaf litter decomposition in wet tropical forests. *Ecology*, 90(12), 3333–3341.

Wilby, A. et al. (2009). Biodiversity, food provision, and human health. In Sala, O. E., L. A. Meyerson, and C. Parmesan (eds.), *Biodiversity Change and Human Health*. Island Press, Washington, DC, pp. 13–39.

Wilde, S. A., Steinbrenner, E. C., Pierce, R. S., Dosen, R. C., and Pronin, D. T. (1953). Influence of forest cover on the state of the ground water table. *Soil Sci. Soc. Proc.*, 17, 65–67.

Williams, E. J., Hutchinson, G. L., and Fehsenfeld, F. C. (1992). NO_x and N_2O emissions from soil. *Global Biogeochemical Cycles*, 6(4), 351–388.

Williams, M. (1990). Forests. In Turner B. L., II, W. C. Clark, R. W. Kates, J. F. Richards, J. T. Mathews, and W. B. Meyer (eds.), *The Earth as Transformed by Human Action*. Cambridge University Press, Cambridge, pp. 179–201.

Williams, M. (2000). Dark ages and dark areas: Global deforestation in the deep past. *Journal of Historical Geography*, 26, 28–46.

Williams, M. (2002). *Deforesting the Earth: From Prehistory to Global Crisis*. University of Chicago Press, Chicago.

Williams, M. R., Fisher, T. R., and Melack, J. M. (1997). Solute dynamics in soil water and groundwater in a central Amazon catchment undergoing deforestation. *Biogeochemistry*, 38(3), 303–335.

Wilson, B. A., Neldner, V. J., and Accad, A. (2002). The extent and status of remnant vegetation in Queensland and its implications for statewide vegetation management and legislation. *Rangeland J.*, 24, 6–35.

Wilson, J. B., and Agnew, A. D. Q. (1992). Positive-feedback switches in plant communities. *Advances in Ecological Research*, 23, 263–336.

Wittemyer, G., Elsen, P., Bean, W. T., Burton, C. O., and Brashares, J. S. (2008). Accelerated human population growth at protected area edges. *Science*, 321, 123–126.

Woodward, C., Shulmeister, J., Larsen, J., Jacobsen, G. E., and Zawadzki, A. (2014). The hydrological legacy of deforestation on global wetlands. *Science*, 346(6211), 844–847.

WRI. (1997). The Last Frontier Forests ecosystems and economies on the edge. World Resources Institute. Washington DC, USA, 49 pp.

WRI. (2007). EarthTrends: Environmental information. World Resources Institute (WRI), Washington, DC available at: http://earthtrends.wri.org.

Wright, J. M., and Chambers, J. C. (2002). Restoring riparian meadows currently dominated by Artemisia using alternative state concepts – above-ground response. *Appl. Veg. Sci.*, 5, 237–246.

Wright, S. J. (2010). The future of tropical forests. *Ann. N.Y. Acad. Sci.*, 1195, 1–27.

Wu, T. H., McKinnel, W. P., and Swanston, D. N. (1979). Strength of tree-roots and landslides on Prince of Wales Island, Alaska. *Can. Geotech. J.*, 16(1), 19–33.

Wu, T. H., and Swanston, D. N. (1980). Risk of landslides in shallow soils and its relation to clearcutting in southeastern Alaska. *For. Sci.*, 26(3), 495–510.

Wunder, S., and Dermawan, A. (2007). Cross-sectoral tropical forest cover impacts: What matters. *Cross-Sectoral Policy Developments in Forestry*, 1–14.

Xue, Y., Sellers, P. J., Kinter III, J. L., and Shukla, J. (1991). A simplified biosphere model for global climate studies. *J. Climate*, 4, 345–364.

Yachi, S., and Loreau, M. (1999). Biodiversity and ecosystem productivity in a fluctuating environment: The insurance hypothesis. *Proceedings of the National Academy of Sciences*, 96(4), 1463–1468.

Yang, X., Post, W. M., Thornton, P. E., and Jain, A. (2013). The distribution of soil phosphorus for global biogeochemical modeling. *Biogsciences*, 10(4), 2525–2537.

Yang, Y. S., Guo, J., Chen, G., Xie, J., Gao, R., Li, Z., and Jin, Z. (2005). Carbon and nitrogen pools in Chinese fir and evergreen broadleaved forests and changes associated with felling and burning in mid-subtropical China. *Forest Ecology and Management*, 216(1), 216–226.

Yasuoka, J., and Levins, R. (2007). Impact of deforestation and agricultural development on anopheline ecology and malaria epidemiology. *Am. J. Trop. Med. Hyg.*, 76(3), 450–460.

Young, A., and Mitchell, N. (1994). Microclimate and vegetation edge effects in a fragmented podocarp-broadleaf forest in New Zealand. *Biological Conservation*, 67, 63–72.

Xiao, H., Ouyang, Z., Zhao, J., and Wang, X. (2000). Forest ecosystem services and their ecological valuation – a case study of tropical forest in Jianfengling of Hainan Island. *The Journal of Applied Ecology*, 11(4), 481–484.

Zak, D. R., Holmes, W. E., Finzi, A. C., Norby, R. J., and Schlesinger, W. H. (2003). Soil nitrogen cycling under elevated CO_2: A synthesis of forest FACE experiments. *Ecological Applications*, 13(6), 1508–1514.

Zavaleta, E. (2000). Valuing ecosystem services lost to Tamarix invasion in the United States. *Invasive Species in a Changing World*, 261–300.

Zegada-Lizarazu, W., and Monti, A. (2011). Energy crops in rotation: A review. *Biomass and Bioenergy*, 35(1), 12–25.

Zen, Z., Barlow, C., and Gondowarsito, R. (2006). Oil palm in Indonesian socio-economic improvement: A review of options. *Industry Economic Journal*, 6, 18–29.

Zeng, N. et al. (1999). Enhancement of interdecadal climate variability in the Sahel by vegetation interaction. *Science*, 286, 1537–1540.

Zeng, N., and Neelin, J. D. (2000). The role of vegetation-climate interaction and interannual variability in shaping the African savanna. *J. Clim.* 13, 2665–2670, doi:10.1175/1520-0442(2000)013,2665:trovci.2.0.co;2.

Zeng, X., Shen, S. S., Zeng, X., and Dickinson, R. E. (2004). Multiple equilibrium states and the abrupt transitions in a dynamical system of soil water interacting with vegetation. *Geophys. Res. Lett.*, 31, L05501, doi:10.1029/2003GL018910.

Zentner, R. P. et al. (2002). Economics of crop diversification and soil tillage opportunities in the Canadian prairies. *Agronomy Journal*, 94(2), 216–230.

Zhang, D., Hui, D., Luo, Y., and Zhou, G. (2008). Rates of litter decomposition in terrestrial ecosystems: Global patterns and controlling factors. *Journal of Plant Ecology*, 1(2), 85–93.

Zhang, L., Dawes, W. R., and Walker, G. R. (2001). Response of mean annual evapotranspiration to vegetation changes at catchment scale. *Water Resources Research*, 37(3), 701–708.

Zhang, Q., Wang, Y. P., Pitman, A. J., and Dai, Y. J. (2011). Limitations of nitrogen and phosphorous on the terrestrial carbon uptake in the 20th century. *Geophysical Research Letters*, 38(22).

Zhang, T., Barry, R. G., Knowles, K., Ling, F., and Armstrong, R. L. (2003). Distribution of seasonally and perennially frozen ground in the Northern Hemisphere. In *Proceedings of the 8th International Conference on Permafrost*. AA Balkema, Vol. 2, pp. 1289–1294.

Zheng, D., Wallin, D. O., and Hao, Z. (1997). Rates and patterns of landscape change between 1972 and 1988 in the Changbai Mountain area of China and North Korea. *Landscape Ecology*, 12, 241–254.

Zhu, Z., Xiong, Z., and Xing, G. (2005). Impacts of population growth and economic development on the nitrogen cycle in Asia. *Science in China Series C: Life Sciences*, 48(2), 729–737.

Zipperer, W. C. (1993). Deforestation patterns and their effects on patches. *Landscape Ecology*, 8, 177–184.

Index

Bold page numbers represent the instance where the term is mentioned specifically.

Absolute humidity, 57
Acid deposition, 88–89
Acid rain, 88
Active layer, 121–124
Afforestation, **4**, 10, 18, **21–22**, 47
Africa, 1–2, 10–11, **18–20**
Aggressive species, 185–186
Agricultural expansion, 15, 20, 24, 27, 149, **151–152**, 161, 171, 174, 176, 188, **192–193**
Agricultural production, **23–24**, 26–30, 102, 154–161, 188–189, 192
Albedo, 41, **57–58**, 61, 64, 68, 70, 108, 135–137
Alpine forest, 5, 63, 65, 67, 135
Alternative states, 103, 133, 137, 141, 144, 177
Amazon, 7, 11, 14, 24, 30, 32, 44, 62, 68, 83–84, 108–111, 133–134, 151, 154, 157, 160–169
Ammonia, 86
Ammonification, 85–86, 91
Ammonium, 85–88, 91–92
Anaerobic conditions, 52, 119–121
Angiosperms, 9–10, **56**, 95, 179
Anisohydric, 56
Apatite, 93, 98
Arable land, 29, 79, 175, 187–188
Arbuscular mycorrhizal fungi, 100, 126
Arctic, 5, 50, 65, 67, 136–137, 182
Ash, 51, 79, 92, 99, 125
Atlantic forest, 14, **104**
Atmospheric deposition, 93, 102, 106, 113
Attractors, 103, 106
Australia, 20–21, 119, 133–134, 139–142
Autocovariance, 143
Autotrophic respiration, 71
Averaging effect, 107

Basin of attraction, 103, 105, 113
Benefits, 145–148, 157–158, **165**, 169–171, **173–177**
Bifurcation, 103–105, 142, 143
Biodiesel, 32
Biodiversity loss, **178**, 183, **184**
Bioenergy, 26, 31–32, 193
Bioethanol, 32
Biofuels, **26**, 31–32, 161, 193
Biogeochemical benefits, 173

Biogeography of forests, **3**, 5
Bistability, 103, 120, 141, 142
Blue water, 39, 40
Boreal forests, **10**, 76, 79, 121, 136
Borlaug hypothesis, 153
Bosque tropical, 9
Boundary layer height, 109, 112
Broadleaf, 5, 7, 9, 42, 87
Broad-leaved rainforests, 9
Buffering effect, 107, 187

Caatinga, 9
Calcium, 71, 97
Calorie demand, 27, 187
Canopy breezes, 63, 68, 111, 174
Canopy interception, 42–43
Canopy storage capacity, 42
Canopy water retention, 42
Carbon assimilation, 4, 56, 66
Carbon balance, 56, 66, 71, 74
Carbon cycle, 71–84
Carbon losses, 33
Carbon starvation, 56
Carbon stock, 2, 33, 35–37, 166, 171
Carboxylates, 95
Carrying capacity, 116
Cattle production, 24
Cattle ranching, 14–15, 30, 150–151
Cavitation, 4, 56, 66
Central America, 15
Civilization, 3, 18
Climate benefits, 173
Climate change, **44–45**, **58**, 142, 183
Climate warming, 44, 59, 69
Cloud base height, 44–45, 108
Cloud condensation nuclei, 62–63, 70
Cloud deposition, 44, 114
Cloud forest, 43–45, 69, 114
Cloud microphysics, 62, 70
Cloudiness, 5, 44, 59
Clouds, 35, 37, 43–45, 62, 113
Cold sensitivity, 66
Cold stress, 4, 66
Collective outcomes, 159

249

Common pool, 159
Community managed forests, 159–160
Community structure, 181
Concentration time, 47
Condensation, 43–45, 62–63, 101
Conifers, 10, 49, 56, 132
Conservation, 84, 156, 162–163, 165–171, 185–186, 191
Conserved state, 149
Consumer behavior, 188
Convective boundary layer, 108–109
Conversion of forest, 4, 14–16, 20–21, 23–24, 31, 80, 82, 171–172, 182
Cooling effect, 41
Corruption, 157
Critical slowing down, 143
Crop breeding, 29
Crop diversification, 192
Crop landraces, 30
Cropland, 21, 23–24, 30–32, 82–83, 85, 176, 187–189
Crown Lands Alienation Act, 21
Cultivation, 15, 25–26, 79, 84–85, 99–100, 127, 128
Cultivation period, 128
Currency devaluation, 156

Decomposition, 72–73, 79–80, 84–86, 88, 96, 99
Deforestation, 4
Deforested state, 103
Degraded forest, 4
Demographic factors, 150, 153
Denitrification, 86, 88, 91–93
Desiccation, 4–5, 56, 66
Dew point, 44–45
Diets, 27, 187
Diffusion, 54–55, 57, 88
Discount rate, 146, 158, 162
Displacement of land use, 30, 32, 149, 155
Displacement uses, 149
Distribution of species, 181
Disturbance, 4, 66–67, 79, 103, 105, 124, 129–142, 147, 182–183, 185–186
Disturbance diversity, 185
Diversity-stability hypothesis, 107
DOC, 83
Domestic costs, 156, 176
Drainage, 52, 122, 141–142
Drivers of deforestation, 23–26, 149–152
Dry deposition, 101, 113
Dry forests, 7–9
Dry tropics, 6

Easter Island, 3
Ecological reserves, 186
Economic development path, 22
Economic factors, 26, 150, **154–156**
Economic modelling, 163–165
Economic uses, 148–149
Ecosystem function, 104, 107, 144, 147–148, 181
Ecosystem services, 147–148, **169–171**, 183
Education, 166
Embolism, 56, 66
Emission, 62, 70, 72, **76–80**, 84–85, 89–90, 161, 170–171, 174, 189

Energy balance, 55, 57, 59, 61, 70, 108–109
Environmental externalities, 145
Environmental factors, 38, 54, 150, 162–163
Erosion, 3, 39–40, 45–46, 50–52, 72, 73, 79, 83–84, 92, 96, 99–101, 125, 130–132, 174–175
Europe, 2, 11, **18**
Evaporation, 40, **54–59**, 60–61, 100–101, 110
Evapotranspiration, 39–42, 44–45, **54–59**, 60–64, 68, 70
Evapotranspiration recycling, 59, 110
Evergreen semi-deciduous, 7, 14
Extensive systems, 85
External benefits, 171, 177, 192
Externalities, 145
Extinction, 180–182, 185
Extracellular phosphatases, 94, 175

Fallow period, 29, 85, 128
Farm abandonment, 16
Feedback, 44, 62, 68, 103, **106–141**
Fertilized soil, 87
Fertilizers, 28–30, 89, 117, 129, 175, 188–191
Fiber, 153, 165
Fire, **50–52**, 62, 69, 73, 79–80, 92, 99–100, **132–135**
Fire frequency, 133–134
Fire suppression, 134
Fixation, 85–86, 96–97, 100, 125
Flammability, 133
Flash flood, 47
Fog, 43–45, 69–70, 101, 113–114
Food, 15, 20, 26–32, 89, 153, 165–166, 173, 184, **187–193**
Food security, 187, **192**
Forest cover change, **10–22**
Forest scarcity path, 22
Forest transition, 22, 155–156
Forested area, 2, 10–11, 13, 15–18, 22, 33, 45, 76, 155
Fossil fuel, 31, 84, 89
Freezing, 65, 66, 106, **135–138**
Frontier forest, 4, 151
Frontier land, 146, 159
Frost, 5, **63–67**, 135–137
Frost damage, 5, 66–67, 135
Frost desiccation, 5, 66
Frost heaving, 65
Frost hollows, 64
Frost stress, 63
Fuel, 15, 16, 26, 31–32, 84, 89, 133, 147, 152, 165
Fuelwood, 146, 150, 152, 167
Functional diversity, 185

Germination, 5, 66
Globalization, 30
Governance, 160
Grassland, 7, 23, 24, 26, 41, 82–83, 122, 133, 135, 140
Grazing, 118, 160, 169, 183
Green revolution, 28, 30, 154, 189–190
Green water, 39–40
Gross primary production (GPP), 71, 73–76, 147
Groundwater, 40, 46, 49, **52**, **70**, 90, 119, 138–142, 175
Growing season, 4–5, 43, 48, 50, 66, 122
Guard cells, 55
Gymnosperm, 56, 179

Habitat, 45, 49, 66, 68, 104, 135, 147, 173, **180–187**
Halophytes, 49
Harvesting, 15–17, 26, 90, 131, 137, **152**, 166, 168
Heliophytes, 149
Herbicides, 188
Hopkins' biogeoclimatic law, 5
Horizontal precipitation, 43
Hotspots of biodiversity, 166, 180, 185
Human disease, 186–187
Human health, 186–187
Human welfare, 169–170
Humus, 72–73, 86, 88, 96, 129
Hydraulic failure, 3, **56**
Hydraulic redistribution, 118
Hydrograph, 47
Hydrological benefits, 173–174
Hydrophobicity, 51, 53
Hydrophytes, 49
Hysteresis, 105, 116

Ice nuclei, 62, 70
Immobilization, 86, 96, 127, 175
Inbreeding depression, 180
Income stability, 192
Indirect values, 149
Infectious disease, 169, 186–187
Infiltration, **45–46**, 47, 50–53, 117–118
Infiltration-excess runoff, 45–47
Infrastructure extension, 149–151
Insecticides, 29, 188
Insecure tenure, 158
Institutional factors, 153, 156–157
Institutional instability, 162
Insurance hypothesis, 107
Intensification, 28–29, 153–155, **188–191**
Interannual variability, 68, 142
Interception, 42–43, 50–52, 54, 121–122
Internal benefits, 192
International trade, 30, 154–155, 188
Intertropical convergence zone, 6
Invasions, 180, 192
Inverse tree lines, 64–65
Investment, 30, 158, 160, 161, 164, 168, 170
Iron Age, 20
Irreversibility, 103
Irrigation, 28–30, 113, 117, 141, 170, 176, **190**
Isohydric, 56

Jobs, 22, 165, 171

Kampfzone, 5
Katabatic flows, 65
Keystone species, 147, 186
Kuznet's curve, 155

Land concessions, 61, 67
Landsat, 35
Landslides, 129–132
Land-use change, 24, 26, 31, 32, 71, 77–78, 80, 82–84, 155, 180–182, 185–187
Large-scale land acquisitions, 161
Latent heat flux, 55, 61, 108–109
Latent heat of vaporization, 55

Leaching, 72–73, 83, 86, 88, 90, 96, 98–100, 125
Leaf area index, 8, 42, 49, 57–58, 65, 115
Leakage, 32, 156, 171
Lidar, 37
Life expectancy, 166
Lifting condensation level, 109, 112
Literacy, 166
Litter, 42–43, 46, 72–76, 86–88, 95–97
Litter interception, 42–43
Local management, 159
Logging, **24–25**, 78–79, 121, 166, 177
Long-wave radiation, 63
Low temperature photoinhibition, 65, 66

Macroeconomic modes, 163
Macropore, 45
Maintenance respiration, 71
Malaria, 169, 187
Mangroves, 6, 7, 49, 67, 135
Manufactured capital, 169
Marginal net benefits, 146, 148
Markets, 14, 145, 150, 152, 154–156, 168–170
Mass wasting, 130, 147
Maximum temperatures, 63
Medical care, 166
Medicine, 148, 166, 173, 184
Mesophytes, 49
Mesoscale circulations, 63, 70, 174
Methane, 73
Microbial activity, 73, 86, 91
Microclimate, 39, 45, 58, 63–67, 135, 137
Microeconomic models, 163–164
Micronutrients, 71
Microwave, 34–35
Migrations, 156
Minimum temperatures, 63, 66, 135
Mismanagement, 157
Modified natural forest, 4
MODIS, 35, 61
Monocultures, 188, 191–192
Monsoon forests, 9
Mycorrhizae, 73, 95, 100, 126
Mycorrhizal fungi, 95, 100, 126

N deposition, 89, 90, 97
Near InfraRed, 34
Negative covariance effect, 107
Net ecosystem carbon balance (NECB), 74
Net ecosystem production (NEP), 73
Net primary production (NPP), 73
Net profit, 146
Nitrate, 85, 86, 88, 90–92, 175
Nitric oxide, 87
Nitrification, 85–86, 88, 90–93, 175
Nitrogen cycle, 85–92
Nitrous oxide, 90, 190
Noise-induced transitions, 142
Non-timber forest products, 145, 149, **166**, 173

Occluded pool, 93
Occult deposition, 113
Occult precipitation, 43–45, 69
Oil palm, 20, 31, 165, 191–192

Open access, 159, 168
Opportunity costs, 145, 170
Optical window, 33
Optimal level of deforestation, 146
Organic aerosols, 59, 62, 70
Orographic clouds, 44, 45, 114
Ostracism, 159
Outsourcing of deforestation, 156
Overconsumption, 27
Overexploitation, 168

P availability, 93–96, 126
P limitations, 96–97, 126
P sorption, 93
Paired watershed, 50
Paper products, 24
Pasture, 22–24, 43, 58, 68, 82 84, 99, 108–109, 146, 170, 174
Pathogens, 187–188
Peak flow, 47, 50, 52
Peat swamp forest, 192–193
Performance enhancing effect, 107
Permafrost, 121–125, 136
Pests, 188, 191–192
Pharmaceutical products, 166, 173
Phosphatase enzyme, 95, 100, 126
Phosphorus cycle, 93–100
Photosynthesis, 55, 56, 65, 66, 71–72, 85, 93–94
Phreatophytes, 49, 52, 54
Plantation, 4, 15, 20, 25, 28, 31–32, 76, 104, 155–156, 192–193
Planted forest, 4, 25
Policy, 150, 156–162
Policy failure, 157
Population, 26–27, 32–33, 150, 152–153
Potential function, 106
Precipitation, 42–44, 106–112
Precipitation recycling, 40, 59–61, 70, 108–110
Predator loss, 184
Price, 22, 31, 145–146, 149–154, 156–157, 164–168
Primary biological aerosols, 62
Primary forests, 13, 17, 161, 186
Private benefits, 145, 166, 171
Private land, 159
Profits, 156, 160
Property rights, 157–160
Protected land, 159, 182–183, 185–186
Proximate causes, 149–152
Public benefits, 145
Public goods, 145–146
Public land, 159
Pyromineralization, 99

Radar, 35–37
Radiative cooling, 63–67, 135–137
Rainfall variability, 68, 124
Rainforests, 7, 9, 20, 74, 83
Random drivers, 142, 143
Recycling ratio, 60, 61, 109, 110
REDD, Reduced Emissions from Deforestation and Degradation, 33, **170–171**

Reforestation, 4, **21–23**, 47, 104, 155–156, 174
Regeneration capacity, 66
Regional models, 164
Relative humidity, 43–46, 50, 57
Remote sensing, 33–38
Repeated deforestation, 127
Replacement cost, 167
Reproduction, 66, 135, 181
Reservoir species, 187
Resilience, 103–107, 177
Resource availability, 105–118
Response diversity, 107, 185
Restoration, 104, 144, 174, 178
Returns, 145, 158, 160, 168
Reversibility, 103
Road construction, 132, 151
Root depth, 58, 61, 106
Root uptake, 86
Runoff, 39–41, **45–46**, 50–53, 59–60, 90, 101, 125, 169
Rural credit, 156
Rural wages, 156, 176

Salinity, 138–142
Salt, 49, 52, 70, 106, 138–142
Salt marshes, 6, 135
Saprolite, 83, 129
SAR, Synthetic Aperture Radar, 35
Saturation vapor pressure, 44, 57
Saturation-excess runoff, 45–47, 53
Sciophytes, 49, 65
Sclerophyllous forests, 9
Seasonal precipitation fluctuations, 68
Second generation biofuels, 193
Secondary forests, 4, 25, 26, 83, 89, 128, 186
Seed production, 66
Sensible heat flux, 55, 68, 108–109, 112
Shear strength, 130
Shelterwood forest regeneration, 137
Shifting cultivation, 25–26, 79, 117, 126–128
Short Wave Infrared, 34
Sink of Carbon, 77–78, 84
Slash and burn, 18, 20, 25, 52, 79, 83–84
Smallholder agriculture, 153, 156
Smoke, 62, 92, 96
Snow interception, 43, 50
Snowmelt, **50**, 69
Snowpack, 39, 50, 69
Social benefits, 146
Soil barrier, 96–97
Soil cohesion, 83, 130
Soil conservation, 84, 191–192
Soil cryoturbation, 65
Soil erosion, *see* "Erosion"
Soil infiltration capacity, 45–46, 51, 117–118
Soil mantle, 130
Soil mass balance, 131
Soil moisture, 51, 57–58, 70, 73, 91, 100–101, **117–118**, 125
Soil pH, 91, 98, 100
Source of Carbon, 77–78, **84**
South Asia, 15–16